Lectures in Astrobiology I
Part 2: From Prebiotic Chemistry to the Origin of Life on Earth
Study Edition

Advances in Astrobiology and Biogeophysics

springer.com

This series aims to report new developments in research and teaching in the interdisciplinary fields of astrobiology and biogeophysics. This encompasses all aspects of research into the origins of life – from the creation of matter to the emergence of complex life forms – and the study of both structure and evolution of planetary ecosystems under a given set of astro- and geophysical parameters. The methods considered can be of theoretical, computational, experimental and observational nature. Preference will be given to proposals where the manuscript puts particular emphasis on the overall readability in view of the broad spectrum of scientific backgrounds involved in astrobiology and biogeophysics.

The type of material considered for publication includes:

- Topical monographs
- Lectures on a new field, or presenting a new angle on a classical field
- Suitably edited research reports
- Compilations of selected papers from meetings that are devoted to specific topics

The timeliness of a manuscript is more important than its form which may be unfinished or tentative. Publication in this new series is thus intended as a service to the international scientific community in that the publisher, Springer, offers global promotion and distribution of documents which otherwise have a restricted readership. Once published and copyrighted, they can be documented in the scientific literature.

Series Editors:

Dr. André Brack
Centre de Biophysique Moléculaire
CNRS, Rue Charles Sadron
45071 Orléans, Cedex 2, France
Brack@cnrs-orleans.fr

Dr. Gerda Horneck
DLR, FF-ME, Radiation Biology
Linder Höhe
51147 Köln, Germany
Gerda.Horneck@dlr.de

Prof. Dr. Michel Mayor
Observatoire de Genève
1290 Sauverny, Switzerland
Michel.Mayor@obs.unige.ch

Prof. Dr. John Baross
School of Oceanography
University of Washington
Box 357940
Seattle, WA 98195-7940, USA
jbaross@u.washington.edu

Dr. Christopher P. McKay
NASA Ames Research Center
Moffet Field, CA 94035, USA

Prof. Dr. H. Stan-Lotter
Institut für Genetik
und Allgemeine Biologie
Universität Salzburg
Hellbrunnerstr. 34
5020 Salzburg, Austria

Muriel Gargaud
Bernard Barbier   Hervé Martin   Jacques Reisse   (Eds.)

Lectures in Astrobiology

Volume I
Part 2: From Prebiotic Chemistry
to the Origin of Life on Earth
Study Edition

With a Foreword by Christian de Duve

and 14 Tables and 88 Figures, 82 in Color

Dr. Muriel Gargaud
Editor-in-Chief
Université Bordeaux 1
Observatoire Aquitain
des Sciences de l'Univers
2 Rue de l'Observatoire
33270 Floirac, France

Dr. Bernard Barbier
Centre de Biophysique Moléculaire
Rue Charles Sadron
45071 Orléans, France

Prof. Hervé Martin
Université Blaise Pascal
Laboratoire Magmas et Volcans
5 Rue Kessler
63038 Clermont-Ferrand, France

Prof. Jacques Reisse
Member of the Belgian Academy
of Sciences
Université Libre de Bruxelles (CP 165/64)
Department of Chemistry
Applied Sciences
Avenue F. Roosevelt, 50
1050 Brussels, Belgium

Library of Congress Control Number: 2005937604

ISSN 1610-8957

ISBN-10 3-540-29004-4 Springer Berlin Heidelberg New York
ISBN-13 978-3-540-29004-9 Springer Berlin Heidelberg New York

This work is subject to copyright. All rights are reserved, whether the whole or part of the material is concerned, specifically the rights of translation, reprinting, reuse of illustrations, recitation, broadcasting, reproduction on microfilm or in any other way, and storage in data banks. Duplication of this publication or parts thereof is permitted only under the provisions of the German Copyright Law of September 9, 1965, in its current version, and permission for use must always be obtained from Springer. Violations are liable to prosecution under the German Copyright Law.

Springer is a part of Springer Science+Business Media
springer.com

© Springer-Verlag Berlin Heidelberg 2006
Printed in Germany

The use of general descriptive names, registered names, trademarks, etc. in this publication does not imply, even in the absence of a specific statement, that such names are exempt from the relevant protective laws and regulations and therefore free for general use.

Typesetting and production: LE-TEX Jelonek, Schmidt & Vöckler GbR, Leipzig
Cover design: Erich Kirchner, Heidelberg

Printed on acid-free paper   54/3141/YL - 5 4 3 2 1 0

"It seems plain and self-evident, yet it needs to be said: the isolated knowledge obtained by a group of specialists in a narrow field has in itself no value whatsoever, but only in its synthesis with all the rest of knowledge and only inasmuch as it really contributes in this synthesis toward answering the demand, 'Who are we?'"

Erwin Schrödinger

Foreword

The twentieth century will be remembered as the century of scientific revolutions. It started with the discoveries of physics, which revealed the fine structure of matter and the fundamental laws of nature. Then came cosmology, which traced the history of the Universe, from the Big Bang to the dizzying recession of myriad whirling galaxies. Finally, with the advances of biochemistry, cell biology, and molecular biology, life itself has disclosed its secrets.

These revolutions, in turn, have spawned new technologies – nuclear power, space travel, informatics, bioengineering – that could not even be conceived one century earlier; they have also opened new fields of inquiry that had been relegated before to the realm of the unknowable, objects only of gratuitous speculation or imaginative fiction. Among these new fields, the origin and evolution of life on Earth have become topics of intense theoretical and experimental research.

The latest offspring of this upheaval is exobiology, the science of extraterrestrial life, also known as astrobiology or bioastronomy. Of all branches of science, it is the most universal and all-encompassing, involving virtually every scientific discipline. It is also the emptiest, being so far without known object. No sign of life beyond our planet has yet been uncovered.

Whether or not its quest will one day be fulfilled, exobiology has already produced many valuable fruits and is bound to produce many more in the future. It has brought together and impelled physicists, chemists, cosmologists, astronomers, planetologists, geologists, paleontologists, biologists of all kinds, and other specialists who had until then labored each in the isolation of their individual disciplines to interact. It has stimulated many investigations that would otherwise have been performed with considerably less vigor, perhaps not at all. It has revealed a number of significant facts on the cosmic properties and interrelationships out of which life and mind emerged on our planet and may, perhaps, have done so elsewhere in our galaxy or in others. It has even alerted philosophers and theologians to a very real possibility that, only 400 years ago, was a heresy punishable by death. It has evoked new dreams in the collective imagination of humans who, ever since their distant ancestors started contemplating the skies, have asked the question: Is there life out there? Are there others like us elsewhere?

The present book is the outcome of a remarkable pluridisciplinary effort initiated by a group of French scientists and materialized already into four summer schools organized under the aegis of the CNRS, Exobio'99, Exobio'03 et Exobio'05 in Propriano, Corsica (1999, 2003 and 2005) and Exobio'01, in La Colle-sur-Loup, South of France (2001). The lectures given during these meetings are an invaluable source of information covering every aspect, from astronomy to molecular biology, likely to illuminate the exobiology problem. This exceptional documentation will now be generally available, thanks to the present work and to volume 2 in progress, in a world-wide accessible form reviewed by international referees. As a participant myself, who, for personal reasons was unable to provide a chapter, I am particularly pleased and honored to have the opportunity to preface this truly unique compendium.

Christian de Duve

Preface

Astrobiology, also known as bioastronomy or exobiology, refers to a vast area of scientific research. The formation of the solar system, its accretion and the formation of the planets, the origin of the molecules out of which living beings are made, the traces of present and past life within the solar system and elsewhere, as well as the search for extra-solar planets, are all part of astrobiology. And the above list is not exhaustive.

For obvious reasons, astrobiology is a field without the traditional barriers between astronomers, chemists, physicists, geologists, and biologists or between experimentalists and theorists, observers and those who model the observations. As such, a single researcher cannot possess all the knowledge necessary to be an "astrobiologist". One can even go a step further and say that while astrobiology clearly exists as a field of scientific research, there are no astrobiologists. Astrobiology exists at a higher level of organization where the knowledge is not that of an individual but that of a research community whose members all share the same interest for the fundamental questions concerning the emergence of life, its evolution, and how life is distributed on Earth and throughout the universe. Each person contributes in piecing together this vast puzzle through their knowledge and their experimental and theoretical tools.

As often, if not always, when treating questions dealing with the past or with a sort of "elsewhere" where one cannot go and that one can only study indirectly, we must be satisfied with plausible scenarios rather than clear proof or other certainties. In this way, the strategy of the astrobiologist is similar to that of an archeologist or a paleontologist. There exists, however, major differences between the path of a chemist interested in the origin of life, and thus in prebiotic chemical evolution, that of a biologist wanting to follow time back starting with current life, and that of a paleontologist searching for the traces of primitive life and its evolution and extinctions.

While paleontologists have some hard data at hand (fossils and other physical traces), the situation is very different for chemists, who are obliged to build a plausible scenario for the appearance of life based on hypotheses developed by specialists in other fields (composition of the primitive terrestrial atmosphere, addition of extraterrestrial organic material, etc.). For the most part, these hypotheses are unverifiable. The biologist, on the other hand, tries to use phylogenetic tools to find and understand LUCA, the first common ancestor who must have been preceded by other micro-organisms with no descendants.

Similarly, there is an important difference between the strategies of a geologist, expert in the transition between the Tertiary and Cretaceous periods, and the planetologist who would like to describe the Earth during the period of intense meteoritic bombardment. The former disposes of observations and measures (iridium content, sediment ashes, shocked quartz, etc.), which provide a reasonable explanation for the great biological Cretaceous Tertiary crisis caused by a major meteorite impact. The latter only has access to indirect data based on observations of lunar craters but also simulations, which are of course based on theoretical models.

Since every scientist has a limited area of expertise, the scenario that he/she proposes can only be validated by the constraints and parameters that he/she knows and masters. Such an individual strategy can thus lead to as many scenarios as there are researchers. A multidisciplinary approach has the advantage of subjecting each individual proposition to a much larger number of constraints. This naturally leads to the rejection of "weak" scenarios and to the emergence of more robust hypotheses. For example, it is pointless for a chemist to invoke the role of a prebiotic chemical reaction if the conditions needed for the reaction to occur are completely incompatible with the primitive Earth conditions determined by the planetologists. This simple example illustrates the importance of interdisciplinary discussion for all those who consider themselves to be astrobiologists. The CNRS summer schools such as Propriano in 1999, 2003 and 2005 and La Colle-sur-Loup in 2001 have contributed to strengthening the dialog within the French scientific community.

The goal of the first summer school, Exobio'99 in Propriano, was to provide participants with an objective image of what we know today about the early Earth conditions – the oceans, the proto-continents, the atmosphere, and even the climate – but also of what we know about the solar system during the first billion years of its history. Some stages in the chemical evolution that may have occurred on the young planet Earth, with a different solar radiation, less intense in the visible part of the spectrum but much more intense in the RX region were also discussed during the first summer school. The discussion then moves towards the biological evolution, the early stages of which are still very poorly understood. The problems related to the exploration of Mars and Titan were then addressed.

The second summer school, Exobio '01 in La Colle-sur-Loup, was more oriented towards the chemistry, molecular biology, biochemistry, and biological evolution of early Earth. Its main theme was the study of organisms referred to as extremophiles, which could provide information on the nature of the first unicellular organisms that populated the young oceans. Among the specific topics addressed were the autoformation of biological membranes, the possible origins of the homochirality of the constituents of living beings, the protometabolisms that may be inferred from the study of metabolisms, and the possible role of ribozymes before the emergence of catalysis by proteinic enzymes.

The texts that follow represent the first volume of the series "Lectures in Astrobiology" and are the result of the first two schools. The chapters were written for readers already familiar with the general topic of the origin of life and life "elsewhere" but not to the extent to which each specialist is in his own discipline. As such, they are meant as much for students as for established scientists seeking to broaden their horizons in the vast field of the origins of life. We hope these texts will initiate vocations and incite researchers and students specialized in one of the individual fields to join the broad forum of astrobiology. It is undeniable that the questions forming the basis of astrobiology are among the big questions that humanity has asked itself since its inception and which recent decades have attempted to answer; answers that seem more and more plausible although necessarily partial.

Acknowledgements

This book is the product of a multidisciplinary community, the members of which all question the knowledge from their disciplines at origin in order to build together the complex structure, which this area of research represents. One needs a very open mind as well as the ability to question ones ideas through the recent discoveries in other fields.

A group of international reviewers and ourselves have read the set of texts that follow. The final versions of these texts, after multiple rewritings and long discussions, sometimes required the opinions of five or six specialists.

We would therefore like to thank all of the authors who have accepted these remarks, criticisms, and multiple discussions warmly, but also all of the "specialist reviewers" who, through their expertise, have contributed to the general coherence of this work.

Last but not least, the editors call upon the reader's indulgence concerning some (or many!) misusages of the English language; English is not the mother tongue of the large majority of authors.

Muriel Gargaud
Bernard Barbier
Hervé Martin
Jacques Reisse

From *left* to *right*: Hervé Martin (geochemist), Muriel Gargaud (astrophysicist), Jacques Reisse (chemist), Bernard Barbier (biochemist)

Contents

General Introduction

From the Origin of Life on Earth to Life in the Universe
André Brack 1
1   The Search for Traces of Primitive Life and Other Imprints 3
     1.1   Microfossils 3
     1.2   Oldest Sedimentary Rocks 4
     1.3   One-handedness of Life 4
2   Reconstructing Life in a Test Tube 5
     2.1   Primitive Earth Atmosphere 6
     2.2   Organic Synthesis 6
     2.3   Delivery of Organics by Comets and Meteorites 7
     2.4   Simulation Experiments 9
     2.5   Recreating the Chemistry of Primitive Life 10
3   Search for Extraterrestrial Life 11
     3.1   The Diversity of Bacterial Life as a Reference
           for Extraterrestrial Life 11
     3.2   The Search for Life in the Solar System 12
     3.3   The Search for Life Beyond the Solar System 15
     3.4   Panspermia, Interplanetary Transfer of Life 17
4   Conclusion 18
References 20

From Prebiotic Chemistry to the Origin of Life on Earth

Introduction 27

1 A Rational Approach to the Origin of Life:
From Amphiphilic Molecules to Protocells. Some Plausible
Solutions, and Some Real Problems
Guy Ourisson, Yoichi Nakatani 29
1.1   From Amphiphilic Molecules to "Protocells"
        by Understandable Processes. Self-Organisation
        and Self-Complexification 29

1.2	Water and Self-Organisation of Amphiphiles		30
	1.2.1	The Structure of Liquid Water	30
	1.2.2	The Structure of Amphiphile-water Mixtures	30
1.3	Properties Ensuing from the Self-organisation of Amphiphiles		33
	1.3.1	Extraction and Orientation	33
	1.3.2	Increased Concentration and Condensation	33
	1.3.3	Vectorial Properties	34
	1.3.4	Coating the Vesicles	35
	1.3.5	Vesicles and Nucleic Acids; Vesicles as Protocells	36
1.4	The Nature of the Primitive Amphiphiles		37
	1.4.1	The Modernity of N-Acyl Lipids	37
	1.4.2	The Archæal Lipids and Their Synthesis	38
	1.4.3	The Terpenoids as Universal Metabolites	41
1.5	Some Remaining Problems		41
	1.5.1	The Problem of the Synthesis of Ingredients	41
	1.5.2	The Problem of Local Concentration	42
	1.5.3	The Problem of the Prevalence of Phosphates	42
	1.5.4	The Problem of Phosphorylation by Phosphoric Anhydrides	43
	1.5.5	The Problem of the C_5 Unit	44
	1.5.6	The Problem of the Cytoskeleton	44
References			44

2 Prebiotic Chemistry: Laboratory Experiments and Planetary Observation

François Raulin, Patrice Coll, Rafael Navarro-González 49

2.1	Simulation Experiments and Photochemical Models		50
	2.1.1	An Historical View of Miller's Experiment and the Development of a New Field: Prebiotic Chemistry	50
	2.1.2	An Overview of Experimental and Theoretical Data	51
	2.1.3	New Scenario for Prebiotic Chemistry	53
2.2	Elementary Prebiotic Chemistry in Aqueous Solution		54
	2.2.1	Prebiotic Chemistry of HCN: Strecker Reaction or Oligomerization (see Box 2.1)	54
	2.2.2	Prebiotic Chemistry of HCHO, Formose Reaction	57
	2.2.3	Prebiotic Chemistry of Tholins	58
2.3	Application of These Laboratory Experimental Data to Space Studies		58
	2.3.1	Telluric Planets	58
	2.3.2	Giant Planets and Their Satellites	60
2.4	Conclusions		67
References			68

3 Chirality and the Origin of Homochirality
John Cronin, Jacques Reisse 73
3.1   Chirality: Basic Concepts 73
3.2   Reactivity of Chiral Molecules 78
3.3   Pasteur and the Discovery of Molecular Chirality 79
3.4   Crystals and Crystallization 81
3.5   Homochirality and Life 82
3.6   The Why and When of Homochirality 84
3.7   Origin of Homochirality
      and Spontaneous Symmetry Breaking 86
3.8   Origin of Homochirality and Parity Violation 89
3.9   Origin of Homochirality and Photochemistry 91
3.10 Amplification of Enantiomeric Excesses 93
      3.10.1 Introduction 93
      3.10.2 Kinetic Resolution 93
      3.10.3 Chiral Catalysis 95
      3.10.4 Asymmetric Autocatalysis: Theoretical Models 95
      3.10.5 Asymmetric Autocatalysis: Experimental Data 97
      3.10.6 On the Possibility to Amplifying Enantiomeric Excesses
             due to Parity Violation 99
3.11 Exogenous Origin of Homochirality 101
3.12 Hypothesis and Summary 104
3.13 Homochirality Analyses in the Solar System
      and Beyond 107
References 108

4 Peptide Emergence, Evolution and Selection on the Primitive Earth. I. Convergent Formation of *N*-Carbamoyl Amino Acids Rather than Free α-Amino Acids?
Auguste Commeyras, Laurent Boiteau, Odile Vandenabeele-Trambouze, Franck Selsis 117
4.1   Introduction 117
4.2   Organic Molecules on the Primitive Earth 118
4.3   Exogenous Amino Acids and Related Compounds 119
      4.3.1 Exhaustive Survey of Exogenous Amino Acids 119
      4.3.2 Formation Mechanisms of Exogenous Amino Acids 120
      4.3.3 Other Meteoritic Compounds Closely Related to Amino Acids 125
      4.3.4 Non-Racemic Exogenous α-Amino Acids 126
      4.3.5 Exogenous Peptides 127
      4.3.6 Conclusion 127
4.4   Endogenous Organic Matter 128
      4.4.1 Endogenous α-Amino Acids 128

4.5 Formation Mechanisms of α-Amino Acids
and N-Carbamoyl Amino Acids
Via Strecker and Bücherer–Bergs reactions 129
    4.5.1 The Set of Reversible Reactions 132
    4.5.2 The Set of Irreversible Reactions 133
    4.5.3 Fate of Primary and Secondary Amines (R^3NH_2, R^3R^4NH) . . 135
    4.5.4 Conclusion. 135
4.6 Prebiotic Formation of α-Amino Amides
and Hydantoins Through Strecker
and Bücherer–Bergs Reactions. 136
    4.6.1 Formation of Exogenous α-Amino Amides and Hydantoins . . . 136
    4.6.2 Endogenous Formation of α-Amino Amides and Hydantoins . 138
4.7 Convergent Evolution Towards N-Carbamoyl
Amino Acids under Prebiotic Conditions. 140
4.8 Conclusions 141
References 142

5 Peptide Emergence, Evolution and Selection on the Primitive Earth. II. The *Primary Pump* Scenario

Auguste Commeyras, Laurent Boiteau, Odile Vandenabeele-Trambouze, Franck Selsis 147
5.1 From N-carbarmoyl Amino Acids (CAA) to Peptides 147
    5.1.1 Introduction 147
    5.1.2 The Primary Pump 149
5.2 Environmental Requirements 151
    5.2.1 Primitive Earth 151
    5.2.2 Primitive Atmosphere 152
    5.2.3 About the pH of Primitive Oceans 156
5.3 Investigation of the Primary Pump 157
    5.3.1 Step-By-Step Experimental Investigation 158
    5.3.2 Integrated Experimental Approach: Chemoselectivity 161
5.4 Energy 163
5.5 Conclusions and Perspectives. 165
References 166

6 The RNA World: Hypotheses, Facts and Experimental Results

Marie-Christine Maurel, Anne-Lise Haenni. 171
6.1 The Modern RNA World 171
    6.1.1 Where in the Living Cell is RNA Found? 171
6.2 An RNA World at the Origin of Life? 177
    6.2.1 Facts. 178
    6.2.2 Hypotheses 178
    6.2.3 But What do We Know about Primitive Replication? 179
6.3 A Pre-RNA World 181

	6.3.1	Evolutive Usurpation 181
	6.3.2	Alternative Genetic Systems 181
6.4	Optimizing the Functional Capacities of Ribonucleic Acids 183	
	6.4.1	Coenzymes and Modified Nucleosides 183
	6.4.2	The Case of Adenine 185
	6.4.3	Mimicking Darwinian Evolution 186
	6.4.4	Other Perspectives 189
6.5	Conclusion 190	
References 190		

7 Looking for the Most 'Primitive' Life Forms: Pitfalls and Progresses
Simonetta Gribaldo, Patrick Forterre 195

7.1	Simpler Doesn't Necessarily Mean Older! 196	
7.2	Hyperthermophiles are not Primitives, but are Remnants from Thermophilic Organisms 197	
	7.2.1	Hyperthermophiles and the Hypothesis of a Hot Origin of Life 197
	7.2.2	Hyperthermophiles are Complex Prokaryotes 198
	7.2.3	Origin of Hyperthermophily 200
	7.2.4	LUCA was Probably not a Hyperthermophile 201
	7.2.5	Temperature and the RNA World 203
7.3	Comparative Genomics: a Novel Approach to Retrace Our Most Distant Past 204	
	7.3.1	Simple or Complex LUCA? A RNA or a DNA Genome? 204
	7.3.2	A Key Step: the Apparition of DNA 206
	7.3.3	Viruses: Essential Players in Evolution 208
	7.3.4	The Origin of the Nucleus: a Further Puzzle 209
7.4	Conclusions and Perspectives 211	
	7.4.1	More Data are Needed 211
	7.4.2	To not Forget Darwin! 211
References 212		

8 The Universal Tree of Life: From Simple to Complex or From Complex to Simple
Henner Brinkmann, Hervé Philippe 217

8.1	Principles of Tree-Reconstruction Methods 217	
8.2	The Universal Tree of Life According to Woese 221	
8.3	Reconstruction Artefacts 224	
	8.3.1	Multiple Substitutions Generate Reconstruction Problems ... 224
	8.3.2	Mutational Saturation Versus Resolving Power 225
	8.3.3	Compositional Bias 228
	8.3.4	Long-branch Attraction 228
	8.3.5	Heterotachy 230
	8.3.6	Rare Genomic Events as an Alternative Approach? .. 232

8.4	Lateral Gene Transfer and the Quest for a Phylogeny of the Organisms 235
8.5	A New Evaluation of the Universal Tree of Life 237
	8.5.1 The Root of the Universal Tree of Life 237
	8.5.2 Prokaryotic Phylogeny 240
	8.5.3 Eukaryotic Phylogeny 241
8.6	The Importance of an Evolution by Simplification and by Extinction 243
8.7	Exobiology, a Procession of Extinctions? 246
References 247

9 Extremophiles
Purificación López-García 257

9.1	Some Concepts About Extremophiles 257
	9.1.1 What is an Extremophile? 258
	9.1.2 Some Extremophile Features 258
	9.1.3 Why Extremophiles are Interesting? 259
9.2	Microbial Diversity 260
9.3	Extreme Environments and Their Inhabitants 261
	9.3.1 Extremophiles and Extremotolerants 261
	9.3.2 Phylogenetic Groups Best Adapted to Extreme Conditions ... 269
	9.3.3 Resistance Forms and Longevity 270
9.4	Extremophiles and Exobiology 271
	9.4.1 Hyperthermophiles 271
	9.4.2 Psychrophiles 273
	9.4.3 Halophiles and Evaporites 274
	9.4.4 The Deep Biosphere 275
9.5	Perspectives 275
References 276

Appendices

1 Earth Structure and Plate Tectonics: Basic Knowledge
Hervé Martin 283

1.1	Earth Internal Structure 283
	1.1.1 Inner Core (from 6378 to 5155 km Depth) 283
	1.1.2 Outer Core (from 5155 to 2891 km Depth) 284
	1.1.3 Lower Mantle (from 2891 to 670 km Depth) 284
	1.1.4 Upper Mantle (from 670 km to 7 km Depth Under Oceans and 30 km Depth Under Continents) 285
	1.1.5 Crusts (from 7 km Depth Under Oceans and 30 km Depth Under Continents to Surface) 285
	1.1.6 Hydrosphere 286
	1.1.7 Atmosphere 286

	1.1.8 Lithosphere and Asthenosphere 286
1.2	Plate Tectonics 287
	1.2.1 Plates on the Surface of the Earth 287
	1.2.2 Margin Definitions 287
	1.2.3 Divergent Margin 288
	1.2.4 Convergent Margin 291
	1.2.5 Hot Spots 293
	1.2.6 Wilson Cycle 294
	1.2.7 Energy for Plate Tectonics 295
References 295

2 Useful Astrobiological Data 297
2.1 Physical and Chemical Data 297
2.2 Astrophysical Data 304
2.3 Geological Data 309
2.4 Biochemical Data 319

3 Glossary 325

4 Authors 375

Index 381

General Introduction

From the Origin of Life on Earth to Life in the Universe

André Brack

On Earth, life probably emerged in water, about 4 billion years ago, with the first chemical systems capable of self-reproduction and also capable of evolution. What did these primitive chemical systems look like? What materials did they use? Is it reasonable to expect similar systems to emerge on other celestial bodies? Recent joint efforts in astronomy, astrophysics, planetology, geochemistry, geology, biology and chemistry bring some preliminary answers to questions that have puzzled humans since Antiquity. Research is presently pursuing three main avenues: (i) the search for fossil traces of life in Archeaen sediments and for other imprints of life, (ii) the reconstitution, in a test tube, of artificial life capable of self-reproduction and evolution, (iii) the search for another example of natural life beyond the Earth.

1 The Search for Traces of Primitive Life and Other Imprints

1.1 Microfossils

The oldest fossils of micro-organisms have been found in greenstone belts in the Pilbara, Australia and in Barberton, South Africa (Schopf, 1993; Westall et al., 2001; see also Westall's Chap. 6, Part I). They are 3.4–3.5 billion years old. From a theoretical point of view, the earliest forms of life would most likely have had a chemolithoautotrophic metabolism, using inorganic materials as a source of both C and energy. Evidence for the presence of chemolithotrophs in the hydrothermal deposits from Barberton and the Pilbara comes from the N and C isotopic data (Beaumont and Robert, 1999). Considering the high temperatures on the early Earth, it is probable that the earliest micro-organisms were thermophilic. Structures resembling oxygenic photosynthetic micro-organisms such as cyanobacteria have been described and carbon isotope data have been interpreted as evidence for their presence (Schopf, 1993). However, these interpretations are strongly disputed (Brasier et al., 2002). Nevertheless, the fact that the micro-organisms inhabiting the Early Archaean environments of Barberton and the Pilbara formed mats in the photic zone suggests that some of them may already have developed anoxygenic photosynthesis (Westall et al., 2001).

1.2 Oldest Sedimentary Rocks

The oldest sedimentary rocks have been collected in southwest Greenland. The Isua sediments are 3.8 Ga old (Schidlowski, 1988), whereas the Akilia sediments formed 3.85 Ga ago (Mozsis et al., 1996). They testify to the presence of permanent liquid water on the surface of the Earth and to the presence of carbon dioxide in the atmosphere. They contain complex organic molecules. The isotopic signatures of the organic carbon found in Greenland metasediments provide indirect evidence that life may be 3.85 billion years old. Taking the age of the Earth as 4.5 billion years, this means that life must have began quite early in Earth's history. This isotopic evidence stems from the fact that the carbon atom has two stable isotopes, ^{12}C and ^{13}C. The ^{12}C/^{13}C ratio in abiotic mineral compounds is 89. In biological material, the process of photosynthesis gives a preference to the lighter carbon isotope and raises the ratio to about 92. Consequently, the carbon residues of previously living matter may be identified by this enrichment in ^{12}C. A compilation has been made of the carbon isotopic composition of over 1600 samples of fossil kerogen (a complex organic macromolecule produced from the debris of biological matter) and compared with that from carbonates in the same sedimentary rocks (Schidlowski, 1988). The enrichment in ^{12}C carbon encompasses specimens right across the geological time scale, including the Isua and Akilia rocks. This offset is now taken to be one of the most powerful indications that life on Earth was active nearly 3.9 billion years ago (Schidlowski, 1988; Mojzsis et al., 1996; Rosing, 1999). Although there are serious reservations concerning the first two studies (contamination by more recent fossilised endoliths, ^{12}C enrichment produced by thermal decomposition of siderite and metamorphic processes), the biogenicity of the ^{12}C enrichment measured by Rosing has been verified by stepped combustion (van Zuilen et al., 2002).

1.3 One-handedness of Life

One-handedness, also called homochirality (from the Greek *kheiros*, the hand; see Cronin and Reisse's Chap. 3, Part II), of the proteins synthesised via the genetic code is another characteristic of all living systems. Each carbon atom of the amino-acid skeleton occupies the centre of a tetrahedron. Except for the amino acid glycine, the four substituents of the central carbon atom are different. The carbon atom is therefore asymmetrical and it is not superimposable to its mirror image. Each amino acid exists in two mirror images called enantiomers (from the Greek *enantios*, opposed). All protein amino acids have the same handedness, they are homochiral. They are left handed and hence known as L-amino acids. Their right-handed mirror images are known as D-amino acids. L-amino acids engaged in a protein chain generate right-handed single-strand α-helices and asymmetrical multistrand β-sheet structures. Nucleotides, the building blocks of nucleic acids DNA and RNA, are also homochiral. Their nomenclature is more complex because each nucleotide possesses four asymmetrical carbon atoms. In this case, the geometry of a selected carbon atom was chosen. Following this mere convention, nucleotides are thus right handed. Right-handed nucleotides generally generate right-handed helical nucleic acids.

Is homochirality a mandatory requisite for life? Pasteur was probably the first person to realize that biological asymmetry could best distinguish between inanimate matter and life. Life that would simultaneously use both right- and left-handed forms of the same biological molecules appears, in the first place, very unlikely for geometrical reasons. Enzyme β-pleated sheets cannot form when both L- and D-amino acids are present in the same chain. Since the catalytic activity of an enzyme is intimately dependent upon the geometry of the chain, the absence of β-pleated sheets would impede, or at least considerably reduce, the activity spectrum of the enzymes. The use of one-handed biomonomers also sharpens the sequence information of the biopolymers. For a polymer made of n units, the number of sequence combinations will be divided by 2^n when the system uses only homochiral (one-handed) monomers. Taking into account the fact that enzyme chains are generally made up of hundreds of monomers, and that nucleic acids contain several million nucleotides, the tremendous gain in simplicity offered by the use of monomers restricted to one handedness is self-evident. Finally, if the biopolymers to be replicated were to contain L- and D-units located at specific sites, the replication process would have to be able to position the right monomers at the right place, but also to select the right enantiomer from among two species that differ only by the geometry of the asymmetrical carbon atoms. For example, the bacteria *Bacillus brevis* are able to synthesise *gramicidine A* constructed on a strict alternation of left- and right-handed amino acids. However, the biosynthesis of this peptide involves a set of complex and sophisticated enzymes that are homochiral. Life on Earth uses homochiral left-handed amino acids and right-handed sugars. A mirror-image life, using right-handed amino acids and left-handed sugars, is perfectly conceivable and might develop on another planet. Thus, homochirality can be a crucial signature for life.

Unfortunately, the direct clues that could help chemists to identify the molecules that participated in the actual emergence of life on Earth about 4 billion years ago have been erased by geological processes, such as plate tectonics and the permanent presence of running water (erosion), as well as by the unshielded solar ultraviolet radiation and by the oxygen produced by life.

2 Reconstructing Life in a Test Tube

Life emerged in water and the first self-reproducing molecules and their precursors were probably organic molecules built up with carbon atom skeletons. Liquid water participated in the early mechanisms of life as a solvent but also by exchanging hydrogen bonds with the organic molecules. Organic molecules were formed from gaseous molecules containing carbon atoms (methane, carbon dioxide, carbon monoxide), nitrogen atoms (nitrogen, ammonia), sulfur atoms (hydrogen sulfide, sulfur dioxide) and energy (electric discharges, cosmic and UV radiation, heat).

2.1 Primitive Earth Atmosphere

In 1924, the Russian biochemist Aleksander Oparin suggested that the small reduced organic molecules needed for primitive life were formed in a primitive atmosphere dominated by methane. The idea was tested in the laboratory by Stanley Miller (Miller, 1953). He exposed a mixture of methane, ammonia, hydrogen and water to electric discharges, to mimic the effects of lightning. Among the compounds formed, he identified four of the twenty naturally occurring amino acids, the building blocks of proteins. Since this historic experiment, seventeen natural amino acids have been obtained via the intermediate formation of simple precursors, such as hydrogen cyanide and formaldehyde. It has been shown that spark-discharge synthesis of amino acids occurs efficiently when a reducing-gas mixture containing significant amounts of hydrogen is used. However, the true composition of the primitive Earth atmosphere remains unknown. Today, geochemists favour a nonreducing atmosphere dominated by carbon dioxide. Under such conditions, the production of amino acids appears to be very limited. Strongly reducing environments capable of reducing carbon dioxide were therefore necessary.

2.2 Organic Synthesis

Deep-sea hydrothermal systems may represent likely reducing environments for the synthesis of prebiotic organic molecules (Baross and Hoffman, 1985; Holm and Andersson, 1995; 1998). The ejected gases contain carbon dioxide, carbon monoxide, sulfur dioxide, nitrogen, and hydrogen sulfide. For instance, high concentrations of hydrogen (more than 40% of the total gas) and methane have been detected in the fluids of the Rainbow ultramafic hydrothermal system of the Mid-Atlantic Ridge (Charlou et al., 1998). The production of hydrogen, a highly reducing agent favouring prebiotic syntheses, is associated with the hydration of olivine into serpentine and magnetite, a reaction known as "serpentinisation". Indeed, hydrocarbons containing 16 to 29 carbon atoms have been detected in these hydrothermal fluids (Holm and Charlou, 2001). Amino acids have been obtained, although in low yields, under conditions simulating these hydrothermal vents (Yanagawa and Kobayashi, 1992). According to Wächtershäuser (1998), primordial organic molecules formed near the hydrothermal systems; the energy source required to reduce the carbon dioxide might have been provided by the oxidative formation of pyrite from iron sulfide and hydrogen sulfide. Pyrite has positive surface charges and bonds the products of carbon dioxide reduction, giving rise to a two-dimensional reaction system, a "surface metabolism" (Wächtershäuser, 1998). Laboratory work has provided support for this new promising hypothesis (Huber and Wächtershäuser, 1997). Iron sulfide, hydrogen sulfide and carbon dioxide react under anaerobic conditions to produce hydrogen and a series of thiols, including methanethiol. Methanethiol and acetic acid have also been obtained from carbon monoxide, hydrogen sulfide, iron and nickel sulfides and catalytic amounts of selenium. Under specific conditions, thioesesters are formed that might have been the metabolic driving force of a thioester world, according to de Duve (de Duve, 1998). Hydrothermal vents are often disqualified (Miller and Bada, 1988) as efficient reactors for the synthesis of bio-organic molecules, because of the high temperature. However, the products that are synthesized in hot vents are rapidly quenched in the surrounding cold water, which may preserve those organics formed. When fluid containing glycine and copper chloride repeatedly circulated through the hot (225°C) and cold (0°C) chambers in a laboratory reactor that simulated a hydrothermal system, glycine peptides up to hexaglycine were obtained (Imai et al., 1999).

2.3 Delivery of Organics by Comets and Meteorites

Comets and meteorites may have delivered important amounts of organic molecules to the primitive Earth (see Despois and Cottin's Chap. 8, Part I). Nucleic acid bases, purines and pyrimidines, have been found in the Murchison meteorite (Stoks and Schwartz, 1982). One sugar (dihydroxyacetone), sugar-alcohols (erythritol, ribitol) and sugar-acids (ribonic acid, gluconic acid) have been detected in the Murchison meteorite but no ribose, the sugar moiety of ribonucleotides, themselves building blocks in RNAs (Cooper et al., 2001). Eight proteinaceous amino acids have been identified in one such meteorite, among more than 70 amino acids found therein. Cronin and Pizzarello found an excess of about 9% of L-enantiomers for some nonprotein amino acids detected in the Murchison meteorite (Cronin and Pizzarello, 1997). The presence of L-enantiomeric excesses in these meteorites points to an extraterrestrial process of asymmetric synthesis of amino acids, asymmetry that is preserved inside the meteorite. These excesses may help to understand the emergence of the biological asymmetry or one-handedness (see Cronin and Reisse's Chap. 3, Part II). The excess of the one-handed amino acids, as found in the meteorites, may result from the processing of the organic mantles of the interstellar grains from which the meteorite was originally formed. This processing could occur, for example, by the effects of circularly polarized synchrotron radiation from a neutron star, a remnant of a supernova. Strong infrared circular polarization, resulting from dust scattering in reflection nebulae in the Orion OMC-1 star-formation region, has also been observed (Bailey et al., 1998). Circular polarization at shorter wavelengths might have been important in inducing this chiral asymmetry in interstellar organic molecules that could be subsequently delivered to the early Earth (Bailey, 2001). Dust collection in the Greenland and Antarctic ice sheets and its analysis by Maurette (Maurette, 1998) show that the Earth captures interplanetary dust as micrometeorites at a rate of about 50–100tons per day. About 99% of this mass is carried by micrometeorites in the 50–500 micrometre size range. This value is much higher than the most reliable estimate of the normal meteorite flux, which is about 0.03tons per day. A high percentage of micrometeorites in the 50- to 100-micrometre size range has been observed to be unmelted, indicating that a large fraction traversed the terrestrial atmosphere without drastic heating. In this size range, the carbonaceous micrometeorites represent 80% of the samples and contain 2% of carbon, on average. This flux of incoming micrometeorites might have brought to the Earth about 10^{23} g of carbon over the period corresponding to the late heavy bombardment, when planetesimals, asteroids and comets impacted the Earth until –3.8 Ga, as inferred from the lunar craters. For comparison, this delivery represents more carbon than that engaged in the present surficial biomass, i.e. about 10^{18} g. These grains also contain a high proportion of metallic sulfides, oxides, and clay minerals that belong to various classes of catalysts. In addition to the carbonaceous matter, micrometeorites might also have delivered a rich variety of catalysts. They may have functioned as tiny chemical reactors when reaching oceanic water.

Micrometeorites, with a composition close to that of the most primitive meteorites, represent very primitive objects of the Solar System. The formation of some of their organic molecules could have occurred early in the protosolar nebulae. These molecules would have escaped the accretion phase within comets in the outer Solar System. Micrometeorites are probably witnesses of a chemical continuum, via the cometary grains, from the interstellar medium, where they form, to terrestrial oceans. Comets are the richest planetary objects in organic compounds known so far. Ground-based observations have detected hydrogen cyanide and formaldehyde in the coma of comets. In 1986, onboard analyses performed by the two Russian missions Vega 1 and 2, as well as observations obtained by the European mission Giotto and the two Japanese missions Suisei and Sakigake, demonstrated that the Halley comet shows substantial amounts of organic material. On average, dust particles ejected from the Comet Halley nucleus contain 14% of organic carbon by mass. About 30% of cometary grains are dominated by light elements C, H, O, and N, and 35% are close in composition to the carbon-rich meteorites. The presence of organic molecules, such as purines, pyrimidines, and formaldehyde polymers has also been inferred from the fragments analyzed by the Giotto Picca and Vega Puma mass spectrometers. However, there was no direct identification of the complex organic molecules probably present in the cosmic dust grains and in the cometary nucleus. Many chemical species of interest for exobiology have been detected in Hyakutake comet in 1996, including ammonia, methane, acetylene (ethyne), acetonitrile (methyl cyanide) and hydrogen isocyanide. In addition, the study of the Hale-Bopp comet in 1997 led to the detection of methane, acetylene, formic acid, acetonitrile, hydrogen isocyanide, isocyanic acid, cyanoacetylene, formamide and thioformaldehyde. It is possible, therefore, that cometary grains might have been an important source of organic molecules delivered to the primitive Earth. Comets orbit on unstable trajectories and sometimes collide with planets. The collision of Comet Shoemaker-Levy 9 with Jupiter in July 1994 gave a recent example of such events. Such collisions were probably more frequent 4 billion years ago, the comets orbiting around the Sun being more numerous. By impacting the Earth, comets delivered probably a substantial fraction of the terrestrial water (about 35% according to the estimation of Tobias Owen based on the relative contents in hydrogen and deuterium and cometary grains could have provided large amounts of organic molecules. The chemistry that is active at the surface of a comet is still poorly understood. Launched in March 2004, the cometary mission Rosetta will analyse the nucleus of a comet. The spacecraft will first study the environment of the comet during a flyby over several months and a probe will land to analyse the surface and the subsurface ice sampled by drilling.

2.4 Simulation Experiments

Ultraviolet irradiation of dust grains in the interstellar medium may result in the formation of complex organic molecules. The interstellar dust particles are assumed to be composed of silicate grains surrounded by ices of different molecules, including carbon-containing molecules. Ices of H_2O, CO_2, CO, CH_3OH, NH_3 were deposited at 12 K under a pressure of 10^{-7} mbar and irradiated with electromagnetic radiation representative of the interstellar medium. The solid layer that developed on the solid surface was analysed by enantioselective gas chromatography and mass spectrometry GC-MS. After the analytical steps of extraction, hydrolysis, and derivatization, 16 amino acids were identified in the simulated ice mantle of interstellar dust particles (Muñoz Caro et al., 2002). These amino-acid identifications confirmed the preliminary amino-acid formation obtained by Mayo Greenberg (Briggs et al., 1992). The chiral amino acids were identified as being totally racemic. Parallel experiments performed with ^{13}C-containing substitutes definitely exclude contamination by biological amino acids. The results strongly suggest that amino acids are readily formed in interstellar space.

Before reaching the Earth, organic molecules are exposed to UV radiation, both in interstellar space and in the Solar System. Amino acids have been exposed in Earth orbit to study their survival in space. The UV flux ($\lambda < 206$ nm) in the diffuse interstellar medium is about 10^8 photons cm^{-2} s^{-1}. In Earth orbit, the corresponding solar flux is in the range of 10^{16} photons cm^{-2} s^{-1}. This means that an irradiation over one week corresponds to 275 000 years in the interstellar medium. Amino acids, like those detected in the Murchison meteorite, have been exposed as free molecules and also as adsorbed molecules on clay minerals to space conditions in Earth orbit onboard the unmanned Russian satellites FOTON 8 and 11. Aspartic acid and glutamic acid not adsorbed on clay minerals were partially destroyed during exposure to solar UV. However, decomposition was prevented when the amino acids were embedded in clays (Barbier et al, 1998; Barbier et al., 2002). Amino acids and peptides have also been subjected to solar radiation outside the MIR station for 97 days as solid films as well as embedded into ground mineral material (montmorillonite clay, basalt and Allende meteorite). Different thicknesses of meteorite powder films were used to estimate the shielding threshold. After three months of exposure, about 50% of the amino acids were destroyed in the absence of mineral shielding. Peptides exhibited a noticeable sensitivity to space vacuum and sublimation effects were detected. Decarboxylation was found to be the main effect of photolysis. No polymerization occurred and no racemization (the conversion of L- or

D-amino acids into a racemic mixture) was observed. Among the different minerals used as 5-μm films, meteoritic powder offered the best protection whereas montmorillonite was least efficient. Significant protection from solar radiation was observed when the thickness of the meteorite mineral was 5 μm or greater.

2.5 Recreating the Chemistry of Primitive Life

By analogy with contemporary living systems, it is tempting to consider that primitive life emerged as a cellular object, requiring boundary molecules able to isolate the system from the aqueous environment (membrane). Also needed would be catalytic molecules to provide the basic chemistry of the cell (enzymes) and information-retaining molecules that allow the storage and the transfer of the information needed for replication (RNA). Starting from the small organic molecules possibly present in the primitive oceans, chemists have tried to reproduce the spontaneous formation of the three families of basic biological molecules in a test tube. They succeeded in reconstructing precursors of membranes (Deamer, 1998; see also Ourisson and Nakatani's Chap. 1, Part II) and of protein enzymes (Brack, 1993). Unfortunately, they have not yet been able to reconstruct precursors of RNA. Synthesis of sugars from formaldehyde leads to a highly complex mixture of compounds, among which ribose represents only a very tiny fraction. Prebiotic synthesis of short strands of RNA faces two main difficulties: synthesis of the first strand of RNA (especially ribose) and replication of small strands fed with racemic nucleotides. RNA analogs and surrogates have been studied. Considering the easy formation of hexose-2,4,6-triphosphates from glycolaldehyde phosphate, $CHO-CH_2O-PO_3H_2$, in a process analogous to the formose reaction, Eschenmoser and coworkers (Eschenmoser et al., 1999) synthesized polynucleotides containing ribose in the form of pyranose (pyranosyl-RNA or p-RNA) in place of the usual "natural" ribose, the furanose form found in RNA. p-RNAs form Watson–Crick-paired double helices that are more stable than RNA. Replication experiments have had marked success in terms of sequence copying but have failed to demonstrate template-catalysis turnover numbers greater than one. p-RNA seems to be an excellent choice as a genetic system if it can be demonstrated that the prebiotic synthesis of pyranosyl nucleotides is much easier than the synthesis of standard isomers.

Chemists are now tempted to consider that primitive replicating systems must have used simpler information-retaining molecules than biological nucleic acids or their analogs. They are looking for simple self-sustaining chemical systems capable of self-replication, mutation, and selection. It has been shown that simple molecules, unrelated to nucleotides, can actually provide exponentially replicating autocatalytic models. Beautiful examples of autocatalytic micelle growth have been published (Bachmann et al., 1992). However, these autocatalytic systems do not really store hereditary information and cannot therefore evolve by natural selection. These autocatalytic primitive living entities must have been simple and robust enough to survive the heavy bombardment and the nonshielded solar UV. Some investigations are also focusing on autocatalytic systems adsorbed on mineral surfaces (Luther et al. 1998; Orgel, 1998).

3 Search for Extraterrestrial Life

3.1 The Diversity of Bacterial Life as a Reference for Extraterrestrial Life

Life on Earth is based upon the chemistry of carbon in water. The temperature limits compatible with the existence of life are thus imposed by the intrinsic properties of the chemical bonds involved in this type of chemistry at different temperatures. Presently, the maximum temperature limit known for terrestrial organisms is around 113°C for the deep-sea microbe *Pyrolobus fumarii* (Stetter, 1998). Important factors preventing life at temperatures well above 110°C are the thermal instability of supramolecular systems involved in biological systems and the temperature dependence of membrane permeability. In some theoretical scenarios, life appeared at very high temperatures. This means that today's hyperthermophiles might be viewed as relics of the last common universal ancestor of all living beings. However, this hot origin of life hypothesis has been seriously disputed, based on the fact that RNA is very unstable at high temperatures (see Gribaldo and Forterre's Chap. 7, Part II). The most attractive hypothesis might be that life appeared in a moderately thermophilic environment, hot enough to boost catalytic reactions, but cold enough to avoid the problem of macromolecule thermal degradation (see Lopez-Garcia's Chap. 9, Part II). This hypothesis seems to be supported by gene-sequencing studies that suggest that hyperthermophiles appeared later than thermophiles (Galtier et al., 1999; Brochier and Philippe, 2002). Life is extremely diverse in the ocean at temperatures of 2°C. Living organisms, especially micro-organisms, are also present in the frozen soils of Arctic and alpine environments. Antarctica has a wide range of extreme habitats, including microbial ecosystems developing in dry valley rocks. The lower limit for bacterial growth published in the literature is –12°C, the temperature at which intracellular ice is formed.

Salt-loving organisms, known as halophiles, have been well studied. They tolerate a wide range of salt concentrations (1–20% NaCl) and some procaryotes, the extreme halophiles, have managed to thrive in hypersaline biotopes (salines, salted lakes). They are, in fact, so dependent on such high salt concentrations that they cannot grow (and may even die) at concentrations below 10% NaCl.

The chemistry of life on Earth is optimised for neutral pH. Again, some micro-organisms have been able to adapt to extreme pH conditions, from pH 0 (extremely acidic) to pH 12.5 (extremely alkaline), albeit maintaining their intracellular pH between pH 4 and 9. As with temperature, the intracellular machinery cannot escape the influence of pressure. However, there are organisms in the deepest parts of the ocean (pressure 1100 bar). The extreme pressure limit for life on Earth is unknown – environments of above 1100 bar have not been explored.

For a long time, it was believed that deep subterranean environments were sterile. An important recent development has been the recognition that bacteria actually thrive in the terrestrial crust. Subterranean micro-organisms are usually detected in subsurface oil-fields or in the course of drilling experiments. For example, recent research has demonstrated that microbes are present much deeper in marine sediments than was previously thought possible, extending to at least 750 m below the sea floor, and probably much deeper (Parkes et al., 2000). To depths of at least 432 m, microbes have been identified as altering volcanic glass. These data provide a preliminary, and probably conservative, estimate of the biomass in this important new ecosystem. It could amount to about 10% of the surface biosphere.

These discoveries have radically changed the perception of marine sediments and indicate the presence of a largely unexplored deep bacterial biosphere that may even rival the Earth's surface biosphere in size and diversity. Clearly, this discovery also has important implications for the probability of life existing in other planets of the Solar System and elsewhere.

3.2 The Search for Life in the Solar System

3.2.1 Life on Mars and the SNC Meteorites

Mapping of Mars by Mariner 9, Viking 1 and 2 and Mars Global Surveyor revealed channels resembling dry river beds. Odyssey's gamma-ray spectrometer instrument detected hydrogen, which indicated the presence of water ice in the upper metre of soil, in a large region surrounding the planet's south pole, where ice is expected to be stable (Boynton et al., 2002). The amount of hydrogen detected indicates 20 to 50 per cent ice by mass in the lower layer beneath the topmost surface. The ice-rich layer is about 60 cm beneath the surface at latitude 60°S, and approaches 30 cm of the surface at latitude 75°S. The inventory of the total amount of water that may have existed at the surface of Mars is difficult to estimate and varies from a total water depth of some metres to several hundred metres. It is generally considered that liquid water has been restricted to the very early stages of Martian history (see Bribing's Chap. 9, Part I). Mars possessed therefore an atmosphere capable of decelerating carbonaceous micrometeorites and chemical evolution may have been possible on the planet. The Viking 1 and 2 lander missions were designed to address the question of extant (rather than extinct) life on Mars. Three experiments were selected to detect metabolic activity such as photosynthesis, nutrition and respiration of potential microbial soil communities. Unfortunately, the results were ambiguous since, although "positive" results were obtained, no organic carbon was found in the Martian soil by gas chromatography-mass spectrometry. It was concluded that the most plausible explanation for these results was the presence at the Martian surface of highly reactive oxidants, like hydrogen peroxide, which would have been photochemically produced in the atmosphere. The Viking lander could not sample soils below 6 cm and therefore the depth of this apparently organic-free and oxidizing layer is unknown. Direct photolytic processes can also be responsible for the degradation of organics at the Martian surface. Although the Viking missions were disappointing for exobiology, in the long run the programme has proved to be extremely beneficial for investigating the possibility of life on Mars.

Prior to Viking, it had been apparent that there was a small group of meteorites, all of igneous (volcanic) origin, called the SNC meteorites (after their type specimens Shergotty, Nakhla and Chassigny) that had comparatively young crystallisation ages, equal to or less than 1.3 billion years. One of these meteorites, designated EETA 79001, was found in Antarctica in 1979. It had gas inclusions trapped within a glassy component. Both compositionally and isotopically, this gas matched, in all respects, the make up of the Martian atmosphere, as measured by the Viking mass spectrometer. The data provide a very strong argument that at least that particular SNC meteorite came from Mars, representing the product of a high-energy impact that ejected material into space. There are now thirty one SNC meteorites known in total (Jet Propulsion Laboratory, 2003). Two of the SNC meteorites, EETA79001 and ALH84001, supply new and highly interesting information (Brack and Pillinger, 1998). A subsample of EETA79001, excavated from deep within the meteorite, has been subjected to stepped-combustion. The CO_2 release from 200 to 400°C suggested the presence of organic molecules. The carbon is enriched in ^{12}C, and the carbon isotope difference between the organic matter and the carbonates in Martian meteorites is greater than that observed on Earth. This could be indicative of biosynthesis, although some as yet unknown abiotic processes could perhaps explain this enrichment. McKay and co-workers (McKay et al., 1996) have reported the presence of other features in ALH84001 that may represent a signature of relic biogenic activity on Mars, but this biological interpretation is strongly questioned (Bradley et al., 1997; Westall et al., 1998).

Because Mars had a "warm" and wet past climate, it should have sedimentary rocks deposited by running and/or still water on its surface as well as layers of regolith generated by impacts. Such consolidated sedimentary hard rocks ought therefore to be found amongst the Martian meteorites. However, no such sedimentary material has been found in any SNC meteorite. It is possible that they did survive the effects of the escape acceleration from the Martian surface but did not survive terrestrial atmospheric entry because of decrepitation of the cementing mineral. The 'STONE' experiment, flown by ESA, was designed to study precisely such physical and chemical modifications to sedimentary rocks during atmospheric entry from space. A piece of basalt (representing a standard meteoritic material), a piece of dolomite (sedimentary rock) and an artificial Martian regolith material (80% crushed basalt and 20% gypsum) were embedded into the ablative heat shield of Foton 12, which was launched on September 9 and landed on September 24, 1999. Such an experiment had never been performed before and the samples, after their return, were analysed for their chemistry, mineralogy and isotopic compositions by a European consortium. Atmospheric infall modifications are made visible by reference to the untreated samples. The results suggest that some Martian sediments could, in part, survive terrestrial atmospheric entry from space (Brack et al., 2002).

Even if convincing evidence for ancient life in ALH84001 has not been established, the two SNC meteorites (EETA79001 and ALH84001) do show the presence of organic molecules. This suggests that the ingredients required for the emergence of a primitive life were present on the surface of Mars. Therefore, it is tempting to consider that micro-organisms may have developed on Mars and lived at the surface until liquid water disappeared. Since Mars probably had no plate tectonics and since liquid water seems to have disappeared from the surface of Mars very early, the Martian subsurface perhaps keeps a frozen record of very early forms of life. NASA has planned a very intensive exploration of Mars and European and Japanese missions are also taking place. Exobiology interests are included, especially in the analysis of samples from sites where the environmental conditions may have been favourable for the preservation of evidence of possible prebiotic or biotic processes. The ESA Manned Spaceflight and Microgravity Directorate has convened a Exobiology Science Team to design a multiuser, integrated suite of instruments for the search for evidence of life on Mars. The priority has been given to the in situ organic and isotopic analysis of samples obtained by subsurface drilling (Brack et al., 1999). A first exobiology lander, called Beagle 2, was launched in June 2003, as part of the ESA Mars Express mission. It successfully separated from the orbiter but all altempts to contact it were unsuccessful, thus compelling the scientists to consider Beagle 2 as lost.

3.2.2 Europa

Europa appears as one of the most enigmatic of the Galilean satellites. With a mean density of about 3.0g cm^{-3}, the Jovian satellite should be dominated by rocks. Ground-based spectroscopy, combined with gravity data, suggests that the satellite has an icy crust kilometres thick and a rocky interior. The Voyager images showed very few impact craters on Europa's surface, indicating recent, and probably continuing, resurfacing by cryovolcanic and tectonic processes. Images of Europa's surface taken by the Galileo spacecraft show surface features – iceberg-like rafted blocks, cracks, ridges and dark bands – which are consistent with the presence of liquid water beneath the icy crust (Carr et al., 1998). Data from Galileo's near-infrared mapping spectrometer show hydrated salts that could be evaporites. The most convincing argument for the presence of an ocean of liquid water comes from Galileo's magnetometer. The instrument detected an induced magnetic field within Jupiter's strong magnetic field. The strength and response of the induced field require a near-surface, global conducting layer, most likely a layer of salty water (Kivelson et al., 2000; Zimmer et al., 2000).

If liquid water is present within Europa, it is quite possible that it includes organic matter derived from thermal vents. Terrestrial-like prebiotic organic chemistry and primitive life may therefore have developed in Europa's ocean.

If Europa maintained tidal and/or hydrothermal activity in its subsurface until now, it is possible that bacterial activity is still present. Thus, the possibility of an extraterrestrial life present in a subsurface ocean of Europa must be seriously considered. The most likely sites for extant life would be at hydrothermal vents below the most recently resurfaced area. To study this directly would require making a borehole through the ice in order to deploy a robotic submersible. On the other hand, biological processes in and around hydrothermal vents could produce biomarkers that would be erupted as traces in cryovolcanic eruptions and thereby be available at the surface for in situ analysis or sample return. Mineral nutrients delivered through cryovolcanic eruption would make the same locations the best candidates for autotrophic life.

3.2.3 Titan

Titan's atmosphere (see Raulin et al. Chap. 2, Part II) was revealed mainly by the Voyager 1 mission in 1980, which yielded the bulk composition (90% molecular nitrogen and about 1–8% methane). Also, a great number of trace constituents were observed in the form of hydrocarbons, nitriles and oxygen-containing compounds, mostly CO and CO_2. Titan is the only other object in our Solar System to bear any resemblance to our own planet in terms of atmospheric pressure (1.5 bar) and carbon/nitrogen chemistry. It represents, therefore, a natural laboratory to study the formation of complex organic molecules on a planetary scale and over geological times. The ISO satellite has detected tiny amounts of water vapour in the higher atmosphere, but Titan's surface temperature (94 K) is much too low to allow the presence of liquid water. Although liquid water is totally absent, the satellite provides a unique milieu to study, in situ, the products of the fundamental physical and chemical interactions driving a planetary organic chemistry. Titan also serves as a reference laboratory to study, by default, the role of liquid water in exobiology.

The NASA/ESA Cassini-Huygens spacecraft launched in October 1997 arrived in the vicinity of Saturn in 2004 and will perform several flybys of Titan, making spectroscopic, imaging, radar and other measurements. A descent probe managed by European scientists will penetrate the atmosphere and will systematically study the organic chemistry in Titan's geofluid. During 150 min, in situ measurements will provide detailed analysis of the organics present in the air, in the aerosols and at the surface (Lebreton, 1997; Raulin, 1998).

3.3 The Search for Life Beyond the Solar System

3.3.1 What Are We Looking for?

Apart from abundant hydrogen and helium, 114 interstellar and circumstellar gaseous molecules are currently identified in the interstellar medium (see Despois and Cotin's Chap. 8, Part I). It is commonly agreed that the catalog of interstellar molecules represents only a fraction of the total spectrum of molecules present in space, the spectral detection being biased by the fact that only those molecules possessing a strong electric dipole can be observed. Among these molecules, 83 contain carbon whereas only 7 contain silicon. Silicon has been proposed as an alternative to carbon as the basis of life. However, silicon chemistry is apparently less inventive and does not seem to be able to generate any life as sophisticated as the terrestrial carbon-based one. Can these molecules survive the violent accretion phase generating a stellar system? The origin and distribution of the molecules from the interstellar medium to the planets, asteroids and comets of the Solar System is presently at the centre of a debate based on isotope ratios. Some molecules could have survived in cold regions of the outer Solar System, whereas others would have been totally reprocessed during accretion. Whatever the case, the interstellar medium tells us that organic chemistry is universal. What about liquid water? New planets have been discovered beyond the Solar System. On October 6, 1995, the discovery of an extrasolar planet was announced. The planet orbits an eight-billion-year-old star called 51 Pegasus, forty-two light years away within the Milky Way (Mayor and Queloz, 1995). The suspected planet takes just four days to orbit 51 Pegasus. It has a surface temperature around 1000°C and a mass about 0.5 the mass of Jupiter. One year later, seven other extrasolar planets were identified. Among them, 47 Ursa Major has a planet with a surface temperature estimated to be around that of Mars −90 to −20°C) and the 70 Virginis planet has a surface temperature estimated of about 70–160°C. The latter is the first known extrasolar planet whose temperature might allow the presence of liquid water. As of July 2004, 123 exoplanets have been observed (Schneider, 2003).

3.3.2 How to Detect Extraterrestrial Life?

Isotope enrichment of ^{12}C and one-handedness are the two most remarkable signatures of biological terrestrial molecules. With the development of exploration planetary missions, almost any body within the Solar System can be subjected to in situ organic, isotope and mineral analysis. Rocks, in particular, may represent telltale biomarkers of biological activity on both macroscopic (stromatolites) and microscopic (microfossils of bacteria) levels.

Identifying extrasolar life will be more difficult. Extrasolar life will not be accessible to space missions in the foreseeable future. The formidable challenge to detect distant life must therefore be tackled by astronomers and radioastronomers. The detection of water and ozone (an easily detectable telltale signature of oxygen) in the atmosphere will be a strong indication but not an absolute proof (see Selsis et al.'s Chap. 10, Part I). Other anomalies in the atmospheres of telluric exoplanets (rocky Earth-like planets), such as the presence of methane, could be the signature of an extrasolar life. European astrophysicists are proposing the construction of an infrared interferometer made of five space telescopes to study the atmospheres of exoplanets. The mission, called Darwin-IRSI, is presently under study at ESA.

Finally, the detection of an unambiguous electromagnetic signal (via the SETI program) would obviously be an exciting event (SETI, http://www.seti-inst.edu/).

3.4 Panspermia, Interplanetary Transfer of Life

Although it will be difficult to prove that life can be transported through the Solar System, the chances for the different steps of the process to occur can be estimated. These include (1) the escape process, i.e. the removal to space of biological material that has survived being lifted from the surface to high altitudes; (2) the interim state in space, i.e. the survival of the biological material over timescales comparable with the interplanetary passage (3) the entry process, i.e. the nondestructive deposition of the biological material on another planet. The identification of some meteorites as being of lunar origin and some others as most probably being of Martian origin, shows that the escape from a planet of material ranging from small particles up to boulder-size, after the planet has suffered a high-energy impact, is evidently a feasible process. In that context, it is also interesting to note that bacterial spores can survive huge accelerations and shockwaves produced by a simulated meteorite impacts (Horneck et al., 2001).

In order to study the survival of resistant microbial forms in the upper atmosphere and free space, microbial samples have been exposed in situ aboard balloons, rockets and spacecraft (Horneck, 1999). The ESA Microgravity Programme has continued to support experiments of that type. A priori, the environment in space seems to be very hostile to life. This is due to the high vacuum, an intense radiation of galactic and solar origin, and extreme temperatures. In the endeavour to disentangle the network of potential interactions of the parameters of space, methods have been applied to separate each parameter and to investigate its impact on biological integrity, applied singly or in controlled combinations.

Space vacuum has been considered to be one of the factors that may prevent interplanetary transfer of life because of its extreme dehydrating effects. However, experiments in space have demonstrated that certain micro-organisms survive exposure in space vacuum for extended periods of time, provided they are shielded against the intense deleterious solar UV radiation. Most of the results concern the spores (a dormant form) of the bacterium *Bacillus subtilis*. Up to 70% of the bacterial spores survive short-term (e.g. 10 days) exposure to space vacuum. The chances of survival in space are increased if the spores are embedded in chemical protectants such as sugars, or salt crystals, or if they are exposed in thick layers. For example, 30% of *subtilis* spores survived nearly 6 years in space when embedded in salt crystals and 80% survived in the presence of glucose. Specimens of *Streptococcus mitis* survived two-and-a-half years inside a television camera landed on the Moon by the unmanned lunar lander Surveyor 3 and recovered by the astronauts of Apollo 12.

Solar UV radiation has been found to be the most deleterious factor of space, as tested with dried preparations of viruses, bacterial and fungal spores, with

DNA being the critical UV target for lethality. However, about 5% of a species of the extreme halophile *Haloarcula*, (a salt-loving bacteria), survived a two-week space environment during a Foton spaceflight. The radiation field of the Solar System is governed by components of galactic and solar origin. It is composed of electrons, alpha particles and cosmic heavy ions, the latter being the most ionising and therefore the most deleterious components. The heavy particles of cosmic radiation are conjectured as setting the ultimate limit on the survival of spores in space because they penetrate even heavy shielding. The maximum time for a spore to escape a hit by a heavy particle has been estimated to be 10^5–10^6 years. Considering the mean sizes and numbers of meteorites ejected from Mars and falling on Earth, radiative transfer models for galactic cosmic rays, and laboratory responses of *B. subtilis* spores and *Deinococcus radiodurans* cells to accelerated heavy ions, Mileikowsky et al. (2000) have calculated that transfer of viable microbes from Mars to Earth via impact ejecta is a possible event.

During the major part of a hypothetical journey through deep space, microorganisms are confronted with the 3-K cold emptiness of space. Laboratory experiments under simulated interstellar medium conditions point to a remarkably less damaging effect of UV radiation at these low temperatures. Treating *Bacillus subtilis* spores with three simulated factors simultaneously (UV, vacuum and a low temperature of 3K), produces an unexpectedly high survival rate, even at very high UV fluxes. From these data, it has been estimated that, in the most general environment in space, spores may survive for hundreds of years (Weber and Greenberg, 1985).

ESA has initiated the development of the exposure facility EXPOSE, to be attached to the Columbus module of the International Space Station. This will allow the extensive study of the survival of bacteria in space.

4 Conclusion

Recent data collected from different disciplines have given important coherence to the search for life in the Universe that is summarised in Table 1.

More precisely,

1. Studies of the origin of life and the exploration of Mars tell us that the conditions, which probably allowed the emergence of life on Earth, were also present on Mars, 4 billion years ago.
2. There are perhaps hydrothermal systems producing organic prebiotic molecules in the ocean of Europa.
3. Terrestrial bacteria are capable of surviving extreme conditions (temperature, pH, pressure, high salinity, and deep sediments). They resist space conditions and are therefore capable of interplanetary transfer, thus favouring their dispersion in the Universe.

If life started on Earth with the self-organisation of a relatively small number of molecules, its emergence must have been fast and the chances for the appearance

Table 1. Life in the Universe

	Terrestrial life as reference			Extraterrestrial life	
Relics/imprints of life	Origins of life	Limits of life	Life in the Solar System	Extrasolar life	Panspermia
• fossil bacteria • sediments • homochirality	• primitive oceans • ingredients • life in test tube	• temperature • high salinity • pH • pressure • deep sediments	• life on Mars and the SNC meteorites • life on Europa • Titan	• exoplanets • signatures of extrasolar life	• interplanetary transfer of life

of life on any appropriate celestial bodies are real. On the contrary, if the process required thousands of different molecules, the event risks being unique and restricted to the Earth. In addition to its immense cultural and societal appeal, the discovery of a second genesis of life, artificial or natural, would be strong support for the simplicity of the processes involved in the emergence of life. The exploration of Mars and Europa and the recent discovery of exoplanets, open real hope. Many scientists are convinced that bacterial life is not restricted to the Earth. Now, they have to prove it!

References

Bachmann P.A., Luisi P.L., and Lang J. (1992). Autocatalytic self-replicating micelles as models for prebiotic structures. *Nature*, **357**, 57–59.
Bailey J., Chrysostomou A., Hough J.H., Gledhill T.M., McCall A., Clark S., Ménard F, and Tamura M (1998). Circular polarization in star formation regions: Implications for biomolecular homochirality. *Science*, **281**, 672–674.
Bailey J. (2001). Astronomical sources of circularly polarized light and the origin of homochirality. *Origins Life Evol. Biosphere*, **31**, 167–183.
Barbier B., Chabin A., Chaput D., and Brack A. (1998). Photochemical processing of amino acids in Earth orbit. *Planet. Space Sci.*, **46**, 391–398.
Baross J.A. and Hoffman S.E. (1985). Submarine hydrothermal vents and associated gradient environment as sites for the origin and evolution of life. *Origins Life Evol. Biosphere*, **15**, 327–345.
Beaumont V. and Robert F. (1999). Nitrogen isotope ratios of kerogens in Precambrian cherts: a record of the evolution of atmospheric chemistry? *Precambrian Res.*, **96**, 63–82.
Boillot F., Chabin A., Buré C, Venet M., Belsky A., Bertrand-Urbaniak M., Delmas A., Brack A., and Barbier B. (2002). The Perseus exobiology mission on MIR: behaviour of amino acids and peptides in Earth orbit. *Origins Life Evol. Biosphere*, **32**, 359–385.
Boynton W.V. et al. (2002). Distribution of hydrogen in the near surface of mars: evidence for subsurface ice deposits. *Science*, **297**, 81–85.
Brack A. (1993). From amino acids to prebiotic active peptides: a chemical reconstitution. *Pure & Appl. Chem.*, **65**, 1143–1151.
Brack A., Baglioni P., Borruat G., Brandstätter F., Demets R., Edwards H.G.M., Genge M., Kurat G., Miller M.F., Newton E.M., Pillinger C.T., Roten C.-A., and Wäsch E. (2002). Do meteoroids of sedimentary origin survive terrestrial atmospheric entry? The ESA artificial meteorite experiment *STONE. Planet. Space Science*, **50**, 763–772.
Brack A. and Pillinger C. (1998). Life on Mars: chemical arguments and clues from Martian meteorites. *Extremophiles*, **2**, 313–319.
Brack A., Fitton F., and Raulin F. (1999). Exobiology in the Solar System & the search for life on Mars. ESA Special Publication SP 1231, ESA Pub. Noordwijk, The Netherlands.
Bradley J.P., Harvey R.P., and McSween H.Y. (1997). No "nanofossils" in martian meteorite, *Nature*, **390**, 454.

Brasier M.D., Green O.R., Jephcoat A.P., Kleppe A.K., van Kranendonk M., Lindsay J.F., Steele A., and Grassineau N. (2002). Questioning the evidence for Earth's oldest fossils. *Nature*, **416**, 76–81.

Briggs R., Ertem G., Ferris J.P., Greenberg J.M., McCain P.J., Mendoza-Gomez C.X., and Schutte W. (1992). Comet Halley as an aggregate of interstellar dust and further evidence for the photochemical formation of organics in the interstellar medium. *Origins Life Evol. Biosphere*, **22**, 287–307.

Brochier C. and Philippe H. (2002). A non-hyperthermophilic ancestor for bacteria. *Nature*, **417**, 244.

Carr M.H. et al. (1998). Evidence for a subsurface ocean on Europa. *Nature*, **391**, 363–365.

Charlou J.L., Fouquet Y., Bougault H., Donval J.P., Etoubleau J., Jean-Baptiste Ph., Dapoigny A., Appriou P., and Rona P.S. (1998). Intense CH_4 plumes generated by serpentinization of ultramafic rocks at the intersection of $15°20'$ N fracture zone and the Mid-Atlantic Ridge. *Geochim. Cosmochim. Acta*, **62**, 2323–2333.

Cooper G., Kimmich N., Belisle W., Sarinana J., Brabham K., and Garrel L. (2001). Carbonaceous meteorites as a source of sugar-related organic compounds for the early Earth, *Nature*, **414**, 879–883.

Cronin J.R. and Pizzarello S. (1997). Enantiomeric excesses in meteoritic amino acids. *Science*, **275**, 951–955.

De Duve C. (1998). Possible starts for primitive life. Clues from present-day biology: the thioester world, in *The Molecular Origins of Life: Assembling Pieces of the Puzzle*, (ed.) A Brack, p. 219–236, Cambridge University Press, Cambridge.

Deamer D.W. (1998). Membrane compartments in prebiotic evolution, in *The Molecular Origins of Life: Assembling Pieces of the Puzzle*, (ed.) A Brack, p. 189–205, Cambridge University Press, Cambridge.

Eschenmoser A. (1999). Chemical etiology of nucleic acid structure. *Science*, **284**, 2118–2124.

Galtier N., Tourasse N., and Gouy N. (1999). A non hyperthermophilic common ancestor to extant life forms. *Science*, **283**, 220–221.

Holm N. and Charlou J.L. (2000). Indications of abiotic formation of hydrocarbons in the Rainbow ultramafic hydrothermal system, Mid Atlantic Ridge, *Earth Planet. Sci. Lett.*, **191**, 1–8.

Holm N.G. and Andersson E.M. (1995). Abiotic synthesis of organic compounds under the conditions of submarine hydrothermal systems: a perspective. *Planet. Space Sci.*, **43**, 153–159.

Holm N.G. and Andersson E.M. (1998). Organic molecules on the early Earth: hydrothermal systems. in *The Molecular Origins of Life: Assembling Pieces of the Puzzle*, (ed.) A Brack, p. 86–99, Cambridge University Press, Cambridge.

Horneck G. (1999). European activities in exobiology in Earth orbit: Results and perspectives. *Adv. Space Res.*, **23**, 381–386.

Horneck G., Stöffler D., Eschweiler U., and Hornemann U. (2001). Bacterial spores survive simulated meteorite impact. *Icarus*, **149**, 285–193.

Huber C. and Wächtershäuser G. (1997). Activated acetic acid by carbon fixation on (Fe, Ni)S under primordial conditions. *Science*, **276**, 245–247.

Imai E.-I., Honda H., Hatori K., Brack A., and Matsuno K. (1999) Elongation of oligopeptides in a simulated submarine hydrothermal system. *Science*, **283**, 831–833.

Jet Propulsion Laboratory/NASA: Mars Meteorites (2003). http://www.jpl.nasa.gov/snc/

Kivelson M.G., Khurana K.K., Russel C.T., Volwerk M., Walker R.J., and Zimmer C. (2000). Galileo magnetometer measurements: a stronger case for a subsurface ocean at Europa. *Science*, **289**, 1340–1343.

Lebreton J.P. (ed.) (1997). Huygens: Science Payload and Mission. *ESA-SP 1177*, ESA Pub. Noordwijk, The Netherlands.

Luther A., Brandsch R., and von Kiedrowski G. (1998). Surface-promoted replication and exponential amplification of DNA analogues. *Nature*, **396**, 245–248.

Maurette M. (1998). Carbonaceous micrometeorites and the origin of life. *Origins Life Evol. Biosphere*, **28**, 385–412.

Mayor M.D. and Queloz D. (1995). A Jupiter-mass companion to a solar-type star. *Nature*, **378**, 355–359.

McKay D.S., Gibson E.K., Thomas-Keprta K.L., Vali H., Romanek C.S., Clemett S.J., Chellier X.D.F., Maechling C.R., and Zare R.N. (1996). Search for past life on Mars: possible relic biogenic activity in martian meteorite ALH 84001. *Science*, **273**, 924–930.

Mileikowsky C., Cucinotta F., Wilson J.W., Gladman B., Horneck G., Lindegren, Melosh J., Rickman H., Valtonen M., and Zheng J.Q. (2000). Natural transfer of viable microbes in space, Part 1: From Mars to Earth and Earth to Mars. *Icarus*, **145**, 391–427.

Miller S.L. (1953). The production of amino acids under possible primitive Earth conditions. *Science*, **117**, 528–529.

Miller S.L. and Bada J.L. (1988). Submarine hot springs and the origin of life. *Nature*, **334**, 609–611.

Mojzsis S.J., Arrhenius G., McKeegan K.D., Harrison T.M., Nutman A.P., and Friend C.R.L. (1996). Evidence for life on Earth before 3800 million years ago. *Nature*, **384**, 55–59.

Munoz Caro G.M., Meierhenrich U.J., Schutte W.A., Barbier B., Arcones Segovia A., Rosenbauer H., Thiemann W.H.-P., Brack A., and Greenberg J.M. (2002). Amino acids from ultraviolet irradiation of interstellar ice analogues. *Nature*, **416**, 403–405.

Orgel L.E. (1998). Polymerization on the rocks: Theoretical introduction. *Origins Life Evol. Biosphere*, **28**, 227–234.

Parkes R.J., Cragg B.A., and Wellsbury P. (2000). Recent studies on bacterial populations and processes in subseafloor sediments: A review. *Hydrogeol. J.*, **8**, 11–28.

Raulin F. (1998). Titan, in *The Molecular Origins of Life: Assembling Pieces of the Puzzle*, (ed.) A Brack, p. 365–385, Cambridge University Press, Cambridge.

Rosing M.T. (1999). ^{13}C depleted carbon microparticles in > 3700 Ma seafloor sedimentary rocks from West Greenland. *Science*, **283**, 674–676.

Schidlowski M. (1988). A 3800-million-year isotopic record of life from carbon in sedimentary rocks. *Nature*, **333**, 313–318.

Schneider J. (2003). http://www.obspm.fr/encycl/f-encycl.html.

Schopf J.W. (1993). Microfossils of the early Archean Apex Chert: new evidence of the antiquity of life. *Science*, **260**, 640–646

SETI http://www.seti-inst.edu/

Stetter K.O. (1998). Hyperthermophiles and their possible role as ancestors of modern life. in *The Molecular Origins of Life: Assembling Pieces of the Puzzle*, (ed.) A Brack, p. 315–335, Cambridge Univ. Press, Cambridge.

Stoks P.G. and Schwartz A.W. (1982). Basic nitrogen-heterocyclic compounds in the Murchison meteorite. *Geochim. Cosmochim. Acta*, **46**, 309–315.

Van Zuilen M., Lepland A., and Arrhenius G. (2002). Abiogenic and biogenic graphite in the Isua Supracrustal belt. In *Isua Multidisciplinary Study Workshop*, Berlin, Jan 2002, Danish Geol. Surv. Copenhagen, p. 73–74.

Wächtershäuser G. (1998). Origin of life in an iron-sulfur world, in *The Molecular Origins of Life: Assembling Pieces of the Puzzle*, (ed.) A Brack, pp. 206–218, Cambridge University Press, Cambridge.

Weber P. and Greenberg J.M. (1985). Can spores survive in interstellar space? *Nature*, **316**, 403–407.

Westall F., Gobbi P., Mazzotti G., Gerneke D., Stark R., Dobrek T., and Heckl W. (1998). Combined SEM (secondary electrons, backscatter, cathodoluminescence) and atomic force microscope investigation of the carbonate globules in Martian meteorite ALH84001: Preliminary results, *SPIE, Instruments, Methods and Missions for Astrobiology*, **3114**, 225–233.

Westall F., De Wit M.J., Dann J., Van Der Gaast S., De Ronde C., and Gerneke D. (2001). Early Archaean fossil bacteria and biofilms in hydrothermally-influenced, shallow water sediments, Barberton greenstone belt, South Africa. *Precambrian Res.*, **106**, 93–116.

Yanagawa H. and Kobayashi K. (1992). An experimental approach to chemical evolution in submarine hydrothermal systems. *Origins Life Evol. Biosphere*, **22**, 147–159.

Zimmer C., Khurana K.K., and Kivelson M.G. (2000). Subsurface oceans on Europa and Callisto: Constraints from Galileo magnetometer observations. *Icarus*, **147**, 329–347.

From Prebiotic Chemistry to the Origin of Life on Earth

Introduction

Chemical evolution, which started very early in our Universe, is not only a preliminary but also an essential step to any biological evolution. All the discussions concerning a hypothetical RNA world evolving towards a DNA world is based on the assumption that deoxyribose derives from ribose or, at least, that the replacement of ribose by deoxyribose was an important step in biological evolution. Many other examples of the interdependence between chemical evolution and biological evolution are known.

Unfortunately, we have no molecular fossils or relics of the prebiotic world and our knowledge about molecular evolution during the prebiotic period is based on models and scenarios. These scenarios themselves are based on hypotheses concerning the physico-chemical conditions present at the surface of the young Earth. They are also supported by experimental simulations and by indirect observations. Indeed, interstellar clouds that can be observed today are probably not very different from the protosolar nebula and chondritic and micrometeoritic matter that we can study in our laboratories today are probably similar to the chondritic or cometary matter that fell on the young Earth in large quantities during the prebiotic period. LUCA (our last universal common ancestor) was most probably a unicellular entity with a membrane, a metabolism, a reproduction capability and therefore a genetic code. LUCA itself was most probably not the most primitive form of life on Earth. Chemists and biologists must introduce constraints into the scenarios they suggest for the transition steps from nonlife to life. These constraints are based on what we know about the chemical elements and their reactivity, about the physicochemical laws, and very importantly, about the conditions prevailing on the young Earth when the transition occurred. The study of extreme biotopes on Earth, such as the ices above the Vostock Lake or the vicinity of oceanic black smokers, yields information about the diversity of life and also about the eventual universal requirements for life.

Together with astronomers, planetologists and geologists, biologists and chemists involved in exobiological studies participate in the elaboration of these models. As already mentioned, the chemist has no relics, no molecular fossils of the prebiotic period but he/she is the only one who can perform experimental simulations to help a model progress. The biochemist and the biologist are the only ones who work with living species, they know how living species have evolved and they are ready to take the risk to extrapolate towards the past: their approach could be described as a top-down approach while the chemists have a bottom-up approach. The hope (or better the dream) is that, one day, chemists and biologists could reach a similar conclusion about what LUCA looked like!

For the chemist, as for all "specialists" involved in exobiology, the most important contribution to the field can only come from their capacity to interact efficiently with other specialists. Exobiology is probably the best example of an interdisciplinary science: all natural sciences but also mathematics and informatics participate to its development. The next ensemble of chapters is a clear example of the work of chemists, biochemists and biologists, working in different fields but searching to contribute to the understanding of what remains an important problem: the emergence of life from nonlife.

1 A Rational Approach to the Origin of Life: From Amphiphilic Molecules to Protocells. Some Plausible Solutions, and Some Real Problems

Guy Ourisson, Yoichi Nakatani

Synopsis Self-organisation of amphiphiles in water into closed vesicles leads automatically to self-complexification into "protocells". However, some real problems are usually not even mentioned in the various theories of the origin of Life. The present discussion is a follow up of our initial publications (Ourisson and Nakatani, 1994, 1999; see Maddox, 1994)

1.1 From Amphiphilic Molecules to "Protocells" by Understandable Processes. Self-Organisation and Self-Complexification

Our thesis will be that the known general universal laws of the physical world can go a long way towards showing how simple molecular systems can spontaneously become self-organised by understandable processes. This is not an original statement. However, we shall also show that self-organisation automatically carries with it the seeds of self-complexification. Without requiring new "steps", it leads to novel properties, all singularly necessary for life to accrue at some later stage. We believe that this is an original set of far-reaching assumptions, even though it does not (yet?) lead to the definitive solution. It will serve us finally to delineate some problems that have seldom or never been raised, and that would require, sooner or later, careful study and eventually a solution to help us understand how life begun on Earth. These "hidden problems" include:

- The need to have had somewhere a concentrated brew, the "primordial soup" of Oparin (Oparin, 1968)
- The need to have had an efficient synthesis of long chain lipids
- The need to have had in the same brew high enough concentrations of phosphoric acid derivatives
- The need to surmount the unexpected and inherent difficulty of phosphorylation with phosphoric anhydrides
- The need to have had sufficient concentrations of all the "bricks of life", however they are produced
- And the need to understand the formation of the cytoskeleton of microtubules

1.2 Water and Self-Organisation of Amphiphiles

1.2.1 The Structure of Liquid Water

The unique structure of water, dominated by an irregular network of hydrogen bonds, is the key to many of its unusual properties (Fig. 1.1). Dissolution of a substance carrying many hydrophilic groups, for instance a sugar, is possible because it leads to the replacement of many homologous hydrogen bonds (water to water) by many heterologous ones (water to OH). Around each O atom, the H atoms of the same molecule, and those of neighbouring ones, are arranged statistically along tetrahedral directions. The energy required is small, and the foreign solute is well accepted (i.e. it is soluble; Fig. 1.1). In some cases (e.g., that of the sugar trehalose), the new network of hydrogen bonds is so well accommodated that the solute in fact better organises the molecular network of water into quasitetrahedral arrangements (see Tanford 1978, Blokzijl 1993, Lemienx 1996). Rather similar molecular properties characterise other solvents, such as ethanediol (Larsen 1984) or formamide (Auvray, 1991); these solvents appear, however, to have no significance for life as we know it. (For an up-to-date discussion of the structure of water, see for instance Westhof 1993).

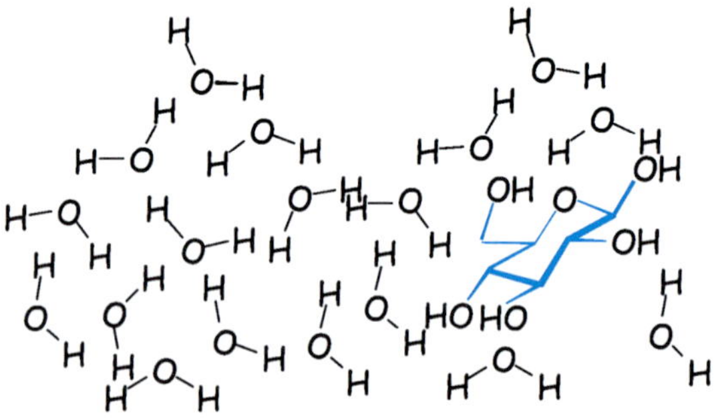

Fig. 1.1. Schematic structure of water and of water containing a molecule of glucose. The hydroxyl groups of the substrate can engage in hydrogen bonding and only little distort the network of H bonds

1.2.2 The Structure of Amphiphile-water Mixtures

By contrast, a molecule like a long-chain alcohol, despite its terminal hydroxyl, is not soluble, because its insertion into the network of hydrogen bonds requires breaking many of them without compensation. It is akin to boring a hole into this compact arrangement, and costs too much energy (Fig. 1.2). Rather than being dissolved in the water phase, the molecules of such an "amphiphile" reach for the surface, and are dissolved only through their polar head-groups (see Wang, 1994; Ron, 2000). At the same time, their long chains are rejected from the water

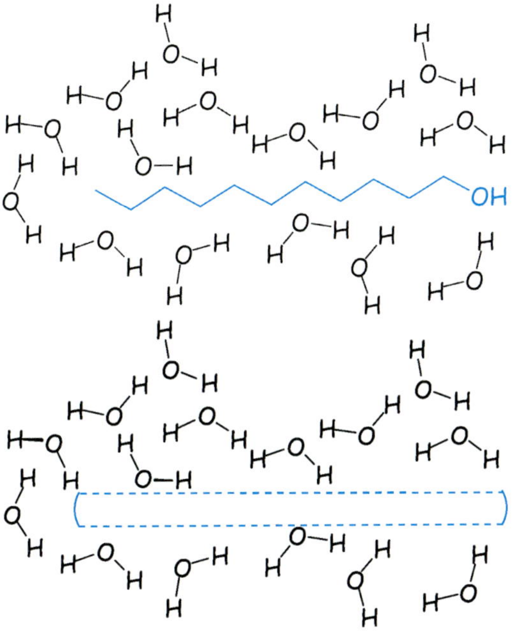

Fig. 1.2. Dissolving a long-chain alcohol in water would require "boring a hole" in the system of H-bonds

to its surface where they self-organise into a monolayer of hydrocarbon chains, more or less compact depending on the side-pressure (Fig. 1.3) (for a detailed review of monolayers, see Kuzmenko et al., 2001; for an analysis of the relevance of monolayers for chiral amplification, a topic not covered in the present chapter but obviously relevant to the origin of life, see Weissbuch et al., 2003).

Amphiphilic molecules of the proper shape (depending on the relative sizes of the polar head-group and the side-chain (Israelachvili et al., 1980), can self-organise in water. It is then possible to obtain a variety of arrangements where the polar heads are grouped together to maximise their contact with the surrounding water, while the lipidic side-chains are segregated. Depending on the particular shape and concentration of the amphiphile, one can then obtain *micelles* (in which the hydrophobic chains are segregated and the hydrophilic heads coat the surface), or larger objects, *vesicles* (syn.: single-walled *liposomes*). These are the systems of particular interest to us. In these, the lipidic chains, all approximately parallel, are ordered into a double layer coating the surface of the vesicle, and separate the inside water from the outside water (Fig. 1.4). They are formed spontaneously by hydration of a lipid film (sonication gives very small vesicles). Vesicles can be of micrometric size, and therefore can be observed by modern optical microscopy; they are then called "giant vesicles". They can be filtered through polycarbonate filters of homogeneous bore size, and have then a narrow diameter distribution, which can be evaluated by light diffusion.

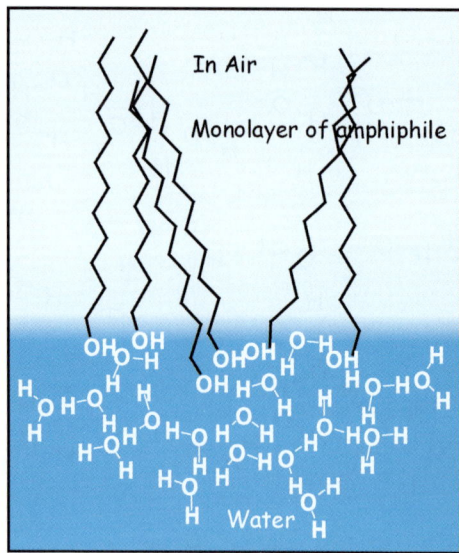

Fig. 1.3. Molecules of a long-chain alcohol self-organise into a monolayer on the surface of water

In these spontaneously formed systems, three distinct phases can be recognised: the water outside, the water inside, and the inside of the bilayer, which is lipidic. This leads, for instance, to the selective extraction into the lipidic membrane of any lipophilic molecule present, for instance a lipophilic dye such as Nile Red. It also leads to the phase separation of racemic and nonracemic amphiphilic α-amino acid derivatives at the air/water interface, a phenomenon that may be important to explain the homochirality of the present peptide amino-acids (Weissbuch et al., 2003).

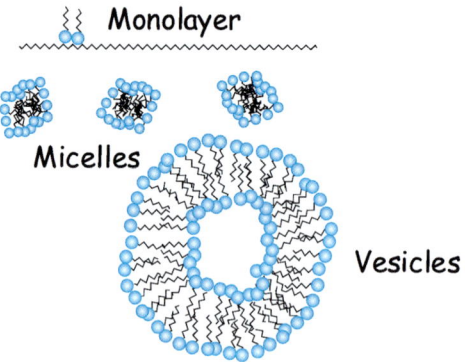

Fig. 1.4. Amphiphiles at the surface of and inside water

1.3 Properties Ensuing from the Self-organisation of Amphiphiles

1.3.1 Extraction and Orientation

The inside of the bilayer is not only lipophilic, but also highly anisotropic: it is a fibrous medium, and any nonisotropic lipophilic molecule will therefore be extracted and oriented. This is obvious, but it is also experimentally proved in at least two cases: that of cholesterol, which has been shown to be oriented parallel to the lipidic chains (with its OH group at the surface) (Nakatani et al., 1996), and that of the hydrocarbon β-carotene, which lies, depending on the lipid used, parallel to the surface, probably in the midlayer space, or embedded in the lipids (Nordén et al. 1977) (Fig. 1.5).

A very interesting discussion of the possibilities offered by this selective extraction is given in Deamer et al. (1994).

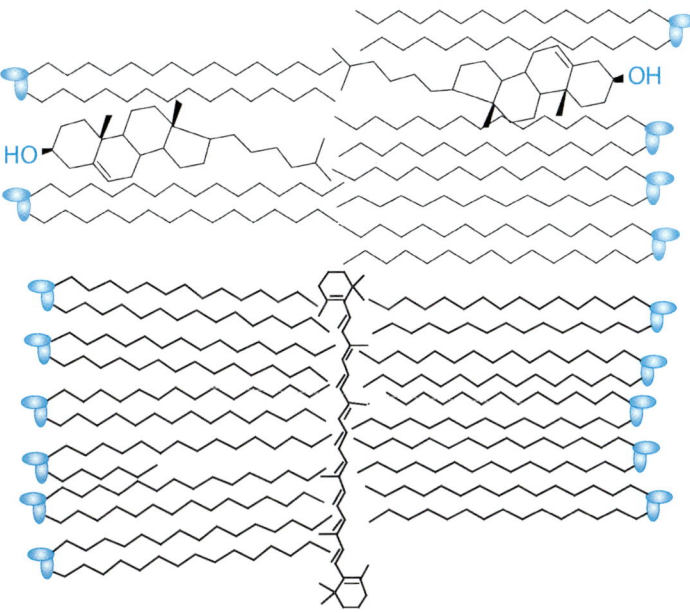

Fig. 1.5. Cholesterol is inserted in lipidic membranes perpendicularly to the surface, whereas β-carotene is oriented parallel to the surface

1.3.2 Increased Concentration and Condensation

Another property induced by the self-organisation of amphiphiles into vesicles is that the selective extraction of lipophilic substances into the double-layer will lead to increased concentrations, and that this opens up the possibility of selective reactions within this double layer. This has been demonstrated in three cases, by Fukuda in Urawa (Fukuda et al. 1981), by Ringsdorf in Mainz (Folda et al., 1982) and by Luisi and Walde in Zurich (Bachmann et al., 1991, Walde et al., 1994; Blocher et al., 1999), with somewhat hydrophobic of amino-acid derivatives that, in the presence of vesicles, underwent nonenzymatic condensation into dipeptides or small peptides (Fig. 1.6).

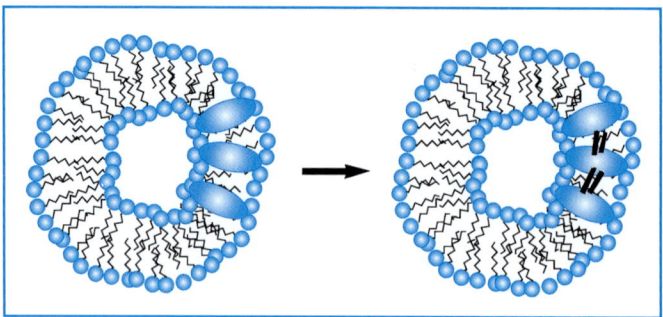

Fig. 1.6. Hydrophobic derivatives of amino-acids are selectively extracted into the double layer of vesicles, where they are spontaneously condensed into derivatives of polypeptides

1.3.3 Vectorial Properties

Self-organisation also leads to vectorial properties: the outer lipids, on a convex surface, are intrinsically different from the inner ones, which are on a concave surface. This had been demonstrated by biochemical arguments already in 1972, in the case of erythrocytes (Bretscher, 1972). It also leads to consequences measurable by NMR, as had been first shown by Chrzeszczyk et al. (1977). We have shown it quite directly in a ^{31}P NMR study of mixed vesicles composed of two different phospholipids, giving each two peaks of different intensities (Fig. 1.7); the ratio of these intensities corresponds to the ratio of the number of molecules in the outside and the inside layers of the vesicle wall, as deduced from the average size of the vesicles (evaluated by light diffusion) (Lee et al., 2002); the same doubling of peaks cam also be observed on vesicles formed of a single phospholipid, at the proper pH. If the NMR signals of the inside and the outside head-groups differ, all their properties must also be sensitive to their location: for instance, they must have different pKa values (Swairjo et al., 1994). Water inside the vesicle must be different from water outside because the outside layer is convex, the inside one concave (and a little more crowded); this is the basis for these "vectorial properties". These differences H must be more marked for small vesicles than for larger ones, but might well be carried over from small ones when they grow. It would be quite interesting to be able to demonstrate whether a measurable pH difference is spontaneously established, and also whether a lipophilic polypeptide is inserted in a predictable orientation (amino-end out?). We have not yet set out to do this.

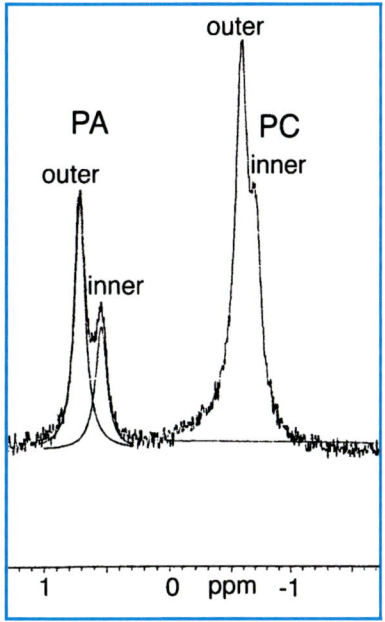

Fig. 1.7. ^{31}P NMR spectrum of vesicles of two phospholipids. Each P-containing headgroup gives rise to a double peak; the more intense doublet is due to the major PC in the mixture, the smaller one to the phytanyl phosphate (here, PA). Each is a doublet, the minor peak being due to the inner head-groups, the major one to the outer ones

1.3.4 Coating the Vesicles

Another novel property arising from self-organisation is the possibility to coat the vesicle with a carbohydrate wrapping, analogous to the bacterial cell wall. This was first demonstrated by Sunamoto with the large fungal polysaccharide pullulane, by labelling it with a fluorescent tag making it visible by optical microscopy, and by linking it to cholesterol, as an anchor, into the outside layer (we shall return below to this concept of anchors) (Ueda et al., 1998) (Fig. 1.8).

We have also shown with Akiyoshi that the same polysaccharide, labelled not with cholesterol but with the highly branched phytol, acts in the same way to cover the outside of vesicles made of geranylgeranyl phosphate (Ghosh et al., 2000).

It still remains to check whether these "coatings" add stability to the vesicles.

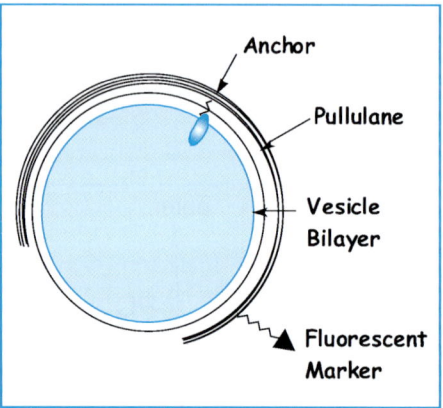

Fig. 1.8. Schematic representation of a double-layer vesicle coated with pullulane, linked to a membrane anchor and labelled with a fluorescent tag

1.3.5 Vesicles and Nucleic Acids; Vesicles as Protocells

Several authors have shown that it is possible to entrap nucleic acid molecules into vesicles, which then act as tight microcontainers. The first success in this field was obtained by enclosing Simian virus 40 (5 kb) into small vesicles of phospholipids ($D = 400$ nm) (Fraley et al., 1980). Next, success was achieved with small (1–20 kb) DNAs, like salmon sperm DNA, enclosed into egg lecithin vesicles (3 ± 5 μm) by dehydration/rehydration (Deamer, Barchfeld, 1982) and with plasmid DNA (3.3 kb) by using the same method, or the reverse-phase method, or the freeze-thaw method (Monnard et al., 1997). Yoshikawa's group in Kyoto has more recently shown that, in the presence of M_g^{2+} ions, one can introduce into vesicles, by several techniques, giant DNA (10^2–10^5 kb) or RNA molecules, large enough to be observed by optical microscopy. The same can be done with their compacted histone complexes. One can use forceful insertion of an individual nucleic acid molecule, under the microscope, pushing it through the membrane by holding it with a laser tweezer, in much the same way as one carries out in vitro fecundation (Nomura et al., 2001). The nucleic acid molecules can be observed undergoing Brownian movement inside the vesicle, "bouncing" on the inner wall. They are then transcribed into m-RNA molecules that are efficiently protected from the action of nucleases introduced in the surrounding medium: the membranes form effective barriers (Tsumoto et al. 2001; Fischer et al., 2002). What is very remarkable is that nucleic acid molecules efficiently find their way into vesicles during the spontaneous formation of these vesicles when a lipid film is left in the presence of a suitable buffer containing the nucleic acid molecules during hydration–dehydration cycles (Deamer, Barchfeld, 1982, Tsumoto et al., 2001). It looks as though they are preferentially entrapped into these newly formed vesicles, as a surprisingly large proportion of them (typically 35–50%) do contain nucleic acid molecules.

Jay and Gilbert showed that this entrapment was facilitated by the presence of a basic protein, lysozyme (Jay, Gilbert, 1987), and the clay montmorillonite facilitates the conversion of fatty acid micelles into vesicles, often encapsulating clay particles which may be of significance for the formation of protocells. (Hanczyc, 2003).

Other very important biochemical reactions have been successfully carried out inside vesicles: the polymerase chain reaction (Oberholzer et al., 1995-1), and the enzymatic RNA replication in self-reproducing vesicles (a non-natural model obtained from oleic acid/sodium oleate) (Oberholzer et al., 1995-2), see also: Walde, Ishikawa, 2001.

Even more spectacular has been the synthesis of proteins inside vesicles: Oberholzer et al., (1999) used non-native RNA (Poly-U) as a template for the synthesis of the non-natural polypeptide polyPhe, and Yu et al., (2001) described the synthesis of a functional protein in giant vesicles (but did not show reliably whether the synthesis was occurring inside or outside the vesicles); finally the synthesis, more efficient inside vesicles than outside, of the green fluorescent protein (GFP) by a plasmid encoding the GFP gene, trapped inside the vesicle, has been described by Nomura et al., (2003).

To sum up, phospholipid vesicles can be considered as model protocells. That they are still very far from being cell models is obvious. That it would still be very worthwhile to study further their spontaneous complexification is, in our view, just as obvious.

This provides additional support, as well as more precise structural detail, for the seminal hypothesis of Deamer and of Morowitz, regarding the role of vesicles as likely precursors of early cells (Deamer, 1986, Morowitz et al., 1988). For more general considerations on this topic, and novel hypotheses, see for instance Szostak et al., (2001).

We shall now turn to a review of a series of questions often hidden when discussing the hypotheses about the origin of life.

1.4 The Nature of the Primitive Amphiphiles

1.4.1 The Modernity of N-Acyl Lipids

So far, we have mentioned "amphiphiles" without really discussing their nature. The best known amphiphiles, those of Bacteria and of Eucaryotes, are quite complex molecules, requiring for their biosynthesis three distinct pathways: the one producing long n-acyl chains, saturated or not, by a series of condensation of C_2 units and reductions, and those required for the C_3 unit of glycerol and for the C_2 unit of choline (Fig. 1.9). The other types of "standard" amphiphiles carry head-groups just as complex or more complex; none can be considered as possibly produced by "prebiotic" mechanisms. Usually, proponents of an early compartmentation into vesicles refer, in an attempt to "explain" a prebiotic formation of the long n-acyl chains, to a remarkable

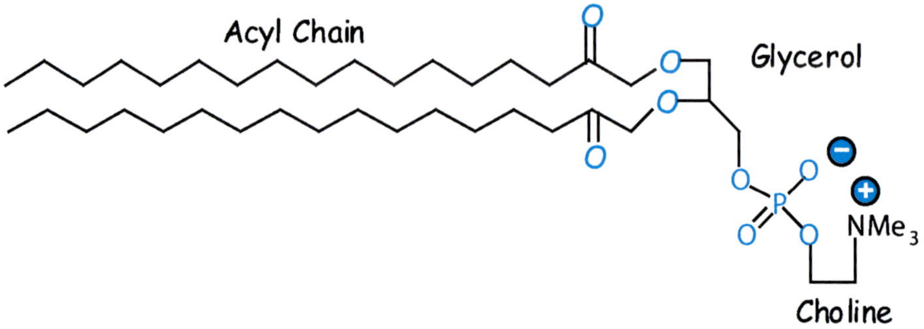

Fig. 1.9. A typical eucaryotic phospholipid: dipalmitoylphosphatidylcholine (DPPC)

process: the Fischer–Tropsch (F–T) reaction, which indeed converts C_1 or C_2 units into longer chains; however, to our knowledge, even the most complex variants of the F–T reaction do not lead to any significant amount of long-chain (C_{14-20}) terminally functionalised n-acids or alcohols. We think that any "explanation" by recourse to this reaction is rather a formal invocation. We also consider as doubtful the contention that meteorites could have been the source of these lipids, as they contain amphiphilic molecules; they do, but in very small amounts, and their complex mixtures containing many aromatic acids are certainly not able to form membranes (Deamer et al., 1994, pp. 112–114).

1.4.2 The Archæal Lipids and Their Synthesis

In recent years, another family of membrane lipids has been discovered in the "new" phylum of Archæa: polyprenyl lipids, usually in C_{20} or, by dimerisation, in C_{40}. These would be just as difficult to obtain under prebiotic conditions; but they display some characteristics much more compatible with a primitive origin: their polyprenyl chains are biosynthesised by C_5 increments (not C_2 as in the biosynthesis of n-acyl chains), which reduces, of course, the number of steps required; these steps are simple alkylations of double bonds (not complex condensations and reductions), and the chemistry involved is "elementary" and has been duplicated (from C_5 to C_{10}) by simple heating of suitable precursors with a montmorillonite clay (Desaubry et al. 2003) (Fig. 1.10).

These steps are, of course, repetitive, and could lead in principle, with the same substrates, from C_{10} to C_{15}, then to C_{20}. Selectivity could be ensured as soon as there is a phase separation at a certain length of the lipophilic chain, with the proper head-group. The complex head-groups present in modern archæal phospholipids call for an independent evolution. We have postulated that the simplest possible polar head, a phosphoric acid salt, might already have the necessary characteristics. We have therefore synthesised a series of phosphate esters containing one or two polyprenyl chains, and have demonstrated that

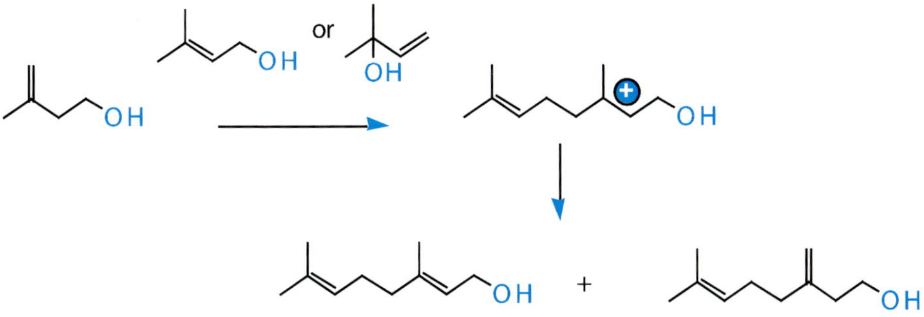

Fig. 1.10. Synthesis of C_{10} terpenic chains from C_5 units

indeed single- or double-chain polyprenyl phosphates do form vesicles, provided their chain(s) contain in toto at least 15 carbon atoms (Birault et al., 1996, Pozzi et al., 1996).

One of the most easily accessible of these "primitive" phosphates is the C_{20} phytyl phosphate (Fig. 1.11), as phytol is a cheap commodity material. We have used this phytyl phosphate extensively, as well as the very similar geranylgeranyl phosphate (Fig. 1.12) to study various properties of polyprenyl phosphate membranes, and retain, as a hypothesis, a scenario involving these phosphates as primitive membrane constituents. It is then possible to postulate a series of hypothetical successive steps, rendering their structures progressively more complex. A complete "genealogical tree" of polyterpenes has thereby be constructed (Ourisson, 1986); however, as it is entirely hypothetical, and relates more to the evolution than to the origin of living organisms, we shall not develop that point here but only present this genealogical tree without comments (Fig. 1.13).

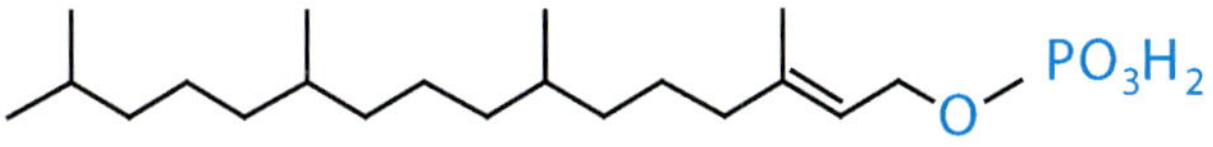

Fig. 1.11. Phytyl phosphate

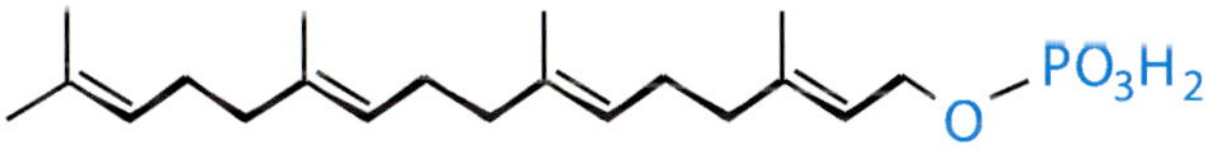

Fig. 1.12. Geranylgeranyl phosphate

Polyprenyl phosphates could be the precursors of *all* polyprenyl derivatives found in memberbranes (Sect. 1.5.4).

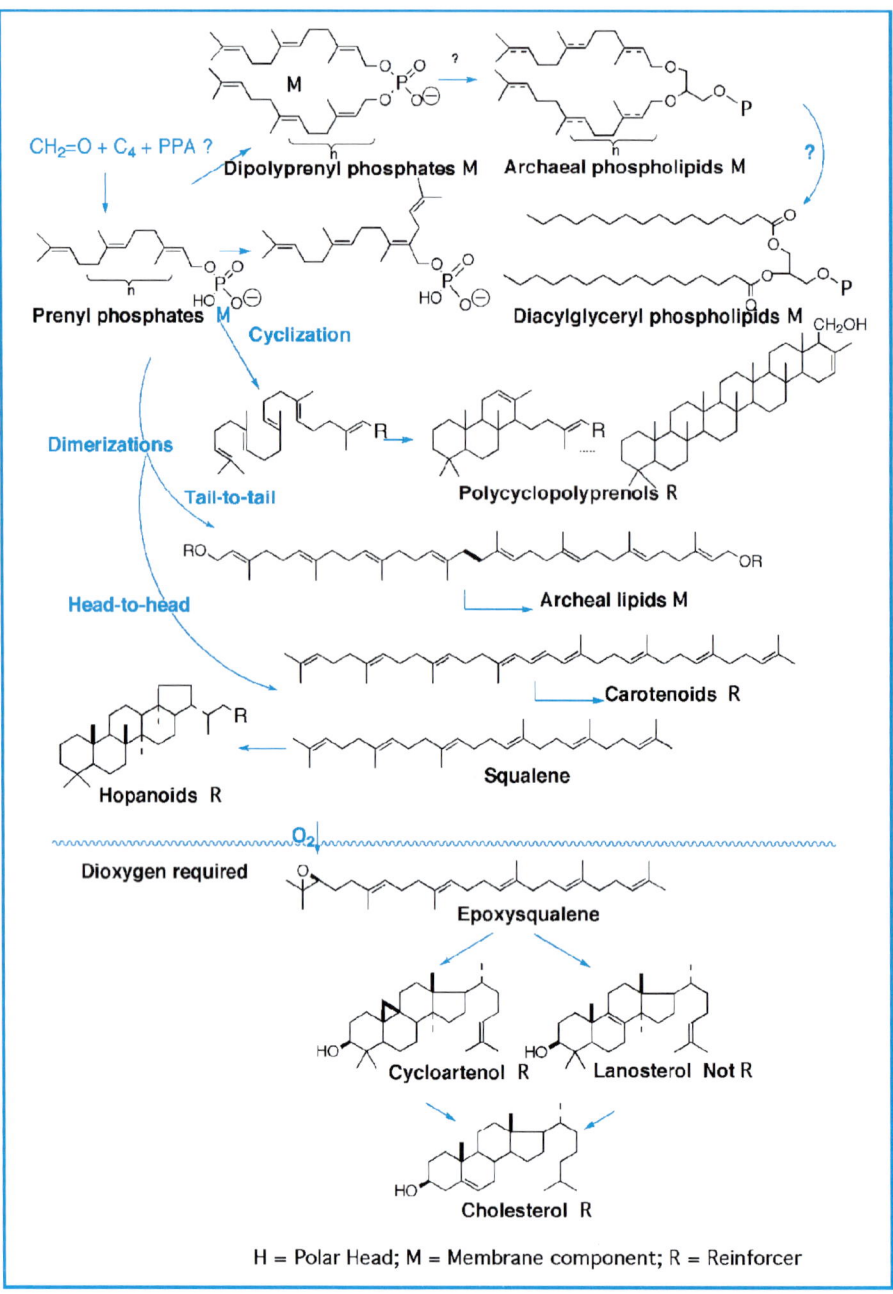

Fig. 1.13. Hypothetical evolution of membrane-related terpenoids (adapted from Ourisson, 1986)

1.4.3 The Terpenoids as Universal Metabolites

We must, however, note that polyterpenes are not "secondary metabolites" as usually considered: they are universal, necessary constituents of all living organisms, and must therefore be promoted to the dignity of essential metabolites. They occur as:

- Constituents of the archæal lipids
- Ubiquitous anchors binding proteins to cell membranes by farnesyl or geranylgeranyl chains
- Ubiquitous constituents of electron transporters across membranes (ubiquinones)
- Ubiquitous membrane constituents involved in the glucosylation of proteins ($C_{\approx 100}$ dolichyl phosphates)
- Light sensors: retinal of the light-driven proton pump of *Halobacterium*, bacteriorhodopsin, and of the light-sensing protein of our eyes, rhodopsin
- Constituents of the photosynthetic systems (the phytyl group of chlorophyll)
- Reinforcers of phospholipidic membranes (cholesterol or equivalents such as phytosterols in Eucaryotes, hopanoids in Bacteria and α,ω-dihydroxylated carotenoids in Bacteria (*St. aureus* and *Halobacterium*)
- Hormones and pheromones of invertebrates and vertebrates
- Vertebrate hormones (vitamin A, calciferol)
- Thousands of pheromones, attractants, repellents, defense substances, toxic alkaloids, etc., in all groups of living organisms.

Thus, a terpenic origin of membrane lipids is a quite tenable hypothesis. However, we shall now list a series of problems that are seldom or never discussed in the present context, and for which we have no solution to propose.

1.5 Some Remaining Problems

1.5.1 The Problem of the Synthesis of Ingredients

From the brief description we have given of the formation of vesicles in water, it is clear that a minimal concentration must be attained locally before any self-organisation sets in (this is the critical micellar concentration, the cmc). Two separate questions arise:

- Where did the necessary ingredients come from?
- How did they reach high enough local concentrations?

The first question has found, in principle, a satisfactory solution in the classical Miller experiment, now 50 years old (Miller, 1953): sparks in a suitable atmosphere (methane, ammonia, dihydrogen, water) lead to the formation of a series of amino-acids and nucleic acids bases; these "bricks of life" are remarkably easy to obtain. With M. Devienne, we have ourselves obtained very similar results, including not only the synthesis of proteogenic amino-acids, but also that of the four nucleic bases and of many other complex substances; these were formed by high-energy bombardment of graphite targets with molecular beams of the desired atoms, simulating plausible interstellar events (Devienne et al., 1998, 2002). Similar results have also been obtained by UV irradiation of ice containing the necessary ingredients (Bernstein et al., 2002; Muñoz Caro et al., 2002). In short, it looks as though any input of enough energy into a set of molecules containing the necessary atoms, followed by quenching, gives rise to a bevy of "bricks of life".

1.5.2 The Problem of Local Concentration

This does not begin to address the question of local concentration. When Oparin spoke of a "primitive brew" (Oparin, 1968), he had in mind a continuous process of formation of the necessary constituents and of accumulation in the primitive oceans, because there was, of course, no biodegradation in the absence of living organisms. This hypothesis may be compatible with Miller's process (Miller, 1953) (despite all the criticisms raised against his choice of composition of a supposedly "primitive", highly reducing atmosphere, already postulated by Oparin). It is, however, hardly compatible with one of the other two mentioned above: interstellar dust or dirty ice crystals may be the site of synthesis of the "bricks of life", but these can hardly have fallen on the primitive Earth in large enough amounts. Anyhow, given the huge volume of the oceans, obtaining large enough concentrations globally appears to be a critical process. We do not see how it could be helped by Darwin's poetic hypothesis of the shallow lagoon by the side of the sea, periodically replenished by waves with dilute ingredients, and constantly concentrated by evaporation. Unless one simultaneously accepts J. Monod's contention that *"L'Univers n'était pas gros de la Vie"* (Monod, 1970), which implies that Life started not because some general conditions were right, but because some contingent coincidence of circumstances made it possible.

1.5.3 The Problem of the Prevalence of Phosphates

Another problem comes from the prevalence of phosphates in all the head-groups of membrane lipids (some sulfate head-groups do exist, but very seldom). In fact, the eminent role of phosphates in many aspects of biochemistry has been emphasised and assigned to the specific properties of this group (Westheimer, 1987).

We should in this respect be conscious that this ubiquitous role of phosphates is in sharp contrast with the low abundance of phosphorus in the universe, and in particular in the Earth's crust, where one evaluates its abundance at only 0.12%. It is well known that, in agriculture, the availability of P is often the growth-limiting factor. Most of the phosphate on Earth is anyway present as calcium phosphate rock, itself derived from the accumulation of fossil debris and therefore not providing an answer to the question of the prebiotic availability of phosphate. Phosphoric and polyphosphoric acids have been reported from volcanic sources (Yamagata et al., 1991) but again these are such localised and limited occurrences that they do not begin to answer the question of the origin of phosphates in prebiotic systems, for which we have no answer.

1.5.4 The Problem of Phosphorylation by Phosphoric Anhydrides

In the preceding paragraph, we have pointed out that phosphates are rather rare on Earth. Furthermore, some aspects of their reactivity are also problematic. On paper, it appears easy to form phosphates from alcohols. Heating phosphoric acid spontaneously gives polyphosphoric acids of various complexities. These polyphosphoric acids are anhydrides, and phosphorylation of an alcohol by any one of them should be easy; however, this is not the case. For instance, a careful study of the phosphorylation of several alcohols with cyclotriphosphoric acid (Fig. 1.14) has shown that this does proceed, but not well, and only at pH 12 (Baba et al., 1990); it requires therefore that the alcohol be present partly as an alkoxide, even though the cyclotriphosphoric acid is then to a large extent anionised.

Not only the phosphorylation of alcohols, but also the simple hydrolysis of these polyphosphates is not an efficient process; the cyclotriphosphates used in domestic detergents, in foods, in tooth pastes, "should" be promptly hydrolysed in water, and precipitated as insoluble calcium phosphate; this is not the case, and complex processes are required to "dephosphate" used waters. The various phosphoric anhydrides involved in biological-energy conversion, in particular adenosinetriphosphate, are also remarkably stable: they require ATP-ase to be hydrolysed. The fact is that all these phosphoric acid anhydrides have, at neutral pH, ionised hydroxyl groups, and are therefore anionic and more difficult to attack by a negatively charged species. Whatever the reason, the fact is that it is not easy to phosphorylate an alcohol! For further discussions on this point, see Baltcheffsky and Baltcheffsky, 1992.

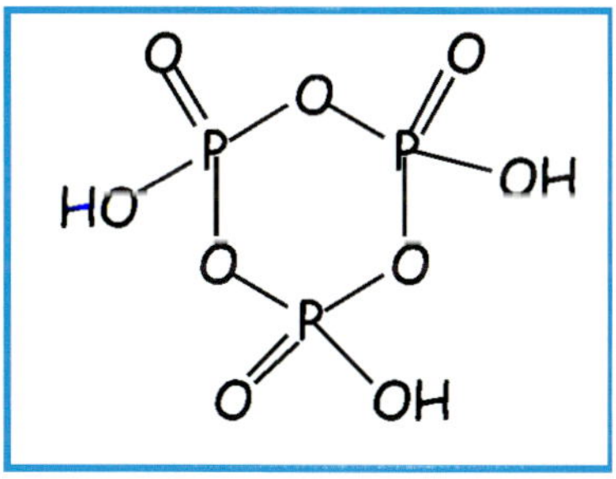

Fig. 1.14. Cyclotriphosphoric acid

1.5.5 The Problem of the C_5 Unit

The isopentenyl, C_5 units postulated to condense into the C_{15+} phosphates involved in membrane formation could be formed by a variety of processes (Fig. 1.15). Biochemically, they are formed either by the classical Lynen–Bloch–Cornforth mevalonic pathway, or by the nonmevalonic pathway more recently discovered by Rohmer (Rohmer et al., 1993, 1996). Both are much too complex to have close nonbiological analogues. However, several reactions could give rise to the branched isopentenyl unit. One is the Prins condensation of formaldehyde and isobutene, to an acetal of isopentanediol (Arundale and Mikeska, 1952; Brace, 1955) or, in acetic acid, to the acetate of isopentenol (Blomquist and Verdol, 1955). Another would be a Paterno–Buchi photochemical condensation of formaldehyde and isobutene, or of acetone and ethylene, to give an acetal of isobutanediol (L. Desaubry, personal communnication). However, none of these appears to be even remotely compatible with prebiotic conditions. We prefer to consider that the abiotic origin of the C_5 unit is still an unsolved problem.

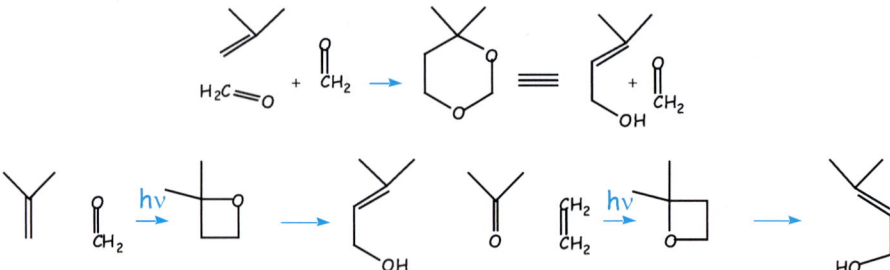

Fig. 1.15. Three possible routes to isopentenol

1.5.6 The Problem of the Cytoskeleton

We consider, of course, that our hypothesis of the early involvement of polyprenyl phosphates to form protocells warrants consideration. We must, however, admit that we cannot make, at the moment, any hypothesis to explain, even in hypothetical terms, the emergence of the cytoskeleton: microtubules in particular. Some such hypotheses are absolutely needed to proceed further.

References

Arundale E., Mikeska L.A., (1952) The olefin-aldehyde condensation, The Prins reaction. *Chem. Rev.*, **51**: 505–555

Auvray X., Perche T., Anthore R., Petipas C., Rico I., Lattes A., (1992) Structure of lyotropic phases formed by sodium dodecyl sulfate in polar solvents, *Langmuir*, **7**: 2385–2393, and many papers by I. Rico and A. Lattes

Bachmann P.A., Walde P., Luisi P.L., Lang J., (1991) *J. Am. Chem. Soc.*, **113**: 8204

Baltcheffsky M., Baltcheffsky H., (1992) Inorganic pyrophosphate and inorganic pyrophosphatases, in Ernster L. (Ed.) *Molecular Mechanisms in Bioenergetics*, New Comprehensive Biochemistry, Elsevier, Amsterdam, 331–348

Bernstein M.P., Dworkin J.P., Sandford S.A., Cooper G.W., Allamandol L.J., (2002) Racemic amino acids from the ultra-violet photolysis of interstellar ice analogues, *Nature*, **416**: 401–403

Birault V., Pozzi G., Plobeck N., Eifler St., Schmutz M., Palanché T., Raya J., Brisson A., Nakatani Y., Ourisson G., (1996) Di(polyprenyl) phosphates as models for primitive membrane constituents: synthesis and phase properties. *Chem. Eur. J.*, **2**: 789–799

Blocher M., Liu D., Walde P., Luisi P.L., (1999) Liposome-assisted selective polycondensation of α-amino acids and peptides, *Macromolecules*, **32**: 7332–7334

W. Blokzijl, J.B.F.N. Engberts (1993). Hydrophobic effects. Opinions and Facts, *Angew. Chem. Int. Ed. Engl.* **32**: 1545–1579

Blomquist A.T., Verdol J.A., (1955) The thermal isobutylene-formaldehyde condensation, *J. Am. Chem. Soc.*, **77**: 78–80

Bretscher M., (1972) Phosphatidyl-ethanolamine: Differential labelling in intact cells and cell ghosts of human erythrocytes by a membrane-impermeable reagent. *J. Mol. Biol.*, **71**, 523–528. – Asymmetrical lipid bilayer structure for biological membranes. *Nature New Biology*, **236**: 11–12

Chrzeszczyk A., Wishnia A., Springer, Jr C.S. (1977) The intrinsic structural asymmetry of highly curved phospholipid bilayer membranes, *Biochim. Biophys. Actan.*, **470**: 161–169

Brace N.O., (1955) The uncatalyzed thermal addition of formaldehyde to olefins, *J. Am. Chem. Soc.*, **77**: 4566–4668

Deamer D.W., (1986) Role of amphiphilic compounds in the evolution of membrane structure on the early Earth, *Origins of Life and Evolution of the Biosphere*, **17**: 3–25

Deamer D.W., Barchfeld G.L., (1982) Encapsulation of macromolecules by lipid vesicles under simulated prebiotic conditions. *J. Molec. Evol.*, **18**: 203–206

Deamer D.W., Harang Mahon E., Bosco G., (1994) in *Early Life on Earth*, Nobel Symposium No 84, Columbia Univesity Press, New York, Self-assembly and function of primitive membrane structures, 107–123

Desaubry L., Nakatani Y., Ourisson G., (2003) Toward higher polyprenols under "prebiotic" conditions. *Tetrahedron Lett.*, **44**: 6959–6961

Devienne F.M., Barnabé C., Couderc M., Ourisson G., (1998) Synthesis of biological compounds in quasi-interstellar conditions. *C.R. Acad. Sci. Paris*, **1, Série IIc**: 435–439

Devienne F.M., Barnabé C., Ourisson G., (2002) Synthesis of further biological compounds in interstellar-like conditions. *C.R. Chimie*, **5**: 651–653

Fischer A., Franco A, Oberholzer T., (2002) Giant vesicles as microreactors for enzymatic mRNA synthesis. *Chem. Bio. Chem.*, **3**: 409–417

Folda T., Gros L., Ringsdorf H., (1983) Formation of oriented polypeptides and polyamides in monolayers and liposomes. *Macromol. Chem. Rapid Commun.*, **3**: 167–174

Fraley R., Subramani S., Berg P., Papahadjopoulos D., (1980) Introduction of liposome-encapsulated SV40 DNA into cells. *J. Biol. Chem.*, **255**: 10431–10435

Fukuda K., Shibasaki Y., Nakahara H., (1981) Polycondensation of long-chain esters of α-amino acids in monolayers at air/water interface and in multilayers on solid surface, *J. Macromol. Sci.-Chem.*, **A(15)**: 999–1014

Ghosh S., Lee S.J., Ito K., Akiyoshi K, Sunamoto J., Nakatani Y., Ourisson G., (2000) Molecular recognition on giant vesicles: coating of phytyl phosphate vesicles with a polysaccharide bearing phytyl chains, *Chem. Commun.*, 267–268

Hanczyk M.M., Fujikawa S.M., Szostak J.W., (2003) Experimental models of primitive cellular compartments: encapsulation, growth, and division, *Science*, **302**: 618–622

Israelachvili J.N., Marcelja S., Horn R.G., (1980) Physical principles of membrane organization, *Quart. Rev. Biophys.*, **13**: 121–200

Jay D.G., Gilbert W. (1987) Basic protein enhances the incorporation of DNA into lipid vesicles: model for the formation of primordial cells *Proc. Natl. Acad. Sci. USA*, **84**: 1978–1980

Kuzmenko I., Rapaport H., Kjaer K., Als-Nielsen J., Weissbuch I., Lahav M., Leiserowitz L., (2001) Design and characterization of crystalline thin film architectures at the air-liquid interface: simplicity to complexity, *Chem. Rev.*, **101**: 1659–1696

Lee S., Desaubry L., Nakatani, Y., Ourisson G., (2002) Vectorial properties of small vesicles. *C.R. Chimie*, **5**: 331–335

R.U. Lemieux, (1996) How water provides the impetus for molecular recongnition in aqueous solution, *Acc. Chem. Res.* **29**: 375–380

Maddox J., (1994) Origin of the first cell membranes? *Nature*, **371**: 101

Miller S.J., (1953) A production of amino-acids under possible primitive Earth conditions. *Science*, **117**: 528–529

Monnard P.A., Oberholzer T., Luisi P.L., (1997) Entrapment of nucleic acids in liposomes. *Biochim. Biophys. Acta*, **1329**: 39–50

Monod J., (1970) *Le Hasard et la Nécessité*. Le Seuil, Paris

Morowitz H.J., Heinz., D., Deamer D.W., (1988) The chemical logic of a minimum protocell. *Origins Life Evol. Biosphere*, **18**: 281–287

Muñoz Caro G.M., Meierhenrich U.J., Schutte W.A., Barbier R., Arcones Segovia A., Rosenbauer H., Thiemann W.H.-P., Brack A., Greenberg J.M., (2002) Amino-acids from ultra-violet irradiation of interstellar ice analogues. *Nature*, **416**: 403–406

Nakatani Y., Yamamoto M., Diyizou Y, Warnock W., Dollé V., Hahn W., Milon A., Ourisson G., (1996) Studies on the totography of biomembranes: regioselective photolabelling in vesicles with the tandem use of cholesterol and a photoactivable transmembrane phospholipidic probe. *Chem. Eur. J.*, **2**: 129–138

Nomura S.M., Yoshikawa Y., Yoshikawa K., Dannenmuller O., Chasserot-Golaz S., Ourisson G., Nakatani Y., (2001) Towards proto: cells: "primitive" lipid vesicles encapsulating giant DNA and its histone complex. *Chem. Bio. Chem.* **2**: 457–459

Nomura S.M., Tsumoto K., Hamada T., Akiyoshi K., Nakatani Y., Yoshikawa K., (2003) Gene expression within cell-sized lipid vesicles. *Chem. Bio. Chem.*, in press

Nordén B., Lindblom G.,Joná I., (1977) Linear spectroscopy as a tool for studying molecular orientation in model membrane systems. *J. Phys. Chem.* **81**: 2086–2093

Oberholzer T., Albrizio, M., Luisi P.L., (1995-1) Polymerase chain reaction in liposomes *Curr. Biol.*, **2**, 677–682

Oberholzer T., Nierhaus K.H., Luisi P.L., (1999) Protein expression in liposomes *Biochem. Biophys. Res. Commun.*, **261**: 238–241

Oberholzer T., Wick R., Luisi P.L., Biebricher C.K., (1995-2) Enzymatic RNA replication in self-reproducing vesicles: an approach to a minimal cell. *Biochem., Biophys. Res. Commun.*, **207**: 250–257

Oparin A.I., (1968) *Genesis and Evolutionary Development of Life*, Academic Press, New York N.Y.

Ourisson G., (1986) Vom Erdöl zur Evolution der Biomembranen (Heinrich-Wieland Lecture) *Nachr. Chem., Tech. Lab.*, **34**: 8–14

Ourisson G., Nakatani Y., (1994) The terpenoid theory of the origin of cellular life: the evolution of terpenoids to cholesterol. *Chem. Biol.*, **1**: 11–23

Ourisson G., Nakatani Y., (1999) Origins of cellular life: molecular foundations and new approaches. *Tetrahedron*, **55**: 3183–3190

Pozzi G., Birault V., Werner B., Dannenmuller O., Nakatani Y., Ourisson G., Terakawa S., (1996) Single-chain polyprenyl phosphates form "primitive" membranes. *Angew. Chem. Int. Ed. Engl.*, **35**: 177–179

Rohmer M., Knani M., Simonin P., Sutter B., Sahm H., (1993) A novel pathway for the early steps leading to isopentenyl diphosphate. *Biochem. J.*, **295**: 517–524

Rohmer M., Seemann M., Horbach S., Bringer-Meyer S., Sahm H., (1996) Glyceraldehyde 3-phosphate and pyruvate as precursors of isoprenic units as an alternative non-mevalonic pathway for terpenoid biosynthesis. *J. Am. Chem. Soc.*, **118**: 2564–2566

Ron E., Huang, J.Y., Popovitz-Biro R., Kjaer K., Bouwman W.G., Howes P.B., Als-Nielsen J., Ron shen Y., Lahav M., Leiserowitz L., (2000) Absolute orientation of molecules of amphiphilic alcohols in crystalline monolayers at the air-water interface. *J. Phys. Chem.*, **104**: 6843–6850

Szostak J.W., Bartel D.P., Luisi P.L., (2001) Synthesizing life. *Nature*, **409**: 387–390

Swairjo M.A., Seaton B.A., Roberts M.F., (1994) Effect of vesicle composition and curvature on the dissociation of phosphatidic acid in small unilamellar vesicles – a ^{31}P-NMR study. *Biochim. Biophys. Acta*, **191**: 354–361

Tanford C., (1978) The hydrophobic effect and the organization of living matter. *Science*, **200**: 1012–1018

Tsumoto K., Nomura S.M., Nakatani Y., Yoshikawa K., (2000) Giant liposome as a biochemical reactor: transcription of DNA and transportation by laser tweezers. *Langmuir*, **17**: 7225–7228.

Ueda T., Lee S.L., Nakatani Y., Ourisson G., Sunamoto J., (1998) Coating of POPC giant liposomes with hydroxylated polysaccharide. *Chem. Lett.*, 417–418

Walde P., Ichikawa S., (2001) Enzymes Inside Lipid Vesicles. Preparation, reactivity and applications. *Biomol. Eng.*, **18**: 143–177

Walde P., Wick R., Fresta M., Mangone A., Luisi P.L., (1994) Autopoietic self-reproduction of fatty acid vesicles. *J. Am. Chem. Soc.*, **116**: 11649–11654

Wang J.-L., Leveiller F., Jacquemain D., Kjaer K., Als-Nielsen J., Lahav M., Leiserowitz L., (1994) Two-dimensional structures of crystalline self-aggregates of amphiphilic alcohols at the air-water interface as studied by grazing incidence synchrotron X-ray diffraction and lattice energy calculations. *J. Am. Chem. Soc.*, **116**: 1192–1204

Weissbuch, I., Zepik, H., Bolbach, G., Shavit, E., Tang, M., Jensen, T.R., Kjaer, K., Leiserowitz L., Lahav, M., (2003) Homochiral oligopeptides by chiral amplification within two-dimensional crystalline self-assemblies at the air-water interface; relevance to biomolecular handedness. *Chem. Eur. J.*, **9**: 1782–1794

Weissbuch I., Rubinstein I., Weygand M.J., Kjaer K., Leiserowitz L., Lahav M., (2003) Crystalline phase separation of racemic and nonracemic zwitterionic α-amino acid amphiphiles in a phospholipidic environment at the air/water interface: a grazing-incidence X-ray diffraction study. *Helv. Chim. Acta*, **86**: 3867–3874

Westheimer F.H., (1987) Why Nature chose phosphates? *Science*, **235**: 1173–1178

Westhof E., (ed.) *Water and Biological Molecules*. MacMillan Press, London 1993, Chap. 1, Savage H.F.J.: "Water structure", Chap. 2 Ludemann, H.D., "Thermodynamic and dynamic properties of water"

Yamagata Y., Watanabe H., Saitoh M., Namba T., (1991) Volcanic production of polyphosphate under primitive Earth conditions. *Nature*, **353**: 516–519

Yu W., Sato K., Wakabayashi M., Nakaishi T., Ko-Mitamura E.P., Shima Y., Urabe I., Yomo T., (2001) Synthesis of functional protein in liposomes *J. Biosc. Bioeng.*, **92**: 590–593

2 Prebiotic Chemistry: Laboratory Experiments and Planetary Observation

François Raulin, Patrice Coll, Rafael Navarro-González

The majority of the current theories on the origins of life are based on the concept of chemical evolution, e.g., a process in which life arises from nonliving matter in accordance with the ordinary laws of physics and chemistry. In this framework it is important to note that the primitive terrestrial environment differed substantially from the present Earth in that there was no free oxygen available in the atmosphere and there was a large abundance of organic compounds that were transformed chemically, by a process referred to as prebiotic chemistry into the first self-replicating systems. This concept was originally introduced by Oparin and Haldane in the 1920s (Oparin, 1924; Haldane, 1929); according to these authors, the evolution of the primordial atmosphere led to the formation of key organic compounds that were reactive intermediates. Upon dissolution in the primitive oceans, these compounds were transformed into amino acids and other molecules relevant for life, constituting the so-called "primordial soup" from which the first heterotrophic cells emerged. This first experimental study in support of the process of chemical evolution was carried out almost five decades later by Miller (1953) whose work came to consolidate this theory.

Later it was realized that the primitive terrestrial atmosphere was not as reducing as originally thought by Oparin and Miller, and under such conditions the synthesis of key organic molecules was not so favorable. Therefore, it was necessary to consider the extraterrestrial inputs of organic material (from comets, meteorites and micrometeorites) to Earth (Oro, 1961) and the synthesis of organic material in the vicinity of hydrothermal sources, from the reduction of carbon dioxide in the presence of hydrogen sulfide and iron sulfide (Baross, Hoffman, 1985). However, the atmospheric synthesis of organic material has always been of great interest, particularly in the exploration of the Solar System such as in the atmosphere of Titan, the largest satellite of Saturn. Even though the environmental conditions of Titan are quite different from our planet, especially in terms of the temperature, there are several analogies to the early terrestrial environment. The study of the chemistry and physical chemistry of these planets offers us an exceptional means of testing the theories and models of the terrestrial primitive environment and prebiotic chemistry that took place approximately 4 billion years ago since most of the geophysical and geochemical evidence of that period was wiped out away.

In this chapter, we discuss three complementary exobiological topics related to the studies of the origin of life and prebiotic chemistry. The first part describes the experimental approach based on the development of what is frequently referred to as laboratory simulations. The first laboratory simulation of the early Earth is not only relevant from a historical point of view but also from exobiology as this type of simulation is routinely used today to model the chemical processes occurring in planetary atmospheres. The second part gives an overview of the elementary steps of prebiotic chemistry: an organic chemistry in liquid water and plausible conditions of the primitive terrestrial environment that likely led to the formation of complex organic compounds that are of biological interest. The last section shows that many of the data obtained in laboratory simulations (as described in Sects. 2.1 and 2.2) are applicable to environments different from the primitive Earth. Combining such laboratory results with astronomical observations and theoretical models, one can then derive relevant information for understanding these astronomical environments, as well as enabling us to test our conclusions in situ.

2.1 Simulation Experiments and Photochemical Models

2.1.1 An Historical View of Miller's Experiment and the Development of a New Field: Prebiotic Chemistry

One of the pioneering experiments specifically designed to test the theory of chemical evolution is the famous celebrated experiment of Miller (also known as the Miller–Urey Experiment; Miller, 1953). Miller studied a model of the primitive Earth's atmosphere in a glass reactor by subjecting it to spark discharges in the presence of liquid water. He initially used a mixture of methane, ammonia, hydrogen, and water vapor, which represented, according to Urey, a model of the early atmosphere of the Earth. These gases are thermochemically stable at ambient temperature and it was thought by Urey in the early 1950s that the Earth accreted at low temperature from an interstellar gas-dust cloud. After several days of sparking, Miller demonstrated the formation of some key organic compounds such as formaldehyde (HCHO) and hydrogen cyanide (HCN) that later were shown to be precursors in the synthesis of several biological and non-biological amino acids. This unique experiment was a pioneering study in the field and was considered as the initiator of a new field of research called prebiotic chemistry, i.e. chemistry of carbonaceous compounds in the absence of life but under conditions relevant to the primitive terrestrial environment. Miller's experiment was conducted with a gas mixture at relatively high pressure (approximately 1 bar), simulating the effect of lightning flashes of thunderstorms in the low terrestrial troposphere.

Lightning activity was certainly not the most abundant energy source in the Earth's atmosphere, compared to the ultraviolet radiation from the Sun. The choice of Miller was most probably guided by experimental constraints. The system of irradiation by electronic impact used by Miller was relatively easy to implement in the laboratory and it allowed the transformation of methane and ammonia (but also of molecular nitrogen if this last molecule is used as a source of nitrogen atoms instead of ammonia). A photochemical system is more complicated to put into practice, and introduces strong constraints in terms of energy levels that can be excited. Indeed, most traditional UV lamps (such as low pressure mercury lamps providing a quasimonochromatic emission to 185 nm), do not allow the photodissociation of methane. However, one can use UV lamps emitting at lower wavelengths (lower than approximately 150 nm) to photodissociate methane. An alternative is to introduce a new compound (such as hydrogen sulfide, H_2S) into the initial gas mixture that can be photodissociated with higher wavelengths and whose products of photodissociation are sufficiently energetic to cause the dissociation of methane.

Since the first publication of Miller's experiment in 1953, this type of experiment has been performed more than several hundred times, by Miller and others, with some variations. The most common modifications are the nature of the initial gas mixture (less reducing or neutral) and the type of energy used to drive the chemical reactions. These studies have shown that electron impact from lightning and solar UV radiation, which are the two most important energy sources in planetary atmospheres, play a complementary role in the synthesis of a large suite of organic compounds.

This experimental approach relies on the use of analytical techniques of chemical analysis, in particular molecular characterization. Their nature and their performances have evolved considerably since the 1950s. During the early 1970s, these experiments became increasingly more complex due to the availability of newly developed analytical techniques with greater sensitivity, allowing in particular the separation and quantification of enantiomers. Thus the experiments carried out by Sagan and his team (Sagan, Khare, 1971) made it possible to show for the first time that amino acids formed during these experiments and later chemically derivatized for their separation by gas chromatography (GC) are produced in racemic form, as one might expect, and thus do not come from an unspecified biological contamination, during the irradiation itself or incorporated during the analytical stage. These results were corroborated later with the advance of new and more powerful analytical techniques (more sensitive and to higher resolution), either by GC or by high-performance liquid chromatography (HPLC). For instance, the work of Kobayashi and his team on the chiral analysis of organic materials, including amino acids (Kobayashi et al., 1989), demonstrated that some amino acids detected at ultralow level in some experiments may actually originate by contamination rather than by synthesis during the preprocessing of the sample, e.g., during hydrolysis.

2.1.2 An Overview of Experimental and Theoretical Data

During these experiments, one could obtain a very broad mixture of compounds of biological interest: a hundred amino acids, including the major proteinic amino acids, more than a dozen purine and pyrimidine bases, including those used in DNA, and many other organics. One can classify them in two categories. The first one corresponds to simple, volatile compounds of low molecular mass such as HCN and other nitriles, HCHO and other carbonyl compounds, carboxylic acids, thiols, amines etc. The second category contains more complex refractory organic compounds of the macromolecular type, whose structures are still not well defined, and are often called "tholins" (see below). Many of these compounds, which may be simple or complex, always lead to compounds of biological interest upon treatment in an aqueous solution.

The experimental evidence indicates that carbon, in the form of methane (CH_4), carbon monoxide (CO) or even carbon dioxide (CO_2), in the presence of various mixing ratios of hydrogen (H_2), can lead to gas phase organic synthesis. Mixtures that mainly contain an oxidized form of carbon (CO_2), in the absence of hydrogen, are not able to yield organic compounds regardless of the type of energy used. Experimental work carried out with intermediate gas mixtures, formed with various mixing ratios of CO_2 and CH_4, leads to the same conclusions: only atmospheres corresponding to a mixing ratio $CH_4/CO_2 > 1$ results in notable outputs of organic compounds (Miller, Schlesinger, 1984; Hattori et al., 1984). This conclusion is also confirmed by photochemical models that were developed in the 1980s (Table 2.1). The first photochemical model of the evolution of the Earth's primitive atmosphere, was published by Pinto et al. (1980), and considered a gas mixture that was mainly composed of carbon dioxide and molecular nitrogen (dinitrogen) with low mixing ratios of molecular hydrogen (dihydrogen) (about 0.01) and carbon monoxide (0.2 ppm). The model included

Table 2.1. Production of formaldehyde in the terrestrial primitive atmosphere according to photochemical models. These models are based on a chemical composition $N_2(0.8)$–H_2O (0.012) (mole fractions), on a total pressure close to the current pressure. Various percentages (expressed in mole fractions) of CO_2 (PAL = present atmospheric level = current mole fraction = 2.4×10^{-4}), CO and H_2 are added. A UV flux= 1 means an intensity equal to the present intesity. T-Tauri means a higher intensity (X1000) considered as similar as the UV-flux of the young Sun during its T-Tauri phase

	Pinto et al. (1980)	Canuto et al. (1983)	Kasting et al. (1984)
No. of reactions	39	47	134
CO_2	1 PAL	1 and 100 PAL	1–1000 PAL
CO	2×10^{-7}	$\sim 10^{-6}$	$\sim 3 \times 10^{-4}$
H_2	8×10^{-3}	17×10^{-6} and 10^{-3}	10^{-5}–10^{-3}
UV flux	1	1 and T-Tauri phase	1
HCHO yield (mole.cm^{-2}.s^{-1})	3×10^8	6×10^3 to 10^{10}	3×10^8 to 9×10^8

approximately 40 chemical reactions, and took into account the possible formation of formaldehyde in the atmosphere with its subsequent transport into the hydrosphere by rain processes. The model predicted that solar UV photons led to an extremely inefficient production of formaldehyde, unable to sustain a moderate concentration of formaldehyde in the oceans that would be required for prebiotic chemistry (see below). More elaborate models (Table 2.1), taking into account the influence of CO_2 pressure and that of solar flux were published in subsequent years (Canuto et al., 1983; Kasting et al., 1984), but were unable to increase the production of formaldehyde. These models, however, do not include the chemistry of nitrogen since molecular nitrogen is photochemically stable in the atmosphere.

2.1.3 New Scenario for Prebiotic Chemistry

It seems relatively well established today that the terrestrial primitive atmosphere was mainly composed of CO_2 and not CH_4, and would have never contained a significant mixing ratio of H_2. Although the experimental data presented above concern mainly the organic synthesis in the gas phase and do not take into account any possible catalytic effects, in particular due to the presence of solid phases (dust or volcanic ash) suitable for inducing heterogeneous processes, the main conclusion drawn from experimental and theoretical studies is that the atmosphere has very likely not played a key role in the formation of organic compounds in the primitive Earth (Navarro-González et al., 1998; 2001).

An alternative mechanism for the production of organics in the early Earth comes from deep hydrothermal vents. These hydrothermal sources are rich in carbon dioxide, hydrogen sulfide and iron sulfide. In the presence of liquid water at high pressures and temperatures, thermodynamical models and laboratory studies (Hennet et al., 1992; Holm, 1992; Marshall, 1994; Brack, 1998; Holm, Andersson, 1998) indicate that carbon dioxide can be reduced into biologically relevant organic compounds. These conditions can also allow the elongation of oligopeptides (Imai et al., 1999). The idea of prebiotic synthesis under the extreme conditions of deep hydrothermal vents is also supported by Günther Wächtershäuser (Wächtershäuser, 1990) who thinks that the first prebiotic systems were capable of replication from an inorganic matrix, namely pyrite, which was quite abundant in the primitive environment. The iron sulfide constituting the pyrite would have the ability to reduce the atmospheric carbon dioxide in the presence of hydrogen sulfide into organic compounds in a process referred to as prebiotic chemiosynthesis. All the requirements of this assumption, in particular the ingredients, are present in the vicinity of the hydrothermal sources.

Finally, if the organic synthesis in the early terrestrial atmosphere were an unefficient process , the origin of the prebiotic molecules could have been essentially exogeneous i.e. deposits of molecules coming from space as minor constituents of meteorities, comets and other extraterrestrial objects falling on the young Earth (Briggs, 1961). A notable fraction of the meteorites that reach the surface of our planet indeed contains many complex organic compounds that could take part in the process of chemical evolution. However, as we will show in the third part of this chapter, the synthesis of organic compounds in the atmosphere is of great interest for exobiology, because they could intervene in a significant way in many extraterrestrial planetary atmospheres.

2.2 Elementary Prebiotic Chemistry in Aqueous Solution

The laboratory simulations of the early Earth are also important to understand the mechanisms by which biologically relevant molecules are made from the raw materials. These chemical pathways are generally referred to as prebiotic chemistry (Miller, Orgel, 1974; Raulin, 1990; Brack, 1998). Important in this regard are the roles of a limited number of key organic molecules such as nitriles (mainly HCN), and aldehydes (mainly formaldehyde). It is the chemical transformations of these compounds in aqueous solution that lead to the building blocks of life.

Today, the knowledge of the prebiotic synthesis of the main building blocks of life (the twenty amino acids of proteins and the five bases of the nucleic acids) is satisfactory. One can assemble prebiotically the amino acids into chains of polypeptides that make up proteins (see, for example, Brack, 1998; Zubay, 2000; and references therein). One can also obtain microstructures, with a membrane similar to those of living cells (see Chap. 1, Part II by Ourisson and Nakatami). On the other hand, neither the prebiotic synthesis of nucleosides (ribose and deoxyribose) nor that of nucleotides and polynucleotides have yet been conclusively demonstrated. The suggestion of a RNA world (see chapter by Maurel) implies that nucleic acids were the first living molecules on the planet; however, the extreme difficulty of their prebiotic synthesis seems to contradict this possibility. The prebiotic synthesis of proteins or at least of polypeptides is more or less much more accessible although there is still a lot of work to realize in this field where new pathways can be unraveled.

2.2.1 Prebiotic Chemistry of HCN:
Strecker Reaction or Oligomerization (see Box 2.1)

The main mechanism of formation of the amino acids in Miller's experiments is the reaction referred as "Strecker raction" (see Box 2.1, well known in organic chemistry for more than one century) (Miller, Orgel, 1974): reaction of HCN with ammonia and an aldehyde in aqueous medium leading to an aminonitrile (step 1), followed by hydrolysis of this aminonitrile into an amino-amide (step 2) and finally into the amino acid (step 3).

Indeed, in Miller's experiment, amino acids are not obtained directly, but are released from a precursor after hydrolysis. However, it has not yet been demonstrated in Miller-like simulation experiments that the precursors to the

Box 2.1. Basic prebiotic chemistry

Prebiotic chemistry is organic chemistry in aqueous solution, under plausible conditions for the primitive terrestrial environment, leading to compounds of biological interest. Elementary prebiotic chemistry uses simple and reactive organic compounds, such HCN, HCHO, HC_3N or their oligomers.

Prebiotic synthesis of amino acids by Strecker reaction

Basic ingredients: aldehyde (RCHO), ammonia (NH_3), hydrogen cyanide (HCN)

* Reaction of HCN with ammonia in aqueous medium leading to an aminonitrile:

$$R-CHO + HCN + NH_3 \rightarrow R-CH(NH_2)-CN + H_2O \quad (2.1)$$

* followed by the hydrolysis of this aminonitrile into an amino-amide then an amino acid:

$$R-CH(NH_2)-CN + H_2O \rightarrow R-CH(NH_2)-CONH_2 \quad (2.2)$$

$$R-CH(NH_2)-CONH_2 + H_2O \rightarrow R-CH(NH_2)-COOH + NH_3 \quad (2.3)$$

Case of the sulfurated amino acids: example of methionine (R = CH_3–S–$(CH_2)_2$)

* step (1) above uses a sulphurated aldehyde obtained prebiotically by reaction between methanethiol and acrolein:

$$CH_3-HS + CH_2=CH-CHO \rightarrow CH_3-S-(CH_2)_2-CHO \quad (2.4)$$

Prebiotic chemistry of HCN

$$HCN + CN^- \stackrel{+H^+}{\rightarrow} \underset{\text{Dimer}}{H-C(=NH)-CN} \stackrel{+HCN}{\rightarrow}$$

$$\underset{\text{Trimer}}{NC-CH(NH_2)-CN} \stackrel{+HCN}{\rightarrow} \underset{\text{tetramer}}{NC-C(NH_2)=C(NH_2)-CN} \stackrel{+nHCN}{\rightarrow}$$

$$\text{oligomers} \stackrel{+H_2O}{\rightarrow} \text{amino acids purines and pyrimidines}$$

Prebiotic chemistry of HC_3N

$$HCC-CN + NCO^-(\text{cyanate}) \rightarrow \text{cytosine} + \text{uracil}$$

$$HCC-CN + H_2O \stackrel{\text{urea}}{\rightarrow} OHC-CH_2-CN \rightarrow \text{cytosine} + \text{uracil}$$

> **Box 2.1. Basic prebiotic chemistry (continued)**
>
> **Prebiotic chemistry of HCHO**
>
> $$HCHO + HCHO \xrightarrow{HCHO} CHO-CH_2OH \xrightarrow{HCHO}$$
>
> $$CHO-CHOH-CH_2OH \xrightarrow{HCHO} C_4$$
>
> $$C_4 + HCHO \rightarrow C_5 \text{ etc...}$$

amino acids are the corresponding aminonitriles. Moreover, such a mechanism implies the (improbable) formation of a broad range of aldehydes required to yield the corresponding amino acids.

Another mechanism involves the polymerization of HCN. This idea was suggested by Matthews (see, for example, Matthews, Moser, 1967; Matthews, 1979; Refs. therein), which caused much controversy in all the conferences on the origin of life and prebiotic chemistry. The idea is tempting: it rests on the assumption that the polymerization of HCN leads to a polymer of well-defined structure, including a side-group imine. This group is suitable for forming, under conditions of soft hydrolysis, a peptide bond, leading thus directly to a polypeptide, without requiring the formation of amino acids. The latter can also be formed starting from polymers of HCN during more severe hydrolysis conditions. This mechanism has the enormous advantage of removing the difficult stage of condensation of the amino acids in the prebiotic synthesis of polypeptides. However, this assumption was constantly criticized and largely disputed, in particular by Ferris (see, for example, Ferris, 1979). This last author claimed that the proposed polymeric structure of HCN has not been supported by analytical means and, furthermore, that its formation depends on the prior formation of a hypothetical HCN dimer that has an improbable regular structure. Ferris, on the other hand, supports the role of the formation of HCN oligomers, with structures still poorly known but most probably much more irregular (Ferris, Hagan, 1984 and Refs. therein). The formation of these oligomers is dependent on the polymerization of the HCN tetramer, diaminomaleonitrile (DAMN, not to be confused with damned, although sometimes this type of compound makes one think of diabolical properties). In any case, without knowing their chemical structure, it is shown that hydrolysis of the HCN oligomers/polymers yields a large variety of amino acids (Ferris, Hagan, 1984 and Refs. therein).

Finally, the last mechanism that should be mentioned calls upon even more complex compounds generically referred to as "tholins" (see below) for the formation of amino acids, but this assumption is not in contradiction with that of Ferris: HCN oligomers can also be regarded as made of tholins.

In any case, it is clearly established that HCN in aqueous solution is easily polymerized into oligomers or "polymers" (actually macromolecules built with nonrepetitive structures) that release a complex suite of organic compounds, including biological amino acids, purines (such as adenine) and pyrimidines upon hydrolysis. The mechanism of HCN polymerization in aqueous solution was studied thoroughly in the 1960s, within the framework of prebiotic chemistry and is now well established (Toupance et al., 1970; Ferris, Hagan, 1984). The first stage is the nucleophilic attack of cyanide (CN^-) into hydrogen cyanide (HCN). This nucleophilic attack can only occur under a narrow range of pH for the reaction to proceed rapidly. Indeed, it was shown (Toupance et al., 1970) that the rate limiting step is precisely the dimerisation occurring at a pH close to the pKa of the acid/base couple: HCN/CN^- that is to say, pH 8 to 9, and at temperatures ranging between 0 and 60°C.

The chemistry of cyanoacetylene, HC_3N, can lead in aqueous solution to the prebiotic synthesis of pyrimidines. However, the direct involvement of this compound does not seem very probable, taking into account its strong reactivity, in particular with ammonia and water. Indeed, this compound is hydrolyzed very easily into cyanoacetaldehyde (Ferris et al., 1968). In fact, it is cyanoacetaldehyde that has an important role in prebiotic chemistry, in spite of pH constraints, taking into account its easy dimerization (Raulin, Toupance, 1975). In particular, the reaction between cyanoacetaldehyde and guanidine provides a relatively simple and efficient pathway for the prebiotic synthesis of cytosine and uracil (Ferris et al., 1974).

2.2.2 Prebiotic Chemistry of HCHO, Formose Reaction

Methanal (commonly called formaldehyde or formol) also undergoes a polymerization reaction in aqueous solution – a reaction known as formose reaction, which is quite relevant in prebiotic chemistry since it leads to the formation of sugars, and among them ribose. This process of polymerization is to some extent similar to the HCN polymerization. It starts by the attack of a monomer molecule, playing the role of nucleophilic reagent, on the carbon of a second monomer molecule, leading, by nucleophilic addition, to a dimer. This dimer can itself undergo the attack of the monomer to give a trimer, etc. One can thus obtain in aqueous solution sugars with $2, 3, \ldots, n$ carbon atoms (C_2, C_3, $\ldots$, C_n), including pentoses (C_5 such as ribose) and hexoses (C_6 such as glucose).

However, for this reaction to proceed it is required to have high concentrations of formaldehyde and very alkaline solutions. These are unlikely conditions for the terrestrial primitive environment. Moreover, this reaction leads to an extremely complex mixture of sugars. Living systems only use one of these sugars in their genetic material: ribose (in the case of RNA). Therefore one of the biggest puzzles in the field of prebiotic chemistry is the problem of selection. In the absence of living systems, what was the natural process that led 1) to the selection of ribose among hundreds of sugars potentially available in the environment of the primitive Earth and 2) to its specific incorporation in polynucleotides? If the prebiotic synthesis of the amino acids, the building blocks of proteins, and polypeptides is well established, this is not the case for nucleotides, the building blocks of nucleic acids and polynucleotides (Shapiro, 1988).

2.2.3 Prebiotic Chemistry of Tholins

In all the experiments simulating the evolution of a gas mixture under an energy-source effect there is always the formation of complex organic material of two types. The first type corresponds to simple volatile organic compounds able to participate in prebiotic reactions, such as HCN, HCHO, HC$_3$N, etc. (see Box 2.1). The second type of products is much less well defined; they are macromolecular products with a bidimensional structure (mainly made of polyaromatic groups) or tridimensional structures, branched or not but still very poorly characterized.

In order to refer generically to this type of material, Sagan and Khare (1979) proposed the word "tholins", derived from the Greek word "tholos", which means muddy. The name "tholins" is assigned to refractory macromolecular organic materials, often obtained in a viscous form, which are systematically formed in simulation experiments. Their composition and chemical structure (which is still very poorly known) depends on parameters such as initial gas composition, energy source, temperature and pressure, etc., among others. However, regardless of their mode of formation, they release biological amino acids upon acid hydrolysis (Khare et al., 1986). Therefore they could be the precursors of these biological compounds in aqueous solution. Many studies were developed during recent years on the chemical and physical properties of "tholins", mainly in the case of Titan, but this topic will be discussed in a next section.

2.3 Application of These Laboratory Experimental Data to Space Studies

The study of the different planetary objects in the Solar System in particular with the possibility of space exploration, enables us today to test the preceding results, which were obtained in the laboratory through simulations (Table 2.2) or those coming from modeling. Thus comparative planetology is an essential approach for exobiology. It obviously makes it possible to seek on other objects for traces of life, biological activity or prebiotic chemistry. But it also makes it possible to observe and study in a real planetary environment, the processes that could have played a key role in the terrestrial chemical evolution process towards the origins of life, and that were until now accessible through numerical modeling or laboratory simulation experiments.

2.3.1 Telluric Planets

Of the three telluric planets in the Solar System that contain an atmosphere, the Earth atmosphere seems for the moment the only one to contain organic compounds in detectable amounts (Table 2.2).

Indeed the systematic exploration of Venus, thanks to the many Soviet missions, and the American Pioneer probes, allowed, among other things, the detailed chemical analysis of its atmosphere. With the exception of COS, which can be regarded as border-line cases, no organic compound was detected in the atmosphere of Venus. The same is true of the exploration of Mars: no organic compound was detected in the Martian atmosphere nor in the first centimeters of the Martian ground. The lack of detection of organic compounds in the atmospheres of Venus and Mars confirms forecasts made starting from laboratory data described in the first part (Table 2.2). Indeed, these atmospheres consist mainly of CO_2 (more than 90%) and N_2 (a few %). Their chemistry is currently dominated by the photochemistry of CO_2 and H_2O, which, in the absence

Table 2.2. Comparison of planetary observations and laboratory synthesis of organics in gas phase (simulation experiments under electron impact)

Gas mixture	Detected products (in particular organics) during the experiment	*Relevant planetary atmosphere* associated with this mixture and detected organics
$CH_4 + NH_3 + H_2O$ $(+H_2)$	RH (saturated & unsaturated), HCN, RCN (mainly saturated) HCHO, other aldehydes, ketones, alcohols, RCO_2H	*Giant planets* hydrocarbons in C_2 & C_3, C_6H_6 (*Jupiter*) HCN (*Neptune*)
	Tholins $\rightarrow$ amino acids, N-heterocycles	Chromophores of the Great Red Spot? (*Jupiter*)
$CH_4 + N_2 + H_2O$	RH (saturated and unsaturated), HCN, RCN (saturated and unsaturated) including HC_3N and C_2N_2 HCHO, other aldehydes, ketones, alcohols, RCO_2H	(Without important fraction of H_2O) Titan & Triton hydrocarbons in C_2–C_4, C_6H_6, nitriles in C_2–C_4 (*Titan*)
	Tholins $\rightarrow$ amino acids, N-heterocycles	Tholins in aerosols? (*Titan*) Tholins on the surface? (*Triton*)
$CO + NH_3 + H_2O$	HCN Tholins $\rightarrow$ amino acids	
$CO_2 + N_2 + H_2O +$ CO/H_2	RH (mainly saturated) HCN, other RCN HCHO, other aldehydes, ketones Tholins $\rightarrow$ amino acids	*Primitive Earth?*
$CO_2 + N_2 + H_2O$	No organic compounds	*Primitive Earth?* Mars, Venus: no organics
$CO_2 + N_2 + H_2O +$ O_2	No organic compounds	*Current Earth:* many organics, including HCN, HCHO, C_2N_2

of significant quantities of CO cannot lead to notable concentration of organic compounds, such as formaldehyde.

Only the terrestrial atmosphere even if strongly oxidizing contains organic compounds: they are in disequilibrium and it is a sign of life on our planet. However, it cannot be excluded that Mars may have had life at some time in the past.

2.3.2 Giant Planets and Their Satellites

All the giant planets have atmospheres with notable concentrations of methane, which induces an organic chemistry (Table 2.2), possibly coupled with the chemistry of ammonia (Jupiter), phosphine (Saturn) or even dinitrogen (Neptune), although the latter is for the moment only a presumption.

Indeed in all these planets, many organic compounds have already been detected (Table 2.2). Leaving aside the case of Neptune's atmosphere where HCN was clearly detected (Rosenquist et al., 1992; Marten et al., 1993; Lellouch et al., 1994), the list is limited for the moment to hydrocarbons, whose formation is easily explained by the photochemistry of methane.

It is fair to say that although the photochemistry of methane has been extensively studied both theoretically and experimentally, there are still significant gaps in our understanding of its chemistry; this is case, for instance, with some primary processes that lead to significant uncertainties in photochemical models (Dobrijevic and Parisot, 1998; Smith, 1999). This is illustrated in Table 2.3, by the great variability in the data reported in the literature for Lyman alpha photons (Yung, deMore, 1999; Mordaunt et al., 1993; Smith and Raulin, 1999). It is thus essential to develop experimental techniques to determine the photochemical parameters of this molecule and of this photo-products (C_2 and C_3 hydrocarbons) as well as the main branching ratios, and their variation with wavelength, temperature, pressure as well as the absorption coefficients among others.

The chemistry of methane is particularly active in Titan's atmosphere, Saturn's largest satellite. With a diameter of more than 5100 km, this satellite is

Table 2.3. Methane photodissociation products by Lyman alpha photons 121.6 nm: variability of the literature data

	Pre-1993 data, Yung and deMore, 1999 (and refs included)	Branching ratio		Smith and Raulin 1999
		Mordaunt et al., 1993 Diagram 1	Diagram 2	
$CH_3 + H$	0	0.49	0.51	0.41
$^1CH_2 + H_2$	0.41	0	0.24	0.53
$^3CH_2 + H + H$	0.51	0	0.25	0
$CH + H_2 + H$	0.08	0.51	0	0.06

in size, the second largest moon in the Solar System. Titan is the only satellite with a dense atmosphere, composed mainly of dinitrogen and methane, with small quantities of dihydrogen. This atmosphere is nearly five times denser than the Earth's, with a surface temperature of 90–100 K and a surface pressure of 1.5 bar. The atmosphere of Titan is very rich in organic compounds. They are present not only in the gas phase but also very likely in the aerosols of which thick layers mask the surface of the satellite.

As in the Earth's atmosphere, with water vapor and carbon dioxide on the one hand and clouds on the other hand, Titan's atmosphere also contains greenhouse gases, (condensed CH_4, equivalent to terrestrial H_2O; noncondensable H_2, equivalent to terrestrial CO_2), and antigreenhouse compounds (aerosols). The thermal profile of the lower atmosphere of Titan is very similar – although the temperatures are much lower there – to that of the Earth, with a troposphere (90–70 K), a tropopause (70 K) and a stratosphere (70–175 K). Moreover, the models of the surface of Titan suggest that it is covered – at least partially – with lakes or seas of methane and ethane. Also, the organic chemistry seems to be present in the three components of what one can call, always by analogy with our planet, the "geofluids" of Titan: air (gas atmosphere), aerosols (solid atmosphere) and surface (oceans) (Fig. 2.1 and Fig. 2.2).

In Titan's atmosphere, the chemistry of methane is coupled with that of dinitrogen, resulting in the formation of many nitrogenated organic compounds in both gas and particulate phases: hydrocarbons, nitriles and tholins. There are complex chemical and physical couplings among the three components of the geofluid of Titan that control the formation and the evolution of the organic matter.

Indeed on Titan there are couplings between the photochemistry of the atmosphere – producing minor constituents and submicrometer particles formed by photooligomerization – and the surface chemistry induced by the cosmic rays.

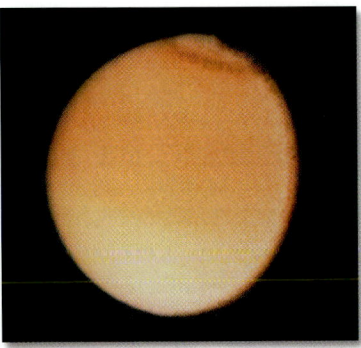

Fig. 2.1. A picture of Titan taken from 2.3 million km by Voyager 2 in 1981. It shows the presence of haze layers covering the whole disk of the satellite. South hemisphere appears brighter than the north and a dark ring is observed near the north pole (credit: NASA)

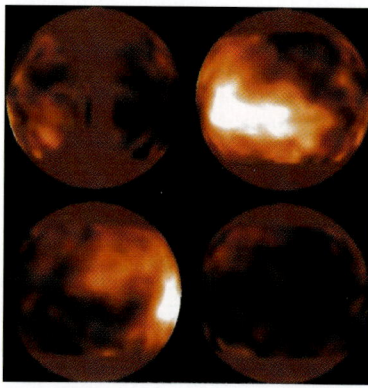

Fig. 2.2. Images of Titan's surface from the Hubble Space Telescope, using near infrared spectral windows. These images obtained by LPL scientists (Univ. Arizona) show bright and dark features over the surface (from the NASA Gallery: http://www.jpl.nasa.gov/cassini/)

Couplings are also involved between the microphysical processes of evolution of aerosols in the upper atmosphere towards the surface, the processes of molecular and turbulent diffusions, and those of condensation in the atmosphere and solubilization in the liquid phase of the lakes or seas which may be present on the Titan surface. Taking into account its many analogies with the Earth (Clarke, Ferris, 1997), the absence of liquid water on its surface, Titan is particularly relevant to understand the process of chemical evolution under anhydrous conditions (Raulin, Owen, 2002). It is interesting to note that all the organic compounds that were detected in the atmosphere of Titan (Table 2.4), were also produced in simulation experiments, including dicyanoacetylene, C_4N_2, an unstable compound at ambient temperature, whose detection in the simulation experiments has been reported (Coll et al., 1999b).

It should be noted that all of these compounds are hydrocarbons and H-, C-, N-containing compounds, mainly organics. The latter are essentialy nitriles, as was also envisaged in laboratory studies (Thompson et al., 1991; De Vanssay et al., 1995; Coll et al., 1998; 1999b; and refs. therein).

Many other organics, including again hydrocarbons and nitriles, are also formed during these experiments and it seems very probable that they are also present in the atmosphere of Titan. This statement is strongly supported by the very recent detection of benzene in Titan's atmosphere through ISO observations (Coustenis et al., 2003; and refs. therein), the presence of which was expected from the results of simulation experiments. Also, the detection of water varpour, in the atmosphere (Coustenis et al., 1998) although at very low concentration, together with CO and CO_2, allows us to consider the possible presence of oxygenated organic compounds, such as methanal and methanol in the atmosphere of Titan, as assumed in the photochemical models of this atmosphere (Yung et al., 1984; Toublanc et al., 1995; Lara et al., 1996). Recent experiments on the modeling of the atmosphere of Titan including traces of CO, show that the main O-containing organic product is oxirane (Coll et al., 2003).

These oxygenated organic compounds that have been identified in the simulation experiments have so far not been detected in the atmosphere of Titan. Their non-detection is probably due to their low concentration and to the observation techniques.

These simulation experiments also lead to the formation of solid, refractory products: macromolecular compounds containing H, C, and N atoms. These "tholins" may be similar to the solid particles of unknown composition that make up fog layers in Titan's upper atmosphere. Tholins have been studied in

Table 2.4. Chemical composition of Titan's stratosphere, and comparison with simulation experiments data (adapted, according to Gautier and Raulin, 1997)

Compounds	Stratosphere mixing ratio (E = Equ.; N = North Pole)	Production in simulation experiments*
• **Main constituents**		
Nitrogen N_2	0.90–0.99	
Methane CH_4	0.017–0.045	
Argon Ar	0–0.06?	
Hydrogen H_2	0.0006–0.0014	
• **Hydrocarbons**		
Ethane C_2H_6	1.3×10^{-5} E	Maj.
Acetylene C_2H_2	2.2×10^{-6} E	Maj.
Propane C_3H_8	7.0×10^{-7} E	++
Ethylene C_2H_4	9.0×10^{-8} E	++
Propyne C_3H_4	1.7×10^{-8} N	+
Diacetylene C_4H_2	2.2×10^{-8} N	+**
Benzene C_6H_6	few 10^{-9}****	+
• **N-Organics**		
Hydrogen cyanide HCN	6.0×10^{-7} N	Maj.
Cyanoacetylene HC_3N	7.0×10^{-8} N	++
Cyanogen C_2N_2	4.5×10^{-9} N	+
Acetonitrile CH_3CN	few 10^{-9}	++
Dicyanoacetylene C_4N_2	Solid phase N	+
• **O-Compounds**		
Carbon monoxide CO	2.0×10^{-5}	
Carbon dioxide CO_2	1.4×10^{-8} E	
Water H_2O	few 10^{-9}***	

* average relative abundance, Maj. = major compound >> ++ >> + ** Coll et al., 1999b; *** Coustenis et al., 1998; **** Coustenis et al., 2003

detail by Sagan and his team (Khare et al., 1981; 1984; 1986; McDonald et al., 1994; and refs. included), their elemental composition determined (McKay, 1996; Coll et al., 1998; 1999a; and refs. therein), and information on the structure was obtained from analysis by pyrolysis coupled to gas chromatography and mass spectrometry (Ehrenfreund, 1995, Coll et al., 1995; 1997, 1999a), and by gel-filtration chromatography of the water soluble fraction (McDonald et al., 1994). The results indicate an elemental composition with a very variable C/N ratio depending on the experimental conditions (Table 2.5). This ratio varies from 1 to 11, and seems to indicate a complex structure including the CN, NH_2, aliphatic and aromatic groups, with an average molecular mass of about 500 to 1000 daltons. Some studies were even carried out to determine the potential assimilation of these materials by micro-organisms, in order to know if they could be used as potential food for hypothetical "Titanic" micro-organisms (Stocker et al., 1990).

Future observational investigations of Titan, which will be carried out within the framework of the Cassini–Huygens mission, either by remote-sensing measurements thanks to the instruments of the Cassini orbiter – which orbits Saturn since July 2004 – or *in situ* measurements from the Huygens probe (Lebreton, 1997; Matson et al., 2002; Lebreton and Matson, 2002), should allow the detection of many compounds identified in the laboratory but not yet observed in Titan's atmosphere (see Box 2.2 and Fig. 2.3).

One of the interests of the experimental simulations is precisely to provide information – at least qualitative – on the nature of the compounds likely to be present in a planetary atmosphere, including those not yet detected. The results of these experiments can thus be used as a guide for the preparation of observational campaigns – by remote sensing or in situ measurements – and for data interpretation.

At LISA a spectroscopic data base in UV and IR has been developed for more than 15 years, for organic compounds of planetological interest (Raulin et al., 1998; Bénilan et al., 2000). The already collected data base relates to hydrocarbons and nitriles as well as to less stable compounds (at ambient temperature),

Table 2.5. C/N ratios in laboratory Titan's tholins according experiments parameters

Experiment	Temperature (K)	Pressure (mbar)	Contamination by oxygen at the time of the sampling?	C/N
Sagan et al., 1984	300	∼1–0.1	Contamination	1.9
Coll et al., 1995	100	∼760	No contamination	11
McKay, 1996	300	∼760	Contamination	5.5
Coll et al., 1997	300	∼0.1	No contamination	1.7
Coll et al., 1999a	100	∼0.1	No contamination	2.8

Box 2.2. The Cassini–Huygens Mission

The Cassini–Huygens mission (Matson et al., 2002), developed jointly by the American (NASA) and European (ESA) space agencies includes the sending of an orbiter (Cassini) around Saturn and of a probe in the atmosphere of Titan. Launch took place in October 1997. After two flybys of Venus, then of the Earth in August 1999, and Jupiter in December 2000, Cassini arrived in the Saturn system in July 2004, and the Huygens probe entered into the atmosphere of Titan on January 14th, 2005. The nominal duration of the Cassini part of the mission is 4 years. The artificial satellite of Saturn, Cassini, contains twelve scientific instruments. Most of them, in particular the IR spectroscopy measurements (CIRS, VIMS) and UV (UVIS), are essential for exobiology. It is the same for most of the six instruments of the Huygens probe, in particular the GC-MS and ACP experiments. The mission also includes 9 interdisciplinary scientists (IDS), who will carry out scientific investigations based on the integrated use of the future Cassini–Huygens data. It is necessary to note the very strong European participation in this mission. (http://www.jpl.nasa.gov/cassini; http://sci.esa.int/huygens)

The Cassini Orbiter

– Scientific instrument	Acronym	Principal Investigator
Optical remote sensing measurements		
– Infrared spectrometer	CIRS	V. Kunde, USA
– Imagery	ISS	C. Porco, USA
– Ultraviolet spectrograph	UVIS	L. Esposito, USA
– Visible/near infrared spectrometer	VIMS	R.H. Brown, USA
Fields of particles and waves		
– Spectrometer of plasma	CAPS	D. Young, USA
– Analyzer of cosmic dusts	CDA	E. Grün, Germany
– Mass spectrometry of ions & neutrals	INMS	J.H. Waite, USA
– Magnetometer	MAG	D. Southwood, GB
– Imagery of the magnetosphere	MIMI	S. Krimigis, USA
– Measurement of radio waves & plasma	RPWS	D. Gurnett, USA
Microwave remote sensing measurements		
– Radar	Radar	C. Elachi, USA
– Radio Measurements	RSS	A.J. Kliore, USA hack

> Box 2.2. The Cassini–Huygens Mission(continued)
>
> - **Interdisciplinary program**   **IDS**
> - Magnetosphere and plasmas   M. Blanc, France
> - Dust and rings   J.N. Cuzzi, USA
> - Magnetosphere and plasmas   T.I. Gombosi, USA
> - Atmospheres   T. Owen, USA
> - Satellites and asteroids   L.A. Soderblom, USA
> - Aeronomy and interaction
>   with solar wind   D.F. Strobel, USA
>
> ## The Huygens Probe
>
> - **Scientific instrument**   **Acronym**   **Principal Investigator**
> - Gas phase chromatograph
>   coupled to a mass spectrometer   GC-MS   H. Niemann, USA
> - Collector and pyrolyser
>   of aerosols   ACP   G. Israel, France
> - Measurement of the atmospheric
>   structure   HASI   M. Fulchignoni, Italy
> - Descent imagery and spectral
>   radiometry   DISR   M. Tomasko, USA
> - Doppler measurement
>   of the winds   DWE   M. Bird, Germany
> - Measuring Instrument of surface   SSP   J.C. Zarnecki, GB
>
> **Interdisciplinary program**   **IDS**
> - Aeronomy   D. Gautier, France
> - Interactions between atmosphere
>   and surface   J. Lunine, USA
> - Chemistry and exobiology   F. Raulin, France

such as polynes and cyanopolyynes. The data set includes frequency and intensity of the bands or the lines, including low temperatures data, which are of particular interest for the atmosphere of Titan.

Experimental simulations also allow the production of laboratory analogues of these atmospheric aerosols. The availability of analogues is also crucial for the preparation of the instruments denoted to in situ analysis and for the interpretation of observed data. They allow testing and calibration of the instrument before and after launching such as the GC-MS and ACP experiments to be performed by the Huygens probe. But they are also essential to determine the physical and chemical parameters of the atmospheric aerosols whose knowledge is essential

Fig. 2.3. Artist view of the Huygens probe entering Titan's atmosphere after release from the Cassini orbiter (from the NASA Gallery: http://www.jpl.nasa.gov/cassini/)

to data processing. The optical properties measured on tholins obtained during laboratory experiments are of major importance to interpret observations made by satellites (Khar, 1984). New determinations of these parameters have recently been made (Ramirez et al., 2002), using Titan tholins obtained under more realistic conditions (in term of gas-mix design, total pressure and temperature) and with an experimental protocol avoiding contamination by the atmosphere of the laboratory.

2.4 Conclusions

Since the first experiment by Miller, in 1953, models of the primitive terrestrial environment and more particularly of its atmosphere have dramatically changed. According to the models currently available and accepted by the scientific community, it seems probable today that the primitive atmosphere of the Earth was never sufficiently reducing to allow atmospheric organic syntheses at least in homogeneous phase, to play a major role in the origin of organics on the prebiotic Earth.

Elemental prebiotic chemistry, based on chemical transformation in aqueous solution of simple organic molecules, such as HCN or HCHO or their oligomers or co-oligomers (equivalent then to what others called tholins) remains however of interest. These basic ingredients could have been formed at the ocean floor, via heterogeneous reactions at the liquid solid interface in submarine vents, or have been imported on the Earth by meteorites, micrometeorites or comets.

Nevertheless, the atmospheric organic syntheses, are still of great importance, because they can occur in many extraterrestrial environments, in particular in the atmosphere of giant planets and especially in Titan's environment. Indeed, chemical processes in Titan atmosphere are the source of many organics, including some of the simple molecules which are key compounds in prebiotic chemistry.

With the development of space exploration, simulation experiments, coupled with theoretical modeling and in situ observations remain an important tool for the understanding of the physical chemistry of planetary environments. The study of atmospheric organic chemistry in extraterrestrial objects is an essential approach to try to understand the primitive terrestrial environment and the organic processes which triggered the emergence of life.

References

Baross J.A., Hoffman, S.E. (1985). Submarine hydrothermal vents and associated gradient environments as sites for the origin and evolution of life. *Origins Life* **15**, 327–345.

Bénilan Y., Smith N.R., Jolly A., Raulin F. (2000). The long wavelength range temperature variations of the mid-UV acetylene absorption coefficient, *Planet. Space Sci.*, **48**, 463–471.

Brack A. ed. (1998). *The Molecular Origins of Life: Assembling Parts of the Puzzle*, Cambridge University Press, Cambridge.

Briggs M.H. (1961). Organic constituents of meteorites. *Nature*, **191(4794)**, 1137–1140.

Canuto V.M., Levine J.S., Augustsson T.R., Imhoff C.L., Giampapa M.S. (1983). The young sun and the atmosphere and photochemistry of the early Earth, *Nature*, **305**, 281–286.

Clarke D.W., Ferris J.P. (1997). Chemical evolution on Titan: comparisons to the prebiotic Earth, *Origins of Life and Evol. Biosphere*, **27** 225–248.

Coll P., Coscia D., Gazeau M.-C., de Vanssay E., Guillemin J.-C., Raulin F. (1995). Organic chemistry in Titan's atmosphere: new data from laboratory simulations at low temperature, *Adv. Space Res.*, **16** (2), 93–104.

Coll P., Coscia D., Gazeau M.-C., Raulin F. (1997). New planetary atmosphere simulations: application to the organic aerosols of Titan, *Adv. Space Res.*, **19** (7), 1113–1119.

Coll P., Coscia D., Gazeau M.-C., Raulin F. (1998). Review and latest results of laboratory investigation of Titan's aerosols, *Origins of Life and Evol. Biosph.*, **28**, 195–213.

Coll P., Coscia D., Smith N.R., Gazeau M.-C., Ramirez S.I., Cernogora G., Israel G., Raulin F. (1999a). Experimental laboratory simulation of Titan's atmosphere: aerosols and gas phase, *Planet. Space Sci.*, **47** (10,11), 1331–1340.

Coll P., Guillemin J.-C., Gazeau M.-C., Raulin F. (1999b). Report and implications of the first observation of C_4N_2 in laboratory simulations of Titan's atmosphere, *Planet. Space Sci.*, **47** (12), 1433–1440.

Coll P., Bernard J.-M., Navarro-González R., Raulin F. (2003). Oxirane: An exotic oxygenated organic compound in Titan? *Astrophys. J.*, **589**, 700–703.

Coustenis A., Salama A., Lellouch E., Encrenaz Th., Bjoraker G.L., Samuelson R.E, De Graauw Th., Feuchtgruber H., Kessler M.F. (1998). Evidence for water vapor in Titan' s atmosphere from ISO/SWS data, *Astron. Astrophys.*, **336**, L85–L89.

Coustenis A., Salama A., Schulz B., Ott S., Lellouch E., Encrenaz Th., Gautier D., Feuchtgruber H. (2003). Titan's atmosphere from ISO mid-infrared spectroscopy, *Icarus*, **161**, 383–403.

De Vanssay E., Gazeau M.-C., Guillemin J.-C., Raulin F. (1995). Experimental simulation of Titan's organic chemistry at low temperature, *Planetary Space Sci.*, **43**, 25–31.

Dobrijevic M., Parisot J.-P. (1998). Effect of chemical kinetics uncertainties on hydrocarbon production in the stratosphere of Neptune, *Planet. Space Sci.,* **46** (5), 491–505.

Ehrenfreund P., Boon J.P., Commander J., Sagan C., Thompson W.R., Khare B.N. (1995). Analytical pyrolysis experiments of Titan aerosol analogues in preparation for the Cassini–Huygens mission, *Adv. Space Res.,* **15** (3), 335–342.

Ferris J.P. (1979). HCN did not condense to give heteropolypeptides on the primitive earth, *Science,* **203**, 1135.

Ferris J.P., Hagan W.J. (1984). HCN and chemical evolution: the possible role of cyano compounds in prebiotic synthesis, *Tetrahedron,* **40** (7), 1093–1120.

Ferris J.P., Sanchez R.A., Orgel L.E. (1968). Studies in prebiotic synthesis III Synthesis of pyrimidine from cyanoacetaldehyde, *J. Mol. Evol.,* **3**, 301–309.

Ferris J.P., Zameck O.S., Altbuch A.M., Freiman H. (1974). Chemical evolution XVIII Synthesis of pyrimidines from guanidine and cyanoacetylene and cyanate, *J Mol. Biol.,* **33**, 693–704.

Gautier D., Raulin F. (1997). Chemical Composition of Titan's atmosphere, *Special ESA Publication,* ***SP* 1177**, 359–364.

Haldane J.B.S. (1929). The Origin of Life, *The Rationalist Annual,* **148**, 3–11.

Hattori Y., Kinjo M., Ishigami M., Nagano K. (1984). Formation of amino-acids from CH_4-rich or CO_2-rich model atmosphere, *Origins of Life,* **14** (1–4), 145–150.

Hennet R.J.-C., Holm N.G., Engel M.H. (1992). Abiotic synthesis of amino acids under hydrothermal conditions and the origin of life: a perpetual phenomenon? *Naturwissenschaften,* **79**, 361–365.

Holm N.G., Guest Editor (1992). Marine hydrothermal systems and the origin of life, *Origins of Life and Evol. Biosph.,* **22** (1–4), 1–191.

Holm N.G., Andersson E.M. (1998). Hydrothermal systems, in *The Molecular Origins of Life: Assembling Pieces of the Puzzle,* ed. A. Brack, p. 86–99, Cambridge University Press, Cambridge.

Imai E-I., Honda H., Hatori K., Brack A., Matsuno K. (1999). Elongation of oligopeptides in a simulated submarine hydrothermal system. *Science,* **283**, 831–833.

Kasting J.F., Pollack J.B., Crisp D. (1984). Effects of high CO_2 levels on surface temperature and atmospheric oxidation state of the early Earth, *J. Atmosph. Chem.,* **1**, 403–428; and refs therein.

Khare, B.N., Sagan C., Zumberge J.E., Sklarew D.S., Nagy B. (1981). Organic solids produced by electrical discharges in reducing atmospheres: tholin molecular analysis, *Icarus,* **48**, 290–297.

Khare, B.N., Sagan C., Arakawa E.T., Suits F., Callicott T.A., Williams M.W. (1984). Optical constant of organic tholins produced in a simulated Titanian atmosphere: from software X-rays to microwave frequencies, *Icarus,* **60**, 127–137.

Khare, B.N., Sagan C., Ogino H., Nagy B., Er C., Schram K.H., Arakawa E.T. (1986). Amino acids derived from Titan tholins, *Icarus,* **68**, 176–184.

Kobayashi K., Oshima T., Yanagawa H. (1989). Abiotic synthesis of amino acids by proton irradiation of a mixture of carbon monoxide, nitrogen and water, *Chem. Letters,* **1989** (9), 1527–1535.

Lara L.M., Lellouch E., Lopez-Moreno J.J., Rodrigo R. (1996). Vertical distribution of Titan's atmospheric neutral constituents *J. Geophys. Res.,* **101** (E10), 23261–23283.

Lebreton J.P., European Space Agency (1997). *Huygens: Science, Payload and Mission,* ESA SP-1177.

Lebreton J.-P., Matson D.L. (2002). The Huygens probe: science, payload and mission overview, *Space Science Rev.*, **104** (1–4), 59–100.

Lellouch E., Romani P.N., Rosenqvist J. (1994). The vertical distribution and origin of HCN in Neptune's atmosphere, *Icarus,* **108**, 112–136.

Marshall W.L. (1994). Hydrothermal synthesis of amino acids, *Geochim. Cosmochim. Acta,* **58** (9), 2099–2106.

Marten A., Gautier D., Owen T., Sanders D.B., Matthews H.E., Owen T.C., Atreya S.K., Tilanus R.P.J., Deane J.R. (1993). First observation of CO and HCN on Neptune and Uranus at millimeter wavelength and their implications for atmospheric chemistry, *Astrophys. J.,* **406**, 285–297.

Matson D.L., Spilker L.J., Lebreton J.-P. (2002). The Cassini/Huygens mission to the Saturnian system, *Space Science Rev.,* **104** (1–4), 1–58.

Matthews C.N. (1979). Reply to Ferris J.P. (1979), *Science,* **203** 1136.

Matthews C.N., Moser R.E. (1967). Peptide synthesis from hydrogen cyanide and water, *Nature,* **215**, 1230–1234.

McDonald G.D., Thompson W.R., Heinrich M., Khare B.N., Sagan C. (1994). Chemical investigation of Titan and Triton tholins, *Icarus,* **108**, 137–145.

McKay C.P. (1996). Elemental composition, solubility, and optical properties of Titan's organic haze, *Planet. Space Sci.,* **44** (8), 741–747.

Miller S.L. (1953). A production of amino-acids under possible primitive earth conditions, *Science,* **117**, 528–529.

Miller S.L., Orgel L. (1974). *The Origins of Life on the Earth,* Prentice Hall, N. Jersey.

Miller S.L., Schlesinger G. (1984). Carbon and energy yields in prebiotic syntheses using atmospheres containing CH_4, CO and CO_2, *Origins of Life,* **14** (1–4), 83–90, and refs. included.

Mordaunt D.H., Lambert I.R., Morley G.P., Ashfold M.N.R., Dixon R.N., Western C.M., Schnieder L., Welge K.H. (1993), Primary product channels in the photodissociation of methane at 121.6 nm, *J. Chem. Phys.,* **98** (3), 2054–2065.

Navarro-González R., Molina M.J., Molina L.T. (1998). Nitrogen Fixation by Volcanic Lightning in the Early Earth. *Geophys. Res. Lett.,* **25**, 3123–3126.

Navarro-González R., McKay C.P., Nna Mvondo D. (2001). A possible nitrogen crisis for archaean life due to reduced nitrogen fixation by lightning, *Nature* **412**, 61–64.

Oparin, A.I. (1924). *Proiskhozhdenie Zhizni,* Izd. Moskovshii. Rabochii, Moscow.

Oro J. (1961). Comets and the formation of biochemical compunds on the primitive earth. *Nature* **190**, 389–390.

Pinto J.P., Gladstone G.R., Yung Y.L. (1980). Photochemical production of formaldehyde in Earth's primitive atmosphere, *Science,* **210**, 183–185.

Ramírez S.I., Coll P., Da Silva A., Navarro-González R., Lafait J., Raulin F. (2002). Complex Refractive index of Titan's aerosol analogues in the 200–900 nm domain, *Icarus,* **156** (2), 515–530.

Raulin F. (1990). Prebiotic syntheses of biologically interesting monomers in aqueous solutions: facts and constraints, *J. British Interplanet. Soc.,* **43**, 39–45.

Raulin F., Coll P., Bénilan Y., Coscia D., Gazeau M.-C., Khlifi M., Bruston P. (1998). Titan's atmosphere: new data of exobiological importance, in *Planetary Systems: The Long View* eds. L.M. Celnikier & J Trân Thanh Vân, p. 435–441, Editions Frontières, Gif/Yvette, France.

Raulin F., Owen T. (2002). Organic chemistry and exobiology on Titan, *Space Science Rev.,* **104** (1–4), 377–394.

Raulin F., Toupance G. (1975). Etude cinétique de l'évolution du cyanoacétaldéhyde en solution aqueuse, *Bull. Soc. Chim.*, 1975 (1–2), 188–195.

Rosenqvist J., Lellouch E., Romani P.N., Paubert G., Encrenaz Th. (1992). Millimeter wave observations of Saturn, Uranus and Neptune: CO and HCN on Neptune, *Astrophys. J.*, **392**, L99–L102.

Sagan C., Khare B.N. (1971). Long wavelength UV photoproduction of amino acids on the primitive earth, *Science*, **173**, 417–420.

Sagan C., Khare B.N. (1979). Tholins: Organic chemistry of interstellar grains and gas, *Nature*, **277**, 102–107.

Sagan C., Khare B.N., Lewis J. (1984). Organic matter in the solar system, in *Saturn* University of Arizona Press, Tucson., pp. 788–807.

Shapiro R. (1988). Prebiotic ribose synthesis: a critical analysis, *Origins Life Evol. Biosphere*, **18**, 71–85.

Smith, N.R. (1999). Sensibilité des modèles théoriques de l'atmosphère de Titan aux incertitudes sur la photochimie des hydrocarbures simples, These de Doctorat, Université Paris 12.

Smith N.R., Raulin F. (1999). Modeling of methane photolysis in the reducing atmospheres of the outer solar system, *J. Geophys. Res.*, **104** (E1), 1873–1877.

Stocker C., Boston P.J., Mancinelli R.L., Segal W.D., Khare B.N., Sagan C. (1990). Microbial metabolism of Tholins, *Icarus*, **85**, 241–256.

Thompson W., Todd H., Schwartz J., Khare B.N., Sagan C. (1991). Plasma discharge in $N_2 + CH_4$ at low pressures: experimental results and applications to Titan, *Icarus*, **90**, 57–73.

Toublanc D., Parisot J.-P., Brillet J., Gautier D., Raulin F., McKay C.P. (1995). Photochemical modeling of Titan's atmosphere, *Icarus*, **13**, 2–26.

Toupance G., Sebban G., Buvet R. (1970). Etape initiale de la polymérisation de l'acide cyanhydrique et synthèses prébiologiques, *J. Chim. Phys.*, **67** (10), 1870–1874.

Wachtershäuser G. (1990). The case for the chemoautotrophic origin of life in an iron-sulfur world. *Origins Life Evol. Biosphere* **20**, 173–176.

Yung Y.L., Allen M., Pinto J.P. (1984). Photochemistry of the atmosphere of Titan: comparison between model and observations, *Astrophys. J. Suppl. Ser.*, **55**, 465–506.

Yung Y.L., DeMore W.B. (1999). *Photochemistry of Planetary Atmospheres*, Oxford Univ. Press, Oxford.

Zubay G. (2000). *Origins of Life on the Earth and in the Cosmos*, Academic Press, San Diego.

3 Chirality and the Origin of Homochirality

John Cronin, Jacques Reisse

3.1 Chirality: Basic Concepts

> "I call any geometrical figure, or group of points, chiral, and say it has chirality, if its image in a plane mirror, ideally realized, cannot be brought to coincide with itself".

This definition of chirality was given by Kelvin in May 1893, at the time of a conference of the Oxford University Junior Scientific Club (Kelvin, 1904). The definition was illustrated by the following comment:

> "Two equal and similar right hands are homochirally similar. Equal and similar right and left hands are heterochirally similar or allochirally similar (but heterochirally is better). These are also called "enantiomorphs", after a usage introduced, I believe, by German writers".

The hand as metaphor is deeply rooted in the history of stereochemistry. (The term chiral derives from the Greek *cheir*, hand.) As Mislow (1996) has pointed out, Kelvin's geometric definition of chirality is equivalent to that given many years later by Vladimir Prelog in his 1975 Nobel Prize lecture:

> "an object is chiral if it cannot be brought into congruence with its mirror image by translation or rotation."

This is now the commonly accepted definition of chirality.

Every object must be either chiral or achiral and, if it is chiral, it can exist a priori in two non-superposable mirror image, i.e., enantiomorphic/enantiomeric forms. Symmetry group theory provides a mathematical criterion by which it can be unequivocally determined whether an object is chiral or achiral. The criterion is expressed as follows: an object is achiral if, and only if, it possesses an improper axis of rotation of the order n. The symmetry operation associated with this element of symmetry corresponds to a rotation of $360°/n$ around the axis, accompanied by reflection in a plane perpendicular to the axis (see Fig. 3.1).

Here S_1 corresponds to a plane of symmetry (σ) and S_2 corresponds to a center of inversion (i). On the other hand, S_4 does not have an equivalent notation. Occasionally one still finds in the chemical literature the formal criterion stated incorrectly by referring only to the absence of a center and plane of symmetry.

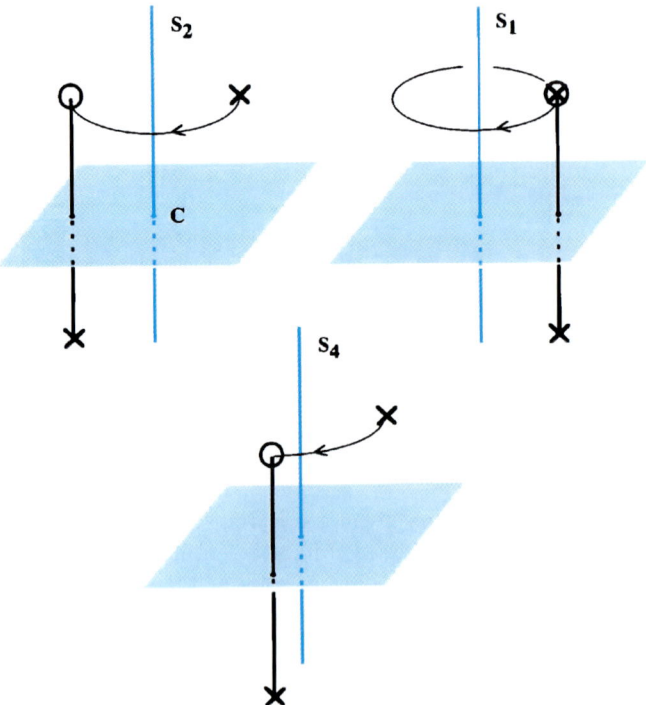

Fig. 3.1. The operations S_1, S_2, and S_4 performed on a material point, X, i.e., rotation about an angle $360°/n$ ($n = 1$ for S_1, $n = 2$ for S_2 and $n = 4$ for S_4) accompanied by reflection in a plane perpendicular to the axis

Nevertheless, it is true that molecular chirality is most commonly due to the absence of a center and (or) planes of symmetry.

A tetrahedron whose four vertices are rendered different, for example by numbering as in Fig. 3.2, is a chiral object of particular interest in chemistry. Indeed, if the center of the tetrahedron is occupied by an atom, for example a carbon atom, and the vertices correspond to different atoms or different groups of atoms bonded to the central atom, the central atom is said to be *asymmetric*. The exchange of two atoms or groups generates the enantiomer, i.e., the nonsuperposable mirror image structure. This is the significant property of the asymmetric tetrahedron from the chirality point of view. The chirality of the molecules found in living things is due to the fact that they contain one, or sometimes many, four-bonded carbon atoms that are asymmetric in this way.

Geometrical chirality can be discussed in spaces of any dimension. The chemist is generally interested in three-dimensional space (E^3) although work of the Mislow group at Princeton has also been directed to quantification of the geometrical chirality of two-dimensional objects such as triangles (Auf Der Heyde et al., 1991; Buda and Mislow, 1991; Buda et al., 1992; Mislow, 1997).

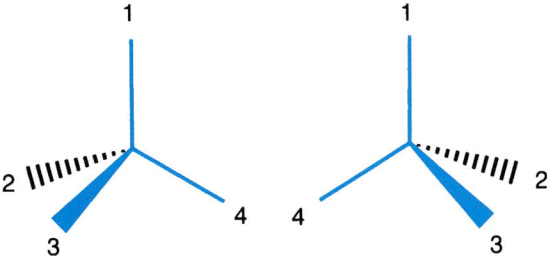

Fig. 3.2. An asymmetric tetrahedron: a chiral object of particular importance in chemistry

Very recently, the problem of the most chiral triangle has been considered by Rassat and Fowler (2003) using a very elegant approach based on quantum mechanics (calculation of eigenfunctions of the Schrödinger equation for the particle confined to an equilateral triangular box).

Two-dimensional space is also of interest because one must often represent a three-dimensional object as a two-dimensional picture, for example, on a page (see Fig. 3.3). Certain rules for handling these representations are imposed by the fact that two enantiomeric objects in two-dimensional space can become congruent (superposable) after passage through three-dimensional space.

Thus it is necessary to be careful in the use of the two-dimensional Fisher projection. For example, if one wishes to determine whether two Fisher projections correspond to the same molecule or, instead, represent two enantiomers, one must test their congruence by translations and rotations only within the plane in which they are represented and not move one or the other through the third dimension.

When the object is a molecule, the Kelvin–Prelog definition remains valid even though molecular objects have some unique properties relating to their dynamics (Mislow, 1997).

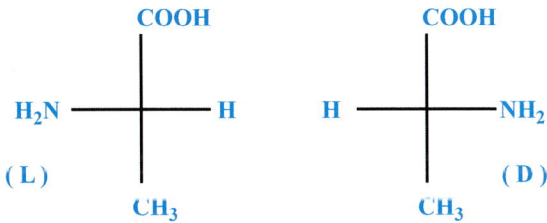

Fig. 3.3. Fisher projection of the two enantiomers of alanine. By convention, the two atoms or groups of atoms (here H and NH_2) at the ends of the *horizontal line* (*bonds*) are viewed as being in front of the plane of the page and the two groups of atoms at the ends of the *vertical line* (here COOH and CH_3) are viewed as being behind the plane of the page. The asymmetric carbon atom is not explicitly shown but is understood to be located at the intersection of the lines

For example, the methane molecule (CH_4) is commonly described as a tetrahedral structure. However, it is sometimes necessary to consider the variations from this ideal structure that can result from motions of the C–H bonds. In reality, the regular tetrahedron is a model that describes the average structure of a great number of methane molecules at any moment (or the average structure of a particular methane molecule taken over time). However, if one were able to photograph, using an infinitely short exposure time, a collection of methane molecules, one would see that only a very small number of them are regular tetrahedrons. The majority of the molecules would be seen to have atomic configurations that are chiral. Of course, one would see among them, in equal numbers, the enantiomeric configurations as well.

An equimolar mixture of enantiomers (molecules that differ only because they are of opposite chiralities) is called a racemic mixture. The term "racemic" was first used by Gay-Lussac to designate an organic acid that lacked optical activity but was of identical composition to optically active tartaric acid (Racemic comes from the Latin, *racemus*, bunch of grapes). Pasteur later showed racemic acid to be composed of equal amounts of the two enantiomers of tartaric acid (see Sect. 3.3) and the term racemic subsequently came to be used as a descriptive term for any such equimolar mixture of enantiomers. This definition is subject to some qualification on the basis of statistical variation, a subject that will be discussed further in Sects. 3.7 and 3.10.

It is important to note that some molecules do not contain asymmetric atoms and owe their chirality to an overall structure that is helical and thereby asymmetric. Although the homochiral molecules isolated from living things owe their chirality primarily to the presence of one or several asymmetric carbon atoms, some biopolymers (proteins, polysaccharides, and DNA) that have asymmetric carbon atoms in their monomeric units (amino acids, sugars, and deoxynucleotides, respectively) also adopt helical conformations. Interestingly enough, at the macromolecular level a hypothetical right helix and a hypothetical left helix made of the same chiral subunits would not be enantiomers but diastereoisomers.

The chirality we have been concerned with up to this point is observed in objects for which the formal interconversion of one enantiomeric form to the other can be carried out by an inversion of space (denoted P), compared to a fixed arbitrary origin. The effect of this inversion is identical to a reflection in a plane mirror. This geometrical chirality does not depend on the overall motion of the object; however, there is another type of chirality associated with movement. An example is a half-cone rotating around its axis. In this case, the inversion of time (operation T), which consists of replacing "t" by "$-t$" and reversing the direction of the movement, transforms the dynamic chiral object into its image. On this basis, Barron (1982) introduced a distinction between "true" chirality and "false" chirality. "False" chirality is associated with objects, like the half-cone example, for which the object-image interconversion can be carried out, not only by inversion of space, but also by inversion of time (followed eventually by rotation in space). Within the framework of this chapter we shall deal only with "true" chirality; however, we must point out that true chirality can be associated with movement (and thus time), for example, in the case of circularly polarized radiation. Indeed, using a wave description, the electric field vector (just like the magnetic field vector) of circularly polarized radiation describes a helix. This helix is palindromic, i.e., the horizontally oriented helix is identical whether traversed left to right or right to left. Thus the inversion of time (which corresponds to the inversion of the direction of the propagation of the light) does not modify the sign of the helicity. On the other hand, the inversion of space transforms right-circularly polarized radiation into left-circularly polarized radiation, i.e., radiation opposite in the sign of its helicity. Thus, one can conclude that circularly polarized radiation possesses true chirality.

In this context, linearly polarized radiation can be described, figuratively, as "racemic" radiation since it is composed of both right-circularly polarized and left-circularly polarized radiation. When such racemic radiation traverses a chiral medium, one of the circularly polarized components is propagated less rapidly than the other and the plane of the polarized light beam is rotated; this phenomenon is called optical rotation (or optical rotatory dispersion when the optical rotation is measured as a function of wavelength). In addition, if the medium is chromophoric, one of the two circularly polarized components is absorbed more than the other, which results in the phenomenon of circular dichroism. The individual enantiomers of chiral compounds differ only with respect to the signs associated with these optical phenomena and have otherwise identical physical and chemical properties except, of course if they interact or react with other homochiral molecules (see Sect. 3.2).

Before ending this section, it may be useful to describe the ways in which enantiomers are specified. In the past, it was usual to designate them with (+) and (−) symbols or with the letters d (dextrorotatory) and l (levorotatory) preceeding the name of the chiral compound. The former was simply the sign of the optical rotation measured with light of the wavelength of the sodium D-line emission. According to a suggestion of Fisher, the D and L nomenclature was based on the relationship between the asymmetric centers of the enantiomers in question and that of a chiral reference compound, the glyceraldehyde of (+) optical rotation (at the D-line wavelength of Na), which Fisher designated D-glyceraldehyde. This assignment depends on the ability to chemically transform the enantiomer into glyceraldehyde. For example, the amino acid serine can be converted to glyceraldehyde by a series of reactions that does not affect the configuration of the groups attached to its asymmetric carbon atom. The product was found to be (−)glyceraldehyde, the levorotatory enantiomer, and thus natural serine is designated L-serine (see Fig. 3.4). The D/L nomenclature is well established for the sugars and amino acids and their derivatives and will be used in this chapter. In Fig. 3.3 the enantiomers of alanine are designated in this way.

In 1950, it was shown by X-ray diffraction carried out by the Bijvoet group that the configuration arbitrarily assigned by Fisher to D-glyceraldehyde was correct (there was a 50% chance!) and since then the D/L nomenclature also

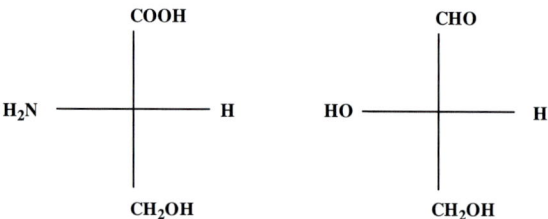

Fig. 3.4. Fisher projection of L-serine and L-glyceraldehyde

correctly denotes absolute configuration. Today, the configurations of chiral molecules are most commonly described by the Cahn–Ingold–Prelog (CIP) system recommended by the IUPAC (International Union of Pure and Applied Chemistry). This nomenclature, which is independent of any measurement of optical rotation, absolutely defines the configuration of every asymmetric center (stereocenter) in a molecule and designates each one as S or R (sinister or rectus).

The reader interested in a concise introduction to stereochemistry and molecular chirality is referred to Mislow (1965). One will find an exhaustive treatment of these subjects in Eliel and Wilen (1994).

3.2 Reactivity of Chiral Molecules

The behavior of chiral molecules when they are not alone is an important consideration. Chemists and biochemists are frequently interested in molecules when they interact (or react) with other identical or different molecules and models are necessary that allow the properties of such sets of interacting molecules to be described. A molecule that is achiral when it is alone can become chiral when it interacts with a chiral molecule. An extreme case is the "chiralisation" of xenon atoms trapped in a chiral molecular cage (Bartik et al., 2000, 2001).

Neglecting for the moment the very small differences in energy due to violation of parity (discussed in Sect. 3.8), two enantiomers have the same internal energy and are characterized by identical reactivity if the second reactant is achiral. What is true for a covalent bond making/breaking chemical reactions is also true for weaker intermolecular interactions if the interaction is with an achiral material. However, for the same reason that a right foot slides easily into a right shoe and with some difficulty into a left shoe, the reaction/interaction will occur at different rates for enantiomers if the molecule with which it reacts/interacts is itself an enantiomer. Enantiomers D1 and L1 will react with reactant D2 with different rates. The interacting pairs (D1–D2) and (L1–D2), whether they are dissociable complexes, reaction transition states, or stable molecules, are called *diastereomers*, meaning that they are stereoisomers (but not enantiomers) and therefore have different internal energies and different properties. The symbols

1 and 2 are not necessarily meant to indicate different molecules: they may be the same or they may be different.

The work of Blount and Idelson (1956) presents an interesting example. They observed that the polymerization of a glutamic acid derivative occurs 20 times faster starting with molecules of the same chirality (only D–D or L–L reactions) rather than with a racemic mixture (D–L and L–D reactions can occur). Matsuura et al. (1965) found that the partial polymerization of a reactive derivative of alanine having a small enantiomeric excess led to an increased enantiomeric excess in the polypeptide products, i.e., they showed that these differences in reactivity could be used to enhance the enantiomeric excess in the product. This is an example of a kinetic resolution (Kagan et al., 1988). The term "resolution" in this context means a process allowing the separation of enantiomers. The possible importance of such processes to the origin of homochirality will be examined in Sect. 3.10. Molecular-recognition processes, which are of enormous importance in biology, are frequently highly chirospecific when both the ligand and the binding molecule are chiral. They represent examples of diastereomeric interactions in which one is generally unfavorable to the point of insignificance. For example, the transport of D-glucose through the membrane of the red cell is facilitated by a chiral transport molecule, a protein.

L-glucose is essentially not recognized by this protein and thus its only mode of penetration into the red blood cell is by unfacilitated diffusion, a much slower process (Rawn, 1989).

3.3 Pasteur and the Discovery of Molecular Chirality

The crystallization of chiral compounds provides many interesting examples of the interaction of homochiral or heterochiral molecules. In fact, it was Pasteur's careful studies of crystals of sodium-ammonium paratartrate that led to the discovery of molecular chirality. Tartaric acid obtained from wine was known to have optical rotation, whereas paratartaric acid lacked optical activity but seemed identical to tartaric acid in all other respects.

In Pasteur's notebooks (Valléry-Radot, 1968), one finds this description of his separation of the enantiomorphic crystals he obtained by crystallization of the sodium-ammonium salt of paratartaric acid.

> "The happy idea came to me to just orient my crystals in a plane perpendicular to the observer, and then I saw that in this confused mass of crystals of paratartrate there were two kinds of them with respect to the distribution of the asymmetric facets. In the one case, the facet of asymmetry close to me was inclined on my line, relative to the plane of orientation of which I spoke, while the others, the asymmetric facet was inclined to my left. In other words, the paratartrate presented itself as formed of two kinds of crystals, one asymmetric to the right, the other one asymmetric to the left. A new idea, very naturally, soon came

to me. The crystals, asymmetric to the right, that I could manually separate from the others, were absolutely identical in form to those of right tartrate".

Pasteur had made a simple discovery with profound implications. He had found that paratartaric acid was a racemic mixture of D- and L-tartaric acids (see Fig. 3.5) but, more importantly, he had shown for the first time that a single organic compound can exist in two forms that differ in the sign of optical rotation and, underlying this, in their molecular asymmetry. The exact nature of this asymmetry was unknown at the time although, according to Mislow (1996), Pasteur explicitly considered that it might be due to an asymmetric tetrahedral arrangement of atoms. It was left to van't Hoff and Le Bel to fully develop the idea that tetrahedral carbon atoms bounded to four different proups are the basis for the chirality of organic molecules such as the tartaric acids and their salts.

The structures of D- and L-tartaric acid are shown in Fig. 3.5. It can be seen that tartaric acid has two asymmetric carbon atoms and, because of this, its stereoisomerism is a little more complicated than we have seen thus far. In the D- and L-tartaric acids, the two asymmetric carbon atoms are of the same configuration (two "right hands" in one case, two "left hands" in the other); however, there exists another form, known as mesotartaric acid, in which the two asymmetric carbon atoms have opposite configurations (one "left hand" and one "right hand"). Consequently mesotartaric acid is optically inactive as a result of internal compensation, that is, the chirality associated with one asymmet-

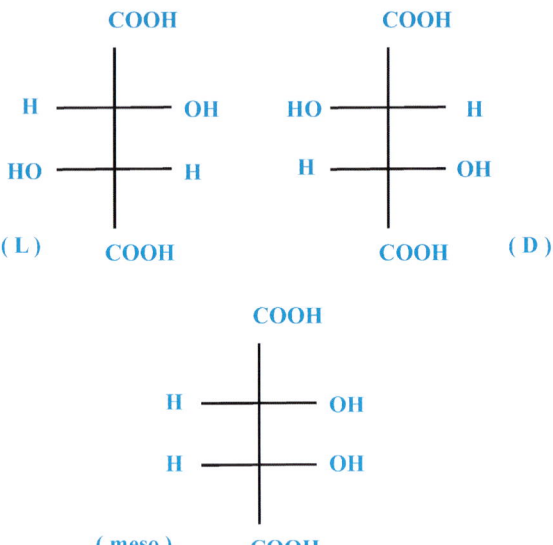

Fig. 3.5. L- and D-tartaric acid and mesotartaric acid in Fisher projection (mesotartaric acid is achiral and is not represented as a mirror image since it is identical)

ric carbon atom is exactly compensated intramolecularly by the chirality of the other asymmetric carbon atom. One might think of mesotartaric acid as being analogous to a right hand and a left hand coupled palm against palm. In contrast to the internal compensation observed in these meso compounds, one can consider a racemic mixture, such as paratartaric acid, as optically inactive due to "external" compensation since the compensation results from the presence of opposed chiralities in different molecules present in equal numbers.

Readers who are interested in the respective roles of Pasteur, van't Hoff, and Le Bel in establishing the foundations of stereochemistry will read with interest the special review volume of Tetrahedron published in 1974 on the occasion of the 100th anniversary of the publication of the work of van't Hoff and Le Bel. The preface to this volume (Robinson, 1974) and an article by Mason (2002) make apparent the pioneering role that Pasteur played in 1860. An article by Lardner et al. (1967) also furnishes information on historical aspects of the chemistry of tetrahedral carbon.

3.4 Crystals and Crystallization

Pasteur's discovery resulted from a combination of exceptional scientific ability and a stroke of good luck. Indeed, the crystallization of a racemic mixture is not always accompanied by enantiomer segregation into enantiomorphic crystals, a process that is unfavorable from the entropic point of view. Generally, molecules of opposite chiralities crystallize together to give a single type of crystal, i.e., a crystal racemate that does not allow separation of the right- and left-handed molecules. Also, the temperature of crystallization is a sensitive factor in determining the type of crystals obtained and this is particularly true in the case of the crystals studied by Pasteur.

Mason (1982), Jacques et al. (1981), Collet (1980), and Collet (1990) thoroughly discuss the question of crystallization of mixtures of enantiomers, the phase diagrams of these mixtures, and the reasons why the crystallization of a racemic mixture more frequently gives crystals of the racemate rather than a conglomerate of crystals composed individually of only one enantiomer. According to Collet (1990), only approximately 10% of racemic mixtures crystallize in the form of conglomerates.

While considering the crystallization of chiral molecules, it is important to note that some achiral molecules can form chiral crystals, which can exist as enantiomorphs. A well known example is quartz, in which the macroscopic chirality of the crystals arises from the helical arrangement of the achiral SiO_2 units of which they are composed (see Fig. 3.6). On a global basis quartz is racemic, a conclusion based on examination of 27,053 quartz crystals (Frondel, 1978).

There are several examples (see Bonner, 1996) of achiral substances that, on crystallization, give enantiomorphic conglomerates. For example, sodium chlorate ($NaClO_3$), when crystallized from supersaturated solutions without agitation, always gives a conglomerate in which there are equal numbers of right and

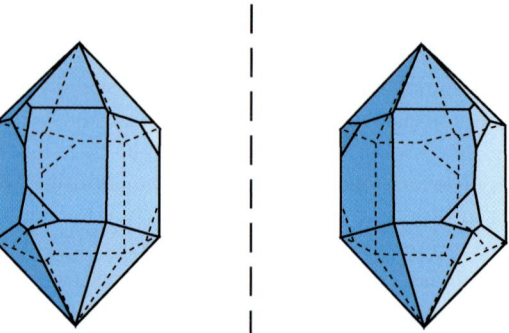

Fig. 3.6. Enantiomorphic crystals of quartz

left crystals. On the other hand, when crystallization is brought about with agitation, sometimes right crystals predominate and sometimes left crystals (Kondepudi et al., 1999). In this case, each experiment gives a spontaneous symmetry breaking, a spectacular phenomenon in which an achiral solution yields a crystalline solid in which the crystals are homochiral! The interpretation of these observations is straightforward: primary nucleation leads to formation of the first chiral crystal and this crystal is then broken by agitation. The fragments, dispersed in the liquid phase, serve as crystallization nuclei and induce secondary crystallization leading, by a cascade effect, to a symmetry breaking (McBride and Carter, 1991). As in the case of quartz, sodium chlorate has been the subject of many studies aimed at showing statistically that the number of right crystals is equal to the number of left crystals. Such crystal counts might at first seem trivial but, as we will see in Sect. 3.8, this is not the case. They aim at eventually obtaining evidence, at the macroscopic level, of an important physical phenomenon called violation of parity.

There are also substances that, in the molten state or in solution, are known to exist as a rapid equilibrium between enantiomers (that is, there is rapid interconversion or racemization) and that, upon crystallization, give homochiral crystals (Havinga, 1954; Kondepudi et al., 1999). In this case there is again a cascade effect in which the separation of the first crystals provides nuclei for further crystallization, effectively displacing the equilibrium toward the first enantiomer to crystallize. These examples of symmetry breaking illustrate the amplification of an initially very slight enantiomeric excess represented here by the first crystal formed. In Sect. 3.10 we shall return to the general problem of the amplification of enantiomeric excesses.

3.5 Homochirality and Life

The importance of Pasteur's discovery to biology was immediately apparent to him. At the time, it was already well known that optical rotation was a property of natural liquids such as turpentine, oil of lemon, and oil of laurel, and therefore must be a property of the molecules, per se. Optical rotation was also known to be a property of solutions of sugars, camphor, tartaric acid and, as we have seen, tartrates. Pasteur's work gave a deeper meaning to these observations by making it clear that molecules, as well as crystals, could exist as asymmetric pairs. Because the direction of optical rotation correlated with a particular asymmetry, and because optical rotation had been observed in so many biological materials, it became clear that the use of molecules of a particular asymmetry was a fundamental property of life. Then the question of the origin of homochirality could be asked for the first time, i.e., how did life choose between the two enantiomeric possibilities with which it was almost always presented? Against the advice of his mentors, the young Pasteur set about attempting to answer this question.

It is ironic that in the second half of the 19th century molecular chirality came to provide one of the last bastions for vitalism, which Pasteur's later research effectively discredited (see Mason, 1982). According to the advocates of this theory, the matter of living things is qualitatively different from that of inanimate matter and one of the differences is that the chiral molecules extracted from living things are always homochiral, whereas, when the chemist synthesizes these same compounds in the laboratory, a mixture of enantiomers in equal amounts, that is, a racemic mixture, is always obtained.

Although Pasteur was unable to answer the fundamental question of the origin of homochirality, he did clearly perceive why chemists were unable to prepare homochiral molecules. At a conference in 1883, he declared:

> "Indeed whenever the chemist in his laboratory combines elements or products born of the elements, he brings into play only nonasymmetric forces. For this reason the syntheses which he carries out never show asymmetry.... I would try asymmetric combinations of elements.... I would make them react under the influence of magnets, solenoids, elliptically polarized light – finally, under the influence of everything which I could imagine to exert asymmetric actions." (Vallery-Radot, 1968)

This statement is similar to, but earlier than, the more general one of Curie, which is now called the symmetry principle:

> "When certain causes produce certain effects, the elements of symmetry of the causes must be found in the produced effects. When certain effects reveal a certain asymmetry, this asymmetry must be found in the causes which gave them birth" (Curie, 1894).

Pasteur made additional comments that are quite interesting in an origin of life context

> "If the immediate principles of life are asymmetric, it is because, in their development, they are governed by asymmetric cosmic forces; therein, in my opinion, is a link between life on the surface of the Earth

and the cosmos, i.e., the entirety of forces spread throughout the universe".

It is remarkable that more than a century ago Pasteur, because of his need to discover "asymmetric forces" that could bring about the symmetry breaking necessary for the homochirality of biomolecules, came to view life on Earth in a cosmic context. In fact, most of the questions that arise today in connection with the origin of the homochirality of life have roots in the work of the 19th century scientists, Pasteur, van't Hoff, and Le Bel.

3.6 The Why and When of Homochirality

Chiral molecules dominate organic chemistry. For example, if we consider the acyclic saturated aliphatic α-amino acids, we see that glycine, the only two-carbon member of the series, is achiral; however, alanine, the C_3 member of the series, is chiral; one of the two C_4 isomers is chiral; all of the three C_5 isomers (including valine) are chiral; and six of the seven C_6 isomers (including leucine and isoleucine) are chiral. In general, the more atoms there are in molecules, the more isomers there are and the larger the number of isomers that are chiral. Clearly, chirality must be a fact of life unless life is somehow constrained to use only the very simplest molecules.

Given that organic chemistry is dominated by chiral molecules, why is biochemistry so rigorously homochiral? The answer to this question lies in understanding the importance of biopolymer structure. Life is based essentially on polymers of two fundamental types, proteins and nucleic acids (DNA and various RNAs). Both of these biopolymers result from the polymerization of chiral monomers and the biological function of both is strongly dependent on their existence in precisely defined three-dimensional structures (conformations). The right-handed double-helical structure of DNA and the unique globular structures of enzymes provide vivid examples of these functionally essential three-dimensional structures.

In the case of proteins, the monomeric units are 19 L-amino acids (to which we could add two very rare L-amino-acids units) plus the achiral amino acid, glycine. In the biosynthesis of a particular protein these amino acids are polymerized to give a chain (polypeptide) having a unique length and amino-acid sequence determined by a particular gene. The resulting polypeptide folds, spontaneously in many cases, to assume the conformation necessary for its biological function. If the amino acids were incorporated in the correct sequence but could vary randomly with respect to their chirality, an enormous number of diastereomeric polypeptides would result (2^{100} for a 100-amino-acid polypeptide). Only a very small fraction of these might be expected to fold correctly to give the exact three-dimensional structure (conformation) required for proper function. Therefore, specification of a particular amino acid as well as its chirality is essential at each step in the biosynthesis of a polypeptide chain in order to insure that all of the products assume the correct three-dimensional structure. The homochirality of amino acids simply precludes chiral variation by providing amino acids of only one configuration (L) for protein synthesis. Of course, this homochirality is now reinforced by the strict chiral specificity displayed by the enzymes of contemporary organisms. The great majority of these enzymes are themselves proteins.

A priori, proteins could have been composed of D-amino acids. If this were the case, the chiralities of all the other constituents of cells would also be of the opposite chirality. This is illustrated by observations made on the enzyme HIV-1 protease when chemically synthesized from D-amino acids. This synthetic enzyme exhibits catalytic activity identical to that of the natural enzyme except that it is specific for the enantiomer of the natural substrate (Milton et al., 1992).

The question has been raised whether molecular homochirality is an absolute requirement for life and many believe that it is. For example, Bonner (1996) finds heterochiral life to be inconceivable. Nevertheless, if, as we have argued above, it is biosynthetic reproducibility of protein structure that life requires, in principle, this does not demand amino-acid homochirality. One can imagine an alternative life form that is both similar and different from terrestrial life: similar in having a triplet nucleotide genetic code, but different in that it specifies both the D and L enantiomers of the 19 chiral amino acids. Its polypeptides could then contain, in precisely ordered sequence, both D and L enantiomers of the chiral amino acids. Each polypeptide would fold and assume a unique three-dimensional structure and would have an active site and display a particular activity just as a homochiral polypeptide does. Such a requirement for both enantiomers of amino acids would, of course, lead to some metabolic duplication of effort. If both D- and L-amino acids were required, it would be necessary, inter alia, to have dual sets of t-RNAs and amino acyl t-RNA synthetases, proteases specific for D–X and L–X peptide bonds, a chirally nonspecific peptidyl transferase, and racemases in order to ensure a balanced supply of both sets of enantiomers.

In the case of the nucleic acids, random heterochirality of the deoxyribonucleotide and ribonucleotide monomeric units of DNA and RNA, respectively, would be just as disastrous for the function of the nucleic acids as is random amino-acid heterochirality for proteins. It would make the regular double-helical structure of DNA impossible as well as the replication and transcription of DNA, during which an extensive regular double-stranded structure is formed transiently (Gol'danski and Kuz'min, 1988). However, in this case, specified heterochirality as suggested for the amino acids of proteins would be equally disastrous because it would also preclude formation of the unique higher-order structure, the antiparallel double helix, that is required for function. Consequently, homochirality is a necessity for a ribose/deoxyribose-nucleotide-based genetic system.

On the other hand, achirality is a possibility for nucleic-acid-like molecules. Achiral nucleic-acid analogues have been synthesized, for example, the peptide nucleic acid (PNA) of Nielsen (1993; 1996). This nucleic-acid analogue, which is completely devoid of asymmetric centers, has a linear array of bases, i.e., the essential feature necessary for encoding and translating information, and can take on the double-helical structure necessary for function (Miller, 1997). If such an informational macromolecule were to provide the genetic memory necessary to reproduce protein amino-acid sequence and, if it specified the use of both D- and L-amino acids, a PNA-protein biochemistry is conceivable in which overall homochirality would not be a necessity. If terrestrial life began on such an achiral-heterochiral basis, the strict homochirality that characterizes life now must have been an outcome of very early evolution and been achieved before the appearance of the Last Universal Common Ancestor (LUCA).

Thus, it is possible to imagine life that is at least partially heterochiral if, but only if, the chiralitry of each amino-acid subunit were fully specified. But how likely is the origin of life on such a basis? If life originated as a PNA-world, catalytic molecules, the analogues of ribozymes, would have been necessary and these would have been asymmetric in their folded forms. Their active sites would have been asymmetric and those that catalyzed key steps, e.g., transamination, in the formation of amino acids would have had a strong tendency to be enantioselective. Therefore, amino-acid heterochirality would have been unlikely. On the other hand, if proteins were necessary for the origin of life, it is difficult to see how the necessary reproducibility of three-dimensional structure could have been achieved without homochirality first having been achieved abiotically. Consequently, we conclude that, although partially heterochiral life is conceivable, its origin and/or evolution on this basis seems unlikely.

The alternative to an evolutionary origin of homochirality is that it was achieved prior to the origin of life. In this case, complex molecules could have been used from the beginning since the problem of chiral variations in monomer sequences that would complicate the production of reproducible biopolymers would have been solved. In the following sections we shall consider some plausible prebiotic processes that might have (a) created a symmetry breaking in chiral molecules and (b) brought about amplification of the possibly small initial enantiomeric excesses that might have resulted.

3.7 Origin of Homochirality and Spontaneous Symmetry Breaking

Many hypotheses have been put forward to explain how homochirality, or at least a significant imbalance in the amounts of enantiomers, might have appeared on the prebiotic Earth as a result of a spontaneous symmetry breaking. For example, a symmetry breaking can occur by the spontaneous separation of a racemic mixture into its constituent enantiomers during crystallization if there is the formation of a conglomerate or a unique L or D crystal (see Sect. 3.4). However, it should be noted that the permissive conditions are tightly constrained and would have a low probability of being met on the primitive Earth. For example, the crystallization of a racemic mixture as a conglomerate followed by separation of the enantiomorphic crystals requires both a high concentration (saturation) as well as the correct temperature for crystallization as a conglomerate and not a racemate. Also, the initial mixture of enantiomers must be relatively pure because the formation of crystals requires that the compound not be a minor constituent of a complex mixture. Finally, some process must be available by which the crystals, once formed, can be spontaneously separated, i.e., an abiotic equivalent of Pasteur's human intervention.

Another possibility is presented by enantiomers in rapid equilibrium in the liquid state that can crystallize as a conglomerate. Here, again, total conversion of the mixture to one enantiomorphic crystal type can be achieved through secondary crystallization caused by strong agitation. Very demanding conditions, even more than in the previous case, must be met in order to observe this behavior. Asakura et al. (2002) continued the work of Kondepudi et al. (1999) in this area. The temperature of crystallization has a narrow range and the excesses of one homochiral crystalline form in comparison with the other varies by an order of magnitude! We concur with the opinion expressed by Bonner in his 1994 review that spontaneous resolution under racemizing conditions, of which there are numerous laboratory examples, probably does not constitute a mechanism of importance for the natural appearance of homochiralty.

Another spontaneous symmetry-breaking scenario is the selective adsorption of one enantiomer from a racemic mixture at the surface of a chiral crystal. As noted in Sect. 3.4, quartz crystals are asymmetric as a consequence of the right- or left-handed helical arrangement of the SiO_2 units and d-quartz and l-quartz selectively adsorb the enantiomers of amino acids. This enantioselective behavior was observed as early as 1935, a role for it in the origin of homochirality suggested by 1938, and the observations more recently confirmed (Bonner, 1996). In verifying this work, Bonner found that the phenomenon required carefully controlled anhydrous conditions, conditions quite unlikely to have been met on the primitive Earth. Furthermore, a survey of terrestrial quartz crystals has shown them to be racemic and, consequently, incapable of giving rise to an overall enantiomeric excess, that is, beyond what might be produced by a single crystal. This limitation to an extremely small locale is a general criticism that can be raised of the hypotheses for the origin of homochirality based on enantioselective adsorption by minerals/crystals. Nevertheless, such hypotheses continue to be suggested as the cause for homochirality on the primitive Earth.

The locality limitation is particularly severe when the adsorptive enantiomorphic surfaces are on the same crystal. For example, enantioselective adsorption has been observed with monoclinic, centrosymmetric and therefore achiral glycine crystals (Weissbuch et al., 1984). Although glycine is achiral, the crystals present enantiotopic faces, that is, their two-dimensional faces are nonsuperposable mirror images of each other. If such crystals float on the surface of a solution containing a racemic mixture of an amino acid and, if only one of the two enantiotopic faces is immersed in the solution, that face can act as a specific adsorbant and enantioselective adsorption approaching 100% can be obtained. More recently, Hazen et al. (2001) have carried out enantioselective adsorption experiments using one of the asymmetric faces of single calcite crystals and gone on to develop a rather comprehensive hypotesis for the origin of self-replicating homochiral peptides.

Clay minerals have been of interest as possible prebiotic catalysts for many years. Many are achiral and thus incapable of playing a role in the origin of homochirality. Others, for example kaolin, are, in principle, chiral, although it has not been possible as yet to demonstrate the chirality of single crystals. Many years ago it appeared that the chirality of kaolin might have been the basis for observations of enantiospecific adsorption and polymerization of amino acids (Jackson, 1971); however, in a careful subsequent study neither effect could be reproduced (Bonner and Flores, 1975). The possibility of such a role for kaolin was later resurrected in a theoretical study that assumed the existence of asymmetric crystals as a result of parity violation (see Sect. 3.8) and showed that enantiospecific amino-acid adsorption is possible (Julg, 1989; Julg et al., 1989).

Following Mills (1932), Dunitz (1996) and Siegel (1998) have suggested that homochirality could have arisen from statistical fluctuations from the equimolar condition that defines the racemic state. As these authors point out, it is incorrect to consider a racemic mixture as consisting of exactly the same number of enantiomeric molecules. If the synthesis of an enantiomeric mixture from achiral starting materials is repeated many times, the products obtained will have different enantiomeric compositions. A plot of frequency vs. enantiomeric composition will give a bell-curve centered on the exactly equimolar composition and with a width proportional to the square root of the variance. The statistics that apply to this situation are the same as those that apply to tossing a coin, that is, an independent event with a binary outcome having a probability $p = 0.5$ for each outcome repeated n times and under the same conditions (Droesbeke, 2001). In this case one obtains a binomial distribution of outcomes with variance proportional to n. If n is equal to 10^{24}, the distribution width will therefore be 10^{12}. According to these authors, a spontaneous symmetry breaking due to statistical fluctuation, followed by its amplification, can be the basis for the origin of homochirality. Mills (1932) explicitly states

> "we might account on the basis of the laws of probability for the existence of an initial minute bias towards one optical system or the other; and this would then, if the principles which I have endeavoured to explain are justified, eventually lead to the complete optical activity of the molecularly dissymmetric components of all living matter."

Furthermore, he stresses the importance of such fluctuations for microscopic prebiotic systems containing, for example, 10^6–10^8 molecules in which the statistical enantiomeric excess, although small in absolute value, can be large relative to the total number of molecules.

In dealing with evolution, Monod (1970) famously employed the dichotomy of chance versus necessity, distinctions that are useful in discussing the origin of homochirality. In the cases described above it is clear that, whatever the cause of the symmetry breaking, the probabilities of formation of both enantiomers are exactly the same and whichever one becomes predominant is a matter of chance. In the section that follows, we will consider an aspect of atomic physics that necessarily gives one particular enantiomer a slight advantage, that is, to use coin tossing as a metaphor, a situation in which the two faces of the coin do not have equal probabilities.

3.8 Origin of Homochirality and Parity Violation

If the homochirality of terrestrial life was a "necessity" and not a matter of "chance" selection, one must ask in what way the biological amino acids of the L-series and monosaccharides of the D-series could have been advantaged. An answer has been suggested by the recognition of a basic asymmetry of matter, the violation of parity that was previously alluded to in regard to the enumeration of quartz crystal enantiomorphs and studies of the handedness of sodium chlorate crystals. This parity violation constitutes a symmetry breaking at the level of the basic laws of atomic physics (Zee, 1999).

Until the 1950s, it was believed that quantum-mechanical operators, wave functions, and the observables that derive from them, preserved parity (P) and were thus independent of the operations of symmetry, that is, inversion with respect to a point or reflection with respect to a plane. In 1956, two young physicists, Tsung-Dao Lee and Chen Ning Yang, suggested that this conservation of parity applied only to quantum systems in which the controlling forces are the intranuclear strong interactions and the electromagnetic interactions. On the other hand, according to Lee and Yang, systems controlled by the weak intranuclear interactions were likely to undergo transformations that did not preserve parity.

Nonconservation of parity was soon experimentally demonstrated in the β-decay of ^{60}Co by Wu and collaborators in 1957. In their experiments, carried out at a temperature close to 0 K and in a static magnetic field in order to create a privileged direction of space, the dissociation of a neutron to a proton, an electron, and an antineutrino was observed. The fact that the intensity of the emission of electrons was not the same parallel and antiparallel with respect to the orientation of the magnetic field constituted proof that the angular momentum of the electrons was preferentially directed antiparallel with respect to the moment associated with the linear movement, i.e., that the electron spin preferentially describes a left-handed helix.

In 1974 Vester reported the first attempts to experimentally induce chirality through the interaction of an achiral reactant with β-radiation from various radioactive sources. He carried out numerous experiments but obtained only false positive and artifactual results. This fruitless work used β-radiation but also the associated circularly polarized light (bremsstrahlung). Bonner (1996) describes later work, including some by his own group. He carried out β-irradiations (exceeding ten years in some cases) using β-radiation from various elements (^{32}P, ^{14}C) and also bremsstrahlung in the hope of obtaining an enantioselective degradation of racemic amino acids but was unable to observe a significant effect.

In addition to the chirality of β-radiation, there is another consequence of the violation of parity that originates in the so-called electroweak interactions that result from unification of the intranuclear weak interactions and electromagnetic interactions. The resulting z force interacts with both the nucleus and electrons and, because of its parity-violating character, distinguishes between left and right. Yamagata (1966) pointed out that because of this coupling, enantiomers cannot have the same energy and that this fact could have relevance to the origin of biological homochirality. The energy difference between enantiomers is called the parity-violating energy difference (PVED). It results from the CPT theorem that states that physics is invariant if, simultaneously, one inverts space (operation P), replaces the elementary particles by their antiparticles (operation C for "loads conjugation") and inverts time, that is, the direction of movement (operation T). To explain the existence of the PVED between enantiomeric molecules, it suffices to take only CP into consideration since in this case time does not play a role.

The apparently equal amounts of enantiomers present in racemic mixtures (i.e., equilibrium mixtures) indicate that K_{eq} must be extremely close to 1.0 and the PVED therefore exceedingly small. To date, it has not been possible to experimentally measure its value; however, there are experimental approaches, in principle, and it may be possible to accomplish this in the future (Quack, 2002). In the meantime, it has been necessary to rely on theoretical methods for determination of PVED. Yamagata's early attempt at such a calculation overestimated its value by several orders of magnitude but, over the last 20 years, increasingly sophisticated ab initio quantum-mechanical calculations have been applied to the problem (reviewed by Quack, 2002). These calculations have generally given PVED values of about 10^{-14} J mol^{-1} corresponding to enantiomeric excesses of about $10^{-15}\%$, although more recent calculations have given larger values for PVED by about one to two orders of magnitude (Quack, 2002).

In addition to the magnitude of the PVED, the question of whether it favors the L- or D-enantiomer of the amino acids and monosaccharides has been of great interest with respect to the origin of terrestrial homochirality. Recent calculations for the amino acids, alanine, valine, serine, and aspartate, and the simplest of the monosaccharides, glyceraldehyde, have shown the PVED to have the "correct" sign, i.e., to favor the biological enantiomers (Zanasi et al., 1999; Mac Dermott, 2002). However, such results are controversial, as their outcome is dependent on both the conformation assumed for the molecule in question as well as the particular calculation method that is used (Buschmann et al., 2000). Bonner

(2000), after a careful recent review of the literature in this area, is unconvinced of a causal link between parity violation and biological homochirality.

3.9 Origin of Homochirality and Photochemistry

As noted in Sect. 3.5, Pasteur included elliptically polarized light among the "asymmetric forces" that he suggested might be capable of inducing enantioselectivity. Nevertheless, it was necessary to wait almost 70 years for an experimental demonstration of this phenomenon. In 1929 Kuhn and Braun showed that partial photolysis of a racemic mixture by irradiation with UV circularly polarized light (UVCPL) gave rise to an enantiomeric excess in the residue. In 1974, Balavoine et al. provided a theoretical treatment of this process that allows calculation of the enantiomeric excess obtained as a function of the degree of photolysis and the differences in the extinction coefficients of the enantiomers for UVCPL. Both Norden (1977) and Flores et al. (1977) demonstrated that aminoacids are subject to enantioselective photolysis by UVCPL. Flores et al. obtained enantiomeric excesses of 1.98% and 2.50% in originally racemic leucine solutions when photolysed to the extent of 59% and 75%, respectively. In this case, the difference between the extinction coefficients relative to their average value (g factor) is about 2%. Recently, Nishino et al. (2001) studied the reaction mechanism and found that acidic conditions are required and that glycine is one of the products of the reaction. Bonner and Bean (2000) have shown that enantioselective photolysis of racemic leucine can also be observed with elliptically polarized light. A role for UVCPL in the origin of the homochirality of terrestrial amino acids has been discussed by Norden (1977; 1978), Rubenstein and Bonner (1983) and by Bonner and Rubenstein (1987). Greenberg (1996) showed that partial UVCPL photolysis of a racemic mixture of the amino acid tryptophan in his laboratory model of low-temperature interstellar-grain chemistry gave an enantiomeric excess in the residue.

Circularly polarized light can also be used to effectuate enantioselective syntheses starting with an achiral reactant. In this case, CPL is the only chiral element. The first example of this was published by Kagan et al. in 1971 and concerned the synthesis of helicene (see also articles by Buchardt (1974), Mason (1982), Rau (1983), and Inoue (1992) which furnish other examples of enantioselective photosyntheses). In Sect. 3.11, we shall discuss the possible role of UV circularly or elliptically polarized light in the formation of the enantiomeric excesses found in the amino acids extracted from certain meteorites.

The results of these experiments are consistent with the Curie principle in that circularly or elliptically polarized light has true chirality and, therefore, a photolysis or a photosynthesis induced by such light must lead to nonracemic products. A recent paper by Rikken and Raupbach (2000; see also the commentary of Barron (2000)) shows that nonpolarized laser light can also lead to enantioselectivity during a photochemical reaction if the irradiated system is placed in a magnetic field and the field is not orthogonal to the propagation direction of the light. Indeed, the combination of a magnetic field B and a wave vector $\boldsymbol{k}$ (along the propagation direction of the light) is endowed with true chirality in so far as $\boldsymbol{k}$ and B are not orthogonal. Reflection in a mirror inverts the propagation direction of the light and therefore the orientation of $\boldsymbol{k}$ but leaves B unchanged since it is an axial vector. The theory of the dielectric constant of an isotropic environment, with or without an external magnetic field, allows quantification of the response of the material environment to the electric field of the electromagnetic radiation. The theoretical expression that follows from this treatment (Jorissen and Cerf, 2002) shows the dielectric constant as an expansion into a series. If one assumes the response of the environment to be linear, one notes that there are four terms that depend on the static magnetic field B in which the sample is placed. These four terms correspond, respectively, to the magnetic rotatory dispersion, the magnetic circular dichroism, the magnetochiral dispersion and the *magnetochiral dichroism*. The results obtained by Rikken and Raupbach (2000) point to the same phenomenon as the one that is responsible for magnetochiral dichroism. There appears a factor $g \cdot B$ that measures the difference between the extinction coefficients of the unpolarized light by the two enantiomers if these are placed in a field B. If B is zero (absence of external field), $g \cdot B$ is therefore also zero and enantioselective photochemistry is not observed. According to Wagnière and Meier (1983), g, in the case of the magnetochiral effect, is several orders of magnitude less than the g characterizing the different response of the two enantiomers to circularly polarized light. This means that to obtain significant enantioselectivity, B must be increased, typically by a few Tesla. (For comparison, the Earth's magnetic field at the surface of the planet is about 10^{-4} Tesla.) Could there exist, at the heart of a protosolar cloud, environments where such large magnetic fields prevailed and where, at the same time, photochemical syntheses produced molecules of biological interest? This remains an open question: the surroundings of a neutron star could be a favorable place in the sense that an enormous magnetic field exists there as well as synchrotron emission of electromagnetic radiation. This light can also be circularly polarized (see Sect. 3.11) but in this case CPL is not necessary for enantioselective syntheses.

Although UVCPL and magnetochiral photochemistry are usually considered to be extrasolar effects, it is necessary to consider the possibility of photosyntheses or enantioselective photolyses on the primitive Earth. In fact, the solar light incident on the Earth is circularly polarized as a result of scattering from atmospheric aerosols (Wolstencroft, 1985; Jorissen and Cerf, 2002). Nevertheless, this effect is canceled, on average, on a daily basis and it is necessary to imagine special topographies such that photosynthesis cannot occur during a part of the day or to invoke light-intensity differences between the morning and evening. The weak circular polarization of the scattered light along with the very constrained conditions that must be satisfied suggests that enantioselective photosynthesis in the atmosphere of the early Earth would not be a very likely occurrence.

3.10 Amplification of Enantiomeric Excesses

3.10.1 Introduction

Thus far we have dealt with several ways in which the symmetry, that is, the perfect enantiomeric equivalence, of racemic compounds can be broken. These symmetry breaking can be nearly quantitative, as in the case of crystallization, or can be minuscule, for example, as predicted for the PVED or allowed by natural statistical deviation from the equivalence of the racemic state. In this section we shall consider mechanisms by which enantiomeric excesses might be amplified, even to homochirality. Bonner's reviews (1991; 1996) critically discuss many of these mechanisms.

3.10.2 Kinetic Resolution

Kinetic resolution (Kagan and Fiaud, 1988), which takes advantage of the different reaction rates of diastereomeric transition states, is particularly pertinent in the case of amino-acid polymerization, a reaction that is commonly assumed to have occurred under prebiotic conditions. As noted previously (Sect. 3.2), transition states for the coupling of chiral amino acids can be diastereomeric (e.g., L L vs. L D) and, as a result, they have different energies and their reactions occur at different rates. Consequently, a homochiral amino-acid pair might react more rapidly than a heterochiral pair. Likewise, the transition states for the next addition can differ similarly (e.g., L L–L vs. D L–L) and their polymerization rates will differ (end-group diasteroisomerism effect).

When about eight amino acids have been joined in this way, the resulting peptide can begin to assume secondary structure, for example, a helical twist. Helixes are themselves chiral and their handedness can be dependent on the chirality of the constituent amino acids. The chirality of such helical coils can reinforce the effect of end-group diasteroisomerism in promoting the preferential addition of a particular amino acid enantiomer (helix effect). The unexpectedly large increase ($20\times$) in the rate of polymerization for derivatives of D- or L-benzyl glutamate compared with that of racemic DL-benzyl glutamate suggested that the helicity of the product might be playing a significant role (Blout and Idelson, 1956). This led Wald (1957) to suggest that the α-helical structure of polypeptides might favor the further addition of amino acids of the same chirality as those that determined the handedness of the helix and that this could have been of significance in the origin of the homochirality of biological amino acids. Later, Spach (1974) demonstrated that the enantioselective effect observed in the polymerization of the benzyl glutamates arises mainly from the helical structure of the product. However, a positive helix effect is not observed with all amino acids. Blair and Bonner (1980) found substantial increases in enantiomeric excess in the polypeptide product when a derivative of leucine was polymerized but enantiomeric decreases with the corresponding valine derivative. Brack and Spach (1981) have pointed out that formation of another type of secondary structure assumed by some polypeptides, that is, the β-pleated sheet, can also give rise to enantiomeric enrichments.

A third effect, the greater stability to hydrolysis of the helical polymer in comparison with the corresponding random coil (stability effect), can further enhance the enantiomeric excesses achieved by polymerization. An example of the amplification that can be achieved by amino-acid polymerization combined with partial hydrolysis has been given by Blair et al. (1981). They found that partial polymerization of a leucine derivative having a 31% L-enantiomeric excess gave polypeptide products with an enantiomeric excess of 45% and that after partial hydrolysis of this material the enantiomeric excess in the surviving residue was 55%. Blair et al. (1981) suggested that repeated wet–dry cycles on the prebiotic Earth might have led to partial polymerization of amino acids followed by partial hydrolysis of the polypeptide products and thus similarly enhanced any initially small enantiomeric excesses.

The work of Eschenmoser et al. (Bolli, 1997) gives an example of what the authors describe as a deracemization process that possibly could occur during polymerization. Activated tetramers of ribonucleotides containing ribose as a pyranose ring were involved in further polymerizations. 8-mers (octamers), 12-mers and even 16-mers are obtained. A strong chiroselectivity was observed and the presence of tetramers containing L-ribose instead of D-ribose had no major influence on the polymerization rates. If we consider one mole of a 50-mer based on 4 different but homochiral ribonucleotides it is impossible to form all the possible sequences because 4^{50} is much larger than the Avogadro number. Following Eschenmoser et al. the final mixture of polymers must necessarily be nonracemic leading therefore to a deracemization process without any chiral external influence. Siegel (1998) also has pointed out the practical impossibility of generating all possible enantiomers in a mixture of polymers with a high (or even moderate) degree of polymerization by starting from racemic mixtures of 20 different monomers.

Recently, Zepik et al. (2002) have given an interesting example of amplification of an enantiomeric excess by polymerization in the two-dimensional environment provided by a water/air interface. The activated monomers of amino acids (lysine and glutamic acid) were functionalized with long-chain hydrocarbons to make them amphiphilic. As a result, they accumulate at the water/air interface where they associate in crystalline two-dimensional aggregates. When the starting mixture was enantiomerically unbalanced, they observed the formation of a racemate along with an enantiomorphic phase that polymerizes, yielding homochiral oligopeptides. In this case intermolecular packing at the interface provides structural order analogous to the helix effect described in the preceding paragraph for homogeneous polymerization. A mechanism for separation of the enantiomeric and racemic phases was not suggested by the authors. Although this phenomenon could be of prebiotic interest with respect to hypothetical chiral amphiphilic compounds it should be noted that amino acids and monosaccharides are quite water soluble and, unless chemically modified, do not accumulate at a water/air interface.

Takats et al. (2003) have shown that clustering of serine molecules is a efficient process for chirality amplification as well as for chirality tranfer to other amino acids or sugars. Although this is not polymerization in the strict sence, it does provide an example of an amino acid based suparmolecular system with possible relevance to prebiotic chemitry.

3.10.3 Chiral Catalysis

Chiral catalysis allows the transfer of chirality from a catalyst to the reaction product and, in many cases, impressive enantiomeric excesses are achieved in the product when the catalyst is enantiomerically pure. Sometimes very significant nonlinear effects are also observed in such catalytic reactions, that is, when the catalyst is not enantiomerically pure, a greater or lesser enantiomeric excess than that of the catalyst may appear in the product (Girard and Kagan, 1998). In the former case, this amounts to an amplification of chirality.

The work of Feringa et al. (1995) describes a physical system in which a small amount of an enantiomer stimulates a large effect. Here an achiral mesophase (achiral liquid crystals), when seeded with a very small quantity of a chiral additive, gives a transition phase leading to a chiral environment at the macroscopic level (transition between an achiral nematic phase and a cholesteric chiral phase). The chiral additive in this case is a photoisomerizable dopant that allows, according to the wavelength of irradiation, the induction of a cholesteric phase of either right or left helicity (chiroptical switching). Of course, the presence of achiral or chiral mesophases on the primitive Earth is an area of sheer speculation.

3.10.4 Asymmetric Autocatalysis: Theoretical Models

Asymmetric autocatalysis can be considered to be a special case of chiral catalysis in which a chiral reaction product acts as the catalyst. Franck (1953) proposed an early theoretical model for such a process, that is, for the amplification of an enantiomeric excess by means of a chemical reaction in which the product is a chiral compound and one enantiomer (for example, the L-enantiomer) catalyzes its own formation while inhibiting the formation of the other enantiomer (the D-enantiomer). The crystallization from solution of an enantiomeric mixture in rapid equilibrium is an example of such a process. In this case, a very slight initial excess of one enantiomer due to a random fluctuation in the racemic mixture can trigger a crystallization cascade that carries the system far from equilibrium, i.e., toward a large enantiomeric excess of the randomly favored enantiomer (see Sect. 3.4).

Following Franck, Decker (1974) systematically studied four different chemical systems all characterized by the presence of autocatalytic feedback. He showed that such systems show bifurcation phenomena that manifest themselves by the sudden amplification of a very small initial enantiomeric excess. Decker explicitly placed this work in the context of the Brussels school of the thermodynamics of open systems far from equilibrium. Kondepudi, based on previous work by Prigogine and Nicolis, has published several papers (Kondepudi and Nelson, 1984; 1985; Kondepudi, 1996; Prigogine and Kondepudi, 1999) devoted to an in-depth study of a system composed of five interdependent chemical reactions likely to show bifurcation phenomena. The system, also a modification of the autocatalytic model of Franck, is as follows:

$$A + B \to X(L)$$
$$A + B \to X(D)$$
$$X(L) + A + B \to 2X(L)$$
$$X(D) + A + B \to 2X(D)$$
$$X(L) + X(D) \to Y$$

The scheme has two autocatalytic steps and an irreversible mutual destruction step, features that appear to be essential in systems that show symmetry breaking. The system is open, that is, A and B, the chiral reactants, are maintained at constant concentrations throughout the simulation experiment and the product Y is eliminated, thereby rendering the reverse reaction impossible and making the evolution of the system irreversible.

Kondepudi et al. are interested in the magnitude of $X(L)-X(D)$, the enantiomeric concentration difference, in relation to the concentration of the perfect racemic mixture ($X(L)-X(D) = 0$). They carried out several simulations corresponding to different values of the molar concentrations product $A \cdot B$ and observe that there exists a critical value of product $A \cdot B$ beyond which the symmetrical solution $X(L) = X(D)$ becomes unstable. This means that a random statistical fluctuation can provoke bifurcation of the system toward a state in which $X(L)$ is no longer equal to $X(D)$ and that corresponds to a spontaneous symmetry breaking. However, if at the start there is a chiral perturbation of the system, for example by the presence of a very small excess of $X(D)$ or of $X(L)$, a bifurcation will be observed when the critical value is attained leading to an amplification of the initial enantiomeric excess (Fig 3.7).

According to Kondepudi et al., this mechanism will allow amplification of initial excesses as small as those that are predicted by the violation of parity (PVED) (see Sect. 3.8) if the accumulation of A and of B can continue in a very large volume (typically a lake of some km^2) for long time periods (typically tens of thousands of years), in this way reaching the critical value of the concentration product $A \cdot B$. Kondepudi and Nelson (1984) note that in their simulations they cannot account for the effect of chance chiral impurities in the environment that would affect the kinetics of the system. They conclude that

"When one finds a real chemical system that breaks chiral symmetry, in order to preserve the effects of weak-neutral current interactions, the

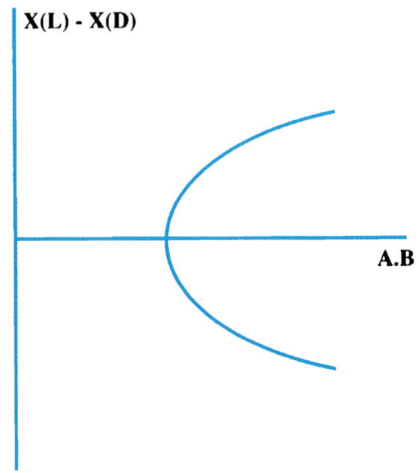

Fig. 3.7. Example of bifurcation predicted by the theoretical model of Kondepudi. A and B are the molar concentrations of the achiral reactants A and B while X(D) and X(L) are the molar concentrations of the chiral products X(L) and X(D) (see text for discussion)

system must be prepared with sufficient purity that the selection occurs due to this interaction and not due to the presence of a chiral impurity."

These condition and constraints raise serious questions as to the significance of such a theoretical model in the context of prebiotic chemistry. However, as will be seen in what follows, the general point is quite significant in evaluating experimental systems of this type.

3.10.5 Asymmetric Autocatalysis: Experimental Data

Recently, asymmetric autocatalysis has been experimentally shown by the group of Soai to give spectacular amplifications (Soai et al., 1995; 2000; Shibata et al., 1998). Soai and his coworkers have studied in detail the addition of an achiral organozinc derivative to an achiral aromatic aldehyde (Soai reaction). The reaction ultimately yields a mixture of enantiomeric alcohols. These alcohols then act as catalysts of their own synthesis. If the reaction is carried out in the presence of a chiral initiator such as leucine with 2% L-enantiomeric excess, the mixture of alcohols obtained at the end of the reaction shows an enantiomeric excess of 21% in one of the enantiomers. If the initiator is enriched in D-leucine, at the end of the reaction the mixture of alcohols is enriched in the other enantiomer and the enantiomeric excess is 23%. The interpretation of these results is as follows. The presence of the chiral initiator leads to very small enantiomeric excess in the first alcohols formed. These then act as the asymmetric catalyst. Since the alcohol is catalyzing its own synthesis an autocatalytic system is operating that amplifies the enantiomeric excess. Soai et al. have observed that an initiator with an enantiomeric excess of only 0.1% can give a mixture of alcohols with an enantiomeric excess greater than 80%! The authors point out that the enantiomeric excess in the leucine used as the initiator is similar to those that can be obtained by enantioselective photodecomposition of a racemic mixture of leucine using UVCPL (Sect. 3.9). They conclude that autocatalysis in the presence of a chiral initiator could have played a role in the appearance of terrestrial homochirality. Furthermore, these researchers (Soai, 1999) have observed that homochiral quartz crystals are also able to play the role of initiator and, starting with the same achiral reactants, give enantiomeric excesses of greater than 80% in the product alcohol. Depending on whether the quartz crystals are l or d, a preponderance of one or the other alcohol is observed.

Subsequently, Singleton and Vo (2002) attempted to use the Soai reaction to test whether the statistically allowed enantiomeric excesses that occur in "racemic" mixtures might be sufficient to initiate the formation of products with a substantial enantiomeric excess. They were unable to demonstrate this for an unexpected reason: the chiral impurities present in even purified solvents are capable of initiating the amplification process in this autocatalytic system and effectively overwhelm any weaker effects. Thus, as Kondepudi and Nelson pointed out (see above), the ubiquity of homochiral molecules in terrestrial environments poses a severe experimental problem for those who would hope to attribute chiral effects to minute enantiomeric excesses such as the statistically allowed variations of racemic mixtures or those that might arise from PVEDs.

Very recently, Soai et al. (2003) appear to have accomplished the amplification of enantiomeric excesses having only a statistical origin. They carried out the synthesis of the chiral pyrimidyl alcohol from only the achiral reagents (the pyrimidyl aldehyde and diisopropyl zinc) in pure solvents and obtained enantiomeric excesses ranging from 29–91% with each enantiomer predominating in approximately half of the cases ($18R$, $19S$). As Mislow (2003) points out in an interesting commentary on this work, this is an example of absolute asymmetric synthesis, i.e., the formation of enantiomerically enriched products without the influence of any pre-existing chirality. Apparently, the random statistical fluctuations in the chirality of the initially formed catalytic zinc complex are amplified and determine the asymmetric outcome of the overall reaction. To be consistent with Curie's symmetry principle "this [product] asymmetry must be found in the causes that gave them birth" and in this case no causal asymmetric influence appears to be available other than random statistical fluctuation.

These results are of obvious importance in the context of the origin of homochirality problem although, as Avalos et al. (2000) have pointed out, the particular reaction studied by the Soai group does not readily proceed under prebiotic conditions. Organometallic chemistry requires rather restrictive non-aqueous conditions but it is hoped that in the near future similar autocatalytic reactions will be discovered that are more compatible with the constraints imposed by prebiotic chemistry. If such are found, it will be possible to agree with

Siegel (2002) when he states that *"in the prebiotic world examples of dominant molecular handedness were already likely to be abundant."*

3.10.6 On the Possibility to Amplifying Enantiomeric Excesses due to Parity Violation

Kinetic resolution and other related processes associated particularly with phase changes like crystallization can lead to amplification only when starting with mixtures having initial enantiomeric excesses that are very much greater (many orders of magnitude) than those that might be provided by the PVED. (see Sect. 3.8.) Consequently, amplification of these extremely small enantiomeric excesses is a subject that has received considerable attention and is treated separately here although, as discussed below, it has much in common with the problem of amplifying statistical excesses in racemic mixtures.

Yamagata (1966), who recognized very early the possible implications of parity violation for the origin of homochirality, also speculated as to how such small enantiomeric excesses might be amplified. He proposed an "Accumulation Principle," that is, the amplification of a very small enantiomeric bias when it is repeatedly multiplied by itself, for example, when it affects each step in a linear reaction sequence composed of a very large number of discrete steps. According to Yamagata, this principle would be applicable to either polymerization or crystallization phenomena. However, the application to polymerization considers only the possibility of formation of two completely homochiral polymers (enantiomers) and fails to consider the inevitable formation of numerous heterochiral polymers (diastereomers). Likewise, the proposal of irreversible consecutive steps is an oversimplification of crystallization. An excess of the l enantiomorph in natural quartz has been cited as possible evidence for the influence of PVED on crystallization (Tranter, 1985); however, as noted in Sect. 3.4, more recent evaluation of the data has failed to confirm a significant enantiomorphic excess (Frondel, 1978). Bonner (1999) has recently evaluated the Accumulation Principle and found it to be incapable of producing amplification by either polymerization or crystallization (see also Avalos, 2000).

Work by Szabo-Nagy et al. (1999) is purported to be the first experimental demonstration of the PVED. The authors prepared a racemic mixture of the *tris*(1,2-ethanediamine) complexes of Co^{3+} and Ir^{3+} by mixing the enantiomers to zero circular dichroism (CD). They then crystallized numerous samples and measured the CD of their solutions. A distribution of CD values was obtained, which, in the case of the metallocomplexes, was shifted from zero. On this basis, the authors claim that undetectably small enantiomeric excesses due to the PVED were amplified by crystallization. The observation of an effect only with the metallocomplexes and not with the racemic mixture of tartaric acids was attributed to the higher atomic number of the atom at the asymmetric center and the resulting larger PVED. The results are controversial (Avalos, 2000). It is not clear that the amplification is attributable to the PVED rather than deviations from the exactly racemic state and/or the effect of chiral impurities on the crystallization process. Circular dichroism is not sufficiently sensitive to detect small enantiomeric exess, even ones much larger than expected taking into account parity violation or statistical fluctuations.

Salam (1993) has proposed another mechanism by which the minuscule enantiomeric excesses due to PVEDs might be amplified. Salam, who received the Nobel Prize for his work on unification of the weak interactions and electromagnetic interactions, envisages an interconversion of the less stable enantiomer to the more stable one in the crystalline phase below some critical temperature. This cooperative effect would occur by a second-order phase transition and is similar to the transition observed when an electrical conducting material becomes a supraconductor. Figureau et al. (1995) attempted to observe a configurational change in racemic cystine over the temperature range 0.6–77 K but failed to obtain a positive result. More recently, Wang et al. (2000) published results that are presented as a partial experimental confirmation of the Salam hypothesis. These authors claim to have observed a phase transition at 270 K by measuring the difference between the specific heats of D- and L-alanine in the solid crystalline phase. Measurements of magnetic susceptibility and by Raman spectroscopy also appear to show unique behavior at this temperature. Wang et al. also found that the specific heat of D-alanine is greater than that of L-alanine. These results are interesting but also surprising, notably because of the high transition temperature reported. One hopes that further studies along these lines will be carried out in order to confirm these results and exclude the possibility of artifacts.

The results of Soai et al. (2003) on the amplification of the enantiomeric excesses that occur in racemic materials as a result of random statistical variation require a rethinking of the possible role of enantiomeric excesses due to the PVED. In the Soai reaction initially 0.1 mMole of the chiral zinc complex was formed and it can easily be calculated that the statistically allowed enantiomeric excess in even this relatively large amount of material would be of the order of 10^{-8}%. Since random enantiomeric excesses of this magnitude will always exist and can, as Soai and his coworkers have demonstrated, be amplified in an autocatalytic reaction to as much as 90%, the presence of much smaller enantiomeric excesses of the order 10^{-15} to 10^{-13}% attributable to the PVED would have no significance.

In summary, it is apparent from this brief review that amplification processes are not lacking and many have been subjected to thorough laboratory testing, in part because of the importance of obtaining pure enantiomers for the pharmaceutical industry (Triggle, 1997). It has also become clear that experimental studies concerning the amplification of extremely small enantiomeric excesses can be subject to misinterpretation due to the presence of chiral contaminants. Finally, it should also be stressed that in the context of research on the origin of the homochiralty of life, amplification mechanisms that require pure crystals, enantiomer solutions near saturation, anhydrous conditions, or reactors on the scale of a lake that are free of chiral impurities for time periods measured in centuries or millenia, are unrealistic.

3.11 Exogenous Origin of Homochirality

The exogenous delivery of organic matter to the primitive Earth is now widely acknowledged to have been an important aspect of its prebiotic chemistry (Chyba and Sagan, 1992). There is little doubt that organic matter of meteoritic, micrometeoritic, and cometary origin accumulated in the hydrosphere of the primitive Earth (for reviews see the chapters by D. Despois, by F. Robert and by A. Morbidelli and D. Benest in Gargaud et al., 2001). Therefore it is understandable that the question of chiral organic matter in meteorites, particularly in carbonaceous chondrites has interested researchers for several decades (see reviews of Mullie and Reisse, 1987; Reisse and Mullie, 1993; Cronin and Pizzarello, 2000). In this regard, it is important to recognize that contamination of meteoritic matter by terrestrial compounds is almost unavoidable and that this can easily give artifacts. The work carried out by Meischein et al. (1966) on optical activity was pioneering in this respect. A meticulous analysis of an extract of the Homestead meteorite led these authors to the conclusion that the optical activity observed in meteoritic extracts by earlier workers was a result of contamination. However, more recently some of the amino acids detected in the Murchison and Murray meteorites have been found to be nonracemic (Cronin and Pizzarello 1997; Engel and Macko, 1997; Pizzarello and Cronin, 1998; 2000; Pizzarello et al., 2003), a discovery of considerable interest with respect to the origin of homochirality. It should be noted that the method of analysis used is no longer measurement of optical rotation or rotatory dispersion but rather gas chromatography (GC) on chiral supports coupled to mass spectrometry (MS) or isotope ratio mass spectrometry (IRMS), more sensitive and specific methods than those that were available in 1966.

There remains a great risk of contamination in the analysis of those meteoritic amino acids that are also present in the biosphere and this cannot be neglected regardless of the analytical methods used. In order to reduce this risk, Cronin and Pizzarello (1997) studied α-methyl amino acids that are either unknown or rarely found in the biosphere. These authors initially reported significant enantiomeric excesses (1 to 9% depending on the amino acid) in favor of the L-enantiomers and more extensive later analyses of isovaline have shown even larger enantiomeric excesses (Pizzarello et al., 2003).

On the basis of these results, it seems very probable that certain of the amino acids present on the primitive Earth had enantiomeric excesses and that these excesses were large enough to be susceptible to amplification by the processes discussed previously, notably by polymerization. It is necessary to add that the α-methyl amino acids, which are relatively abundant in carbonaceous chondrites, are particularly good candidates for amplification by polymerization.

As discussed in Sect. 3.10, the formation of secondary structures of the alpha-helical or beta-pleated-sheet type enhances amplification by polymerization and it has been shown (Altman et al., 1988; Formaggio et al., 1995) that the α-methyl amino acids give polyp eptides with very stable secondary structure of a helical type.

The origin of the enantiomeric excesses found in meteoritic amino acids is an important question. These amino acids have general characteristics that clearly indicate their production by abiotic processes (Kvenvolden et al., 1970; Cronin and Chang, 1993). They are nearly racemic, structurally diverse, with all possible isomeric forms represented, and are found in amounts that decrease exponentially within homologous series. Furthermore, they are isotopically distinct from their terrestrial counterparts, being substantially enriched in the heavier stable isotopes of C, H, and N (Epstein et al., 1987; Engel et al., 1990; Pizzarello et al., 1991; 2003; Engel and Macko, 1997). The isotopic enrichment, particularly in deuterium, is suggestive of chemistry at very low temperature, e.g., the ion–molecule reactions that occur in cold interstellar clouds (Wannier, 1980; Penzias, 1980). Thus, it is possible that meteoritic amino acids are of presolar origin (Epstein et al., 1987), although it is has been suggested that appropriate low-temperature conditions may have also existed within the solar nebula (Aikawa and Herbst, 2001).

As discussed previously (Sect. 3.9) UV circularly polarized light (UVCPL), when absorbed by a racemic organic compound, leads to photolysis with preferential survival of increasing amounts of one enantiomer as the photolysis proceeds. According to Rubenstein et al. (1983) and Bonner and Rubenstein (1987), a neutron star can produce UVCPL as synchrotron radiation resulting from the circulation of an electron plasma in the equatorial plane of the star. In addition, circular polarization has been observed within interstellar clouds and attributed to Mie scattering of light by aligned interstellar grains (Bailey et al., 1998). Bonner and Rubenstein go on to suggest that the UVCPL generated in these ways might affect the organic matter of interstellar clouds or the organic matter at or near the surface of a primitive planet thus giving rise to enantiomeric excesses in its chiral components. This hypothesis was adopted as a possible explanation for the asymmetry observed in meteoritic amino acids (Cronin and Pizzarello, 1997; Engel and Macko, 1997). Greenberg (1996) determined that it is probable that interstellar grains were, at some point in their existence, subjected to circularly or elliptically polarized radiation emitted by a neutron star and therefore concluded that a *"significant fraction of Solar Systems started off with significant enantiomeric excesses."* The possibility of obtaining UVCPL from a neutron star has been questioned by Roberts (1984), and Mason (1997) has expressed doubt, based on the Kuhn–Condon effect, whether the broad band UVCPL emitted by a neutron star could achieve preferential photolysis. The latter objection has been effectively countered by Bonner et al. (1999). Mie scattering is not subject to these objections and a strong case can be made for its greater significance as a source of interstellar UVCPL (Bailey, 2001). Cerf and Jorissen (2000) and

Jorissen and Cerf (2002) have also discussed the pros and cons of a UVCPL scenario for the origin of enantiomeric excesses.

It is also necessary to consider quantitative aspects of the UVCPL photolysis hypothesis in evaluating its relevance to the origin of the enantiomeric excesses observed in meteorites. L-enantiomeric excesses up to 9% were initially observed in the isovaline from the Murchison meteorite and an excess as high as 15% has been measured for this amino acid more recently (Pizzarello et al., 2003). Although not impossible, enantiomeric excesses of this magnitude are difficult to achieve by UVCPL photolysis. The extinction coefficients of the enantiomers of aliphatic amino acids for UVCPL are not very different ($g \cong 0.02$), both enantiomers are subject to photolysis, and substantial enantiomeric excesses are achieved only as the reaction approaches completion. For example, it can be seen from the theoretical treatment of Balavoine et al. (1974) that a practical limit for ee of about 9% is reached for a fixed population of aliphatic amino acid molecules subjected to irradiation with UVCPL. (For $g \cong 0.02$, $ee = 9.2\%$ at 99.99% decomposition.) It seems likely that in any natural setting UVCPL irradiation would occur under conditions inferior to the optimized conditions used in laboratory studies and, consequently, that enantiomeric excesses values smaller than maximal, perhaps much smaller, would be observed if asymmetric photolysis of a racemate were the operative mechanism. From this perspective, some of the L-excesses measured for the meteoritic amino acids seem quite large, which raises doubt about UVCPL as the cause if the process is conceived of in terms of the laboratory model, i.e., a single exposure of a fixed population of molecules. In principle, UVCPL can effect an enantiomeric excess in an amino acid not only by the asymmetric photolysis of its racemate but also by synthesis; however, such photosyntheses are also limited to enantiomeric excess values governed by the g-value, i.e., approximately 2%.

However, two possibilities envisioned by Balavoine et al. (1974) seem worth considering in this context. If a mechanism existed whereby the residual fraction of photolyzed amino acids could accumulate from a large volume and experience a second exposure to UVCPL, this time starting with the small enantiomeric excess achieved previously, a further enhancement in the excess could be attained, although again at the expense of the total amount of amino acid surviving. Such a process of accumulation and re-exposure is conceivable as an interstellar cloud collapses in the process of nebula formation. The second possibility is the secondary formation of amino acids by an asymmetric catalyst. In this case one must imagine the formation of an asymmetric catalyst by the preferential photolysis of a racemic compound and this enantiomerically enriched catalyst then acting to promote the formation of amino acids with substantial enantiomeric excesses. As described in Sect. 3.10.5, in some autocatalytic reactions even small enantiomeric excesses in a chiral initiator can have dramatic nonlinear effects on the chirality of the reaction products.

Magnetochiral photochemistry (Rikken and Raupbach, 2000), which was discussed in Sect. 3.9, is possibly another explanation for the enantiomeric excesses observed in meteoritic amino acids. In this case, photochemistry with unpolarized light propagated parallel to a magnetic field would have been the asymmetric agent that influenced the organic chemistry of the interstellar medium in a region that contributed to meteoritic matter. The enantiomeric excesses achieved would have depended on the magnetic field strength. Presumably, magnetic fields of widely varying field strength exist near stars embedded in interstellar clouds.

3.12 Hypothesis and Summary

As we have seen, life on Earth is inextricably linked to homochirality and the origin of the latter remains a key unanswered question within the larger origin of life problem. In the foregoing sections we have attempted to provide the reader with a basic understanding of chirality as it pertains to organic compounds and then to review some of the more significant ideas that have been put forward with respect to the origin of the homochirality of life. In this section we describe a scenario based on our own evaluation of the many mechanisms that have been suggested. We assume that homochirality was achieved by prebiotic chemistry, i.e., by purely physical/chemical processes operating before the origin of life. We include within "prebiotic chemistry", chemistry that took place in the presolar cloud, as well as chemistry that might have unfolded from it within the solar nebula and on the early Earth. This may encompass reactions of molecules that eventually, in the course of evolution, completely or largely disappeared, for example, the α-methyl amino acids. In evaluating the suggested mechanisms we have considered whether they seem to have a reasonable chance of having operated within the constraints set by the conditions, to the extent that we can know them, that obtained on the prebiotic Earth. We have also been guided by Occam's Razor, the dictum that the best hypothesis is the one that minimizes the number of required assumptions.

Hypothesis: Terrestrial homochirality is a consequence of the postaccretional provision of organic compounds, particularly nonracemic amino acids, by the fall of meteorites, micrometeorites, and possibly comets on the primitive Earth. The enantiomeric excesses that initially characterized the amino acids were amplified at the time of pre-biotic polymerizations alternating with partial hydrolyses. The resulting polypeptides (not necessarily constituted of the contemporary protein amino acids) were necessarily asymmetric and some were endowed with catalytic properties. These polypeptides (protoenzymes) influenced the chirality of the products formed in their catalytic reactions by either the production of a specific enantiomer from achiral reactants or by the selection of a particular enantiomer in the case of chiral reactants.

This hypothesis is based on the following observations and experimental results described in more detail previously:

- Enantiomeric excesses ranging up to 15% are observed in α-methyl-α-amino acids isolated from carbonaceous chondrites impacting the Earth now. Such

amino acids were very likely present and probably more abundant on the primitive Earth when meteoritic, micrometeoritic and cometary infalls were more intense than they are today.
– Partial polymerization of amino acids coupled with partial hydrolysis is an effective amplification mechanism of small enantiomeric excesses. Polymerization reactions are a necessary feature of prebiotic chemistry.
– The α-methyl amino acids form unusually stable helical structures and are thus particularly subject to amplification of their initial enantiomeric excesses by polymerization. Furthermore, they do not readily racemize and the gains achieved in their enantiomeric purity are not susceptible to loss as in the case of their α-hydrogenated counterparts.
– Random polypeptides are sometimes endowed with catalytic activity (Fox and Krampitz, 1964).
– Folded helical polypeptide conformations are inherently asymmetric and the binding of chiral substrates will necessarily be chirospecific and lead to pervasive chiral specificity and eventual homochirality.

The symmetry breaking postulated in this hypothesis is presumed to be due to irradiation of the organic matter of the presolar cloud by UVCPL produced by either a stellar source or by scattering. In this respect, the hypothesis is a more specific version of the one suggested by Pasteur more than a century ago, according to which terrestrial homochirality could be the consequence of *"des actions dissymétriques dont nous pressentons l'existence enveloppante et cosmique"* (Valéry-Radot, 1968). The production of enantiomeric excesses in amino acids by UVCPL is well documented by laboratory studies carried out even under low-temperature interstellar-grain conditions. In this case, the terrestrial preference for L-amino acids rather than D-amino acids would have been a matter of chance and life on extrasolar planets could have the same or the opposite chirality with equal probability as nothing requires that circularly polarized light have the same helicity in all protostellar clouds. The magnetochiral effect is also an attractive possibility; however, it has been subject to laboratory verification only with respect to the displacement of an equilibrium mixture and has not, as yet, been shown to affect stable enantiomers, i.e., enantiomers like those of amino acids that are not ordinarily in rapid equilibrium.

The hypothesis proposes the prebiotic amplification of the small initial enantiomeric excesses by polymerization. Insofar as polymerization leads to even weakly catalytic polypeptides, it also has the potential for providing, at the culmination of the prebiotic period, not only homochiral amino acids, but also a more extensive organic milieu with not only chiral specificity but also specificity with respect to the compounds represented. This latter specificity would arise as a consequence of the statistical impossibility of the formation of all polypeptide sequences, i.e., all catalytic activities, as well as the improbability of the formation of their enantiomeric forms. Although this hypothesis provides an explanation for the origin of both chiral and molecular specificity, it should be noted that it is silent with respect to the vexing problem of how certain catalytic polypeptides, formed at first by chance, could be reproduced, that is, in regard to the essential event that marked the transition from a prebiotic to a biotic world.

The possibility that the small enantiomeric excesses in the amino acids present on the early Earth acted as chiral initiators in Soai-type autocatalytic reactions is an attractive idea. In this way, large enantiomeric excesses could conceivably have been rapidly achieved. We expect that further research in this area will lead to the demonstration of similar effects in prebiotically more realistic aqueous phase reactions but, unless/until this is achieved, Occam's Razor causes us to favor the well-established polymerization–hydrolysis amplification mechanism.

Insofar as this hypothesis postulates "polypeptides first" with respect to the establishment of homochirality as well as early catalytic activities, it is incompatible with an exclusively RNA world, although it would not be incompatible with an RNA world in which, once homochirality was established, polypeptides played a subsidiary role. One might imagine that polypeptides in which α-methyl amino acids were important had, for reasons unknown, a limited future and that the rise of the more familiar nucleic acid-protein world awaited the exclusive use of the more easily (bio)synthesized α-hydrogen amino acids for protein synthesis. As a corollary to this hypothesis, one might postulate that an early catalytic polypeptide had the ability to polymerize formaldehyde to D-ribose and thus solved one of the two chirality problems that must be dealt with on the way to an RNA-world.

With regard to other possible symmetry-breaking mechanisms, we would note that hypotheses invoking specific adsorption on natural mineral surfaces to break symmetry suffer from the fact that, insofar as is known, the enantiomorphic forms of minerals occur equally; thus, unless the origin of homochirality was spatially confined to a very small area/volume, any enantiomeric excess achieved would be countered by a nearby equal and opposite excess. As discussed previously, symmetry breaking by crystallization effects requires a concentration and degree of purity that seem unrealistic under prebiotic conditions.

The violation of parity is difficult to dismiss because of the intuition that such fundamental asymmetry might have far-reaching consequences. However, enantiomeric excesses in chiral molecules deriving from parity violation have not been demonstrated. Furthermore, it is not possible to see how such very small excesses could have emerged from the "noise" of random statistical fluctuation to a degree that would allow their amplification. The possibility that this "noise" itself, i.e., the statistically allowed excursions away from the perfect racemic state, provided the symmetry breaking for the origin of homochirality is an interesting possibility made more plausible by the demonstration of their apparent amplification in the "Soai reaction." If a similar reaction can be shown to occur under aqueous, prebiotic conditions a statistical symmetry breaking must be seriously considered for the origin of homochirality.

3.13 Homochirality Analyses in the Solar System and Beyond

Exobiologists have taken an interest in homochirality as a signature of life and attention has been given to the analysis of chirality (or, more precisely, enantiomeric excesses) beyond the terrestrial environment by the SETH (search for extraterrestrial homochirality) project. SETH could be accomplished remotely on various Solar System bodies by landers equipped with detectors of enantiomeric excesses such as miniaturized polarimeters (Mac Dermott, 1996) or chromatographs with chiral support materials (Meierhenrich et al., 1999). The Rosetta mission to the Churyumov-Gerasimenko comet which was launched in February 2004 and that will reach the comet in November 2014 involves a SETH. SETH also encompasses the broader goal of finding the remains of life and could, for example, contribute to the validation of chemical fossils on other bodies of the Solar System such as Mars. However, diagenesis is accompanied by partial and ultimately complete racemization of chiral molecules and this poses a problem for the use of chirality as a signature of past life. This record is written with ink that inevitably fades with the passage of time. It inevitably fades in terrestrial sediments and must in all sediments (Bada and Miller, 1987; Bada and McDonald, 1995). SETH assumes the necessity of homochirality for life and as long as we know only one form of life it is not possible to know with certainty whether homochirality is an absolute requirement for life although it seems highly probable that it is. The enantiomeric excesses found in meteoritic amino acids add another complication as they are of abiotic origin and widely distributed. It seems clear that drawing the correct inferences from extraterrestrial chirality measurements will require their interpretation within a broad chemical, morphological, and geological context. It is also necessary to remember that, if extant or extinct life on Mars were discovered to have the same molecular homochirality as that of terrestrial life, the meaning of this discovery would not be unambiguous. Transfers of matter between Mars and the Earth are relatively frequent on a time scale of the age of these planets and such transfers may allow the insemination of one planet by the other (Mileikowsky, 2000).

The study of homochirality outside the Solar System, *SEXSOH* (search for extrasolar homochirality) will probably be a goal of future missions but at present, it is too early to predict when (and even how) observations might be carried out.

Dedication

This chapter is dedicated to Kurt Mislow in recognition of his 80th birthday and with appreciation for his many experimental and theoretical contributions to the modern understanding of chirality and stereochemistry in general.

Bibliography

General

Avalos M., Babiano R., Cintas P., Jimenez J.L., Palacios J.C. (2000). From parity to chirality: Chemical implications revisited. *Tetrahedron Asymmetry*, **11**, 2845–2874.
Bouchiat M.A., Pottier L. (1984). An atomic preference between left and right. *Scientific American*, 76–86.
Brack A. (July, 1998), L'asymétrie du vivant in "Pour la Science – Dossier Hors Série," Paris (France) "Les symétries de la nature" (this special issue contains many interesting articles for those who wish to integrate chirality into the general problem of symmetry and the breaking of symmetry in biology, chemistry, physics, and cosmology.)
Cline B.C. (editor) (1996). *Physical Origin of Homochirality on Earth*, American Institute of Physics, Woodbury, New York (USA).
Feringa B.L., van Velden A. (1999). Absolute asymmetry synthesis: the origin, control and amplification of chirality. *Angew. Chem. Int. Ed.*, **38**, 3418–3438.
Jacques J. (1992). *La Molécule et son Double*. Hachette, Paris.
Mason S.F. (1982). *Molecular Optical Activity and the Chiral Discriminations*. Cambridge University Press, Cambridge (UK).
Mislow, K. (199). *Molecular Chirality*. Chapter 1 in "Topics in stereochemistry," vol 22, S.E. Denmark, ed. J. Wiley and Sons, New York (USA).
Nordén B. (1978). The Asymmetry of Life. *J. Mol. Evol.*, **11**, 313–332.
W.J. Lough and I. Wainer (Eds.) (2002). *Chirality in Natural and Applied Science*. Blackwell Science Ltd-CRC Press, USA and Canada.

References

Aikawa Y., Herbst, E. (2001). Two-dimensional distribution and column densities of gaseous molecules in protoplanetary disks. II. Deuterated species and UV-shielding by ambient clouds. *Astron. Astrophys.*, **372**, 1107–1117.
Altman E., Altman K.H., Nebel K., Mutter M. (1988). Conformational studies on host-guest peptides containing chiral alpha-methyl-alpha-amino acids. *Int. J. Pept. Protein Res.*, **32**, 344–351.
Asakura S., Soga T., Uchida T., Osanai S., Kondepudi D.K. (2002). Probability distribution of enantiomeric excess in unstirred and stirred crystallization of binaphthyl melt. *Chirality*, **14**, 85–89.
Auf Der Heyde T.P.E., Buda A.B., Mislow K. (1991). Desymmetrization and degree of chirality. *J. of Mathematical Chemistry*, **6**, 255–265.
Avalos M., Babiano R., Cintas P., Jimenez J.L., Palacios J.C. (2000). From parity to chirality: Chemical implications revisited. *Tetrahedron: Asymmetry*, **11**, 2854–2874.
Bada J.L., McDonald G.D. (1995). Amino acid racemization on Mars: Implications for the preservation of biomolecules from an extinct Martian biota. *Icarus*, **114**, 139–143.
Bada J.L., Miller S.L. (1987). Racemization and the origin of optimalised active organic compounds in living organisms. *Biosystems*, **20**, 21–26.
Bailey J., Chrysostomou A., Hough J.H., Gledhill T.M., McCall A., Clark S., Ménard F., Tamura M. (1998). Circular polarization in star formation regions: Implications for biomolecular homochirality. *Science*, **281**, 672–674.

Bailey J. (2001). Circularly polarized light and the origin of homochirality. *Origins Life Evol. Biosphere*, **31**, 167–183.
Balavoine G., Moradapour A., Kagan H.B. (1974). Preparation of chiral compounds with high optical purity with circularly polarized light, a model for the prebiotic generation of optical activity. *J. Am. Chem. Soc.*, **96**, 5152–5158.
Barron L.D. (1982). *Molecular Light Scattering and Optical Activity*, Cambridge University Press, Cambridge.
Barron L.D. (2000). Chirality, Magnetism and Light. *Nature*, **405**, 895–896.
Barron L.D. (2002). Chirality at the sub-molecular level: true and false chirality, p. 53–84 in Chirality in *Natural and Applied Science*, Eds. W.J. Lough and I.W. Wainer, Blackwell Science Ltd. CRC Press, USA and Canada.
Bartik K., El Haouaj M., Luhmer M., Collet A., Reisse J. (2000). Can monoatomic xenon become chiral? *Chem. Phys.Chem.*, **4**, 221–324.
Bartik K., Luhmer M., Collet A., Reisse J. (2001). Molecular polarization and molecular chiralization: The first example of a chiralized xenon atom. *Chirality*, **13**, 2–6.
Blair N.E., Bonner W.A. (1980). Experiments on the amplification of optical activity. *Origins of Life*, **10**, 255–263.
Blair N.E., Dirbas F.M., Bonner W.A. (1981). Stereoselective hydrolysis of leucine oligomers. *Tetrahedron*, **37**, 27–29.
Blout E.R, Idelson M. (1956). Polypeptides VI. Poly-alpha-L-glutamic acid. Preparation and helix-coil conversions. *J. Am. Chem. Soc.*, **78**, 497–498.
Bolli M, Micura R., Eschenmoser A. (1997). Pyranosyl-RNA/ Chiroselective self-assembly of base sequences by ligative oligomerization of tetranucleotide-2', 3'-cyclophosphates (with a commentary concerning the origin of biomolecular homochirality), *Chemistry and Biology*, **4**, 309–320.
Bonner W.A., Flores J.J. (1975). Experiments on the origin of optical activity. *Origins of Life*, **6**, 187–194.
Bonner W.A., Rubenstein E. (1987). Supernovae, neutron stars and biomolecular chirality. *Biosystems* **20**, 99–111.
Bonner W.A (1991). The origin and amplification of biomolecular chirality. *Orig. Life Evol. Biosphere*, **21**, 59–11.
Bonner W.A. (1994). Enantioselective autocatalysis - spontaneous resolution and the prebiotic generation of chirality. *Orig. Life Evol. Biosphere*, **24**, 63–78.
Bonner W.A. (1996). The quest for chiralitry in *Physical Origin of Homochirality on Earth*, ed. D.B. Cline, p. 17–49, American Institute of Physics, Proc. 379, Woodbury, New York.
Bonner W.A., Rubenstein E., Brown G.S. (1999). Extraterrestrial handedness: A reply. *Orig. Life Evol. Biosphere*, **30**, 329–332.
Bonner W.A. (1999). Chirality amplification – the accumulation principle revisited. *Orig. Life Evol. Biosphere*, **29**, 615–623.
Bonner W.A., Bean B.D. (2000). Asymmetric photolysis with elliptically polarized light. *Orig. Life Evol. Biosphere*, **30**, 513–517.
Bonner W.A. (2000). Parity violation and the evolution of biomolecular homochirality. *Chirality*, **12**, 114–126.
Brack A., Spach G. (1981). Enantiomer enrichment in early peptides. *Origins of Life*, **11**, 135–142.
Buda A.B., Mislow K. (1991). On geometric measure of chirality. *J. of Molecular Structure (Theochem)*, **232**, 1–12.

Buda A.B, Auf der Heyde T., Mislow K. (1992). On quantifying chirality. *Angew. Chem. Int. Ed. Engl.*, **31**, 989–1007.

Buchardt O. (1974). Photochemistry with circularly polarized light. *Angew. Chem. Int. Ed. Engl.*, **13**, 179–185.

Buschmann H., Thede R., Heller D. (2000). New developments in the origins of the homochirality of biologically relevant molecules. *Angew. Chem. Int. Ed.*, **39**, 4033–4036.

Cerf C., Jorissen A. (2000). Is amino-acid homochirality due to asymmetric photolysis in space? *Space Science Reviews*, **92**, 603–612.

Chyba C.F., Sagan, C. (1992). Endogeneous production, exogeneous delivery and impact-shock synthesis of organic molecules: An inventory for the origins of life. *Nature*, **355**, 125–132.

Collet A., Brienne M-J., Jacques J. (1980). Optical Resolution by Direct Crystallization of Enantiomer Mixtures. *Chem. Rev.*, **80**, 215–230.

Collet A. (1990). The Homochiral versus heterochiral packing dilemma in *Problems and Wonders of Chiral Molecules*, ed. M. Simonyi, Akademia Kiado, Budapest.

Cronin J.R., Chang S. (1993). Organic matter in meteorites: Molecular and isotopic analyses of the Murchison meteorite in *The Chemistry of Life's Origins*, eds. J.M. Greenberg, et al., p 209–258, Kluwer Acad. Pub., Netherlands.

Cronin J.R., Pizzarello S. (1997). Enantiomeric exesses in meteoritic amino acids. *Science*, **275**, 951–955.

Cronin J.R., Pizzarello S. (2000). Chirality of meteoritic organic matter: A brief review. in *Perspectives in Amino Acid and Protein Geochemistry*, eds. Goodfriend G. et al., p 15–22, Oxford University Press, Oxford, New York.

Curie P. (1894). Sur la symétrie dans les phénomènes physiques, symétrie d'un champ électrique et magnétique. *J. Chim. Phys.*, 3ème série, t.III, 393–402.

Decker P. (1974). The origin of stochastic information (noise) in bioids: Open systems which can exist in several steady states. *J. Mol. Evol.*, **4**, 49–65.

Droesbeke J.J. (2001). In *Eléments de Statistique*, 4ème édition, p 206–207, Editions de l'Université de Bruxelles.

Dunitz J.D. (1996). Symmetry arguments in chemistry. *Proc. Natl. Acad. Sci. USA*, **93**, 14260–14266.

Engel M.H., Macko S.A., Silfer J.A. (1990). Carbon isotope composition of individual amino acids in the Murchison meteorite. *Nature*, **348**, 47–49.

Engel M.H., Macko S.A. (1997). Isotopic evidence for extraterrestrial non-racemic amino acids in the Murchison meteorite. *Nature*, **389**, 265–268.

Eliel E.M., Wilen S.H.(1994). *Stereochemistry of Organic Compounds*. J. Wiley and Sons, New York (USA).

Epstein S., Krishnamurthy R.V., Cronin J.R., Pizzarello S., Yuen G.U. (1987). Unusual stable isotope ratios in amino acids and carboxylic extracts from the Murchison meteorite. *Nature*, **326**, 477–479.

Feringa B.L., Huck N.P., van Doren H.K. (1995). Chiroptical switching between liquid crystalline phases. *J. Am. Chem. Soc.*, **117**, 9929–9930.

Figereau A., Duval E., Boukenter A. (1995). Can biological homochirality result from a phase transition? *Orig. Life Evol. Biosphere*, **25**, 211–217.

Flores J.J., Bonner W.A., Massey G.A. (1977). Asymmetric photolysis of (R, S)-leucine with circularly polarized light. *J. Am. Chem. Soc.*, **99**, 3622–3625.

Formaggio F., Crisma M., Bonora G.M., Pantano M., Valle G., Toniolo C., Aubry A., Bayeul D., Kamphuis J. (1995). (R)-Isovaline homopeptides adopt the left-handed helical structure. *Peptide Research*, **8**, 6–14.

Fox S.W., Krampitz G. (1964). Catalytic decomposition of glucose in aqueous solution by thermal proteoids. *Nature*, **203**, 1362–1364.

Franck F.C. (1953). On spontaneous asymmetric synthesis. *Biochim. Biophys. Acta*, **11**, 459–463.

Frondel C. (1978). Characters of quartz fibers. *Am. Mineral.*, **63**, 17–27.

Gargaud M., Despois D., Parisot J.-P. (Eds.), (2001). *L'environnement de la Terre Primitive*, Presses Universitaires de Bordeaux, France.

Girard C., Kagan H.B. (1998). Nonlinear effects in asymmetric synthesis and stereoselective reactions: Ten years of investigations. *Angew. Chem. Int. Ed.*, **37**, 2922–2959.

Gol'danski V.I., Kuz'min V.V. (1988). Spontaneous mirror symmetry breaking in nature and origin of life. *Z. Phys. Chem.*, **269**, 216–274.

Greenberg J.M. (1996). Chirality in interstellar dust and in comets: Life from dead stars, in *Physical Origin of Homochirality on Earth*, Ed.D.B. Cline, p. 185–186, American Institute of Physics; Proc. 379, Woodbury, New York.

Havinga E. (1954). Spontaneous formation of optically active substances. *Biochem. Biophys. Acta*, **38**, 171–174.

Hazen R.M., Filley R.F, Goodfriend G.A. (2001). Selective adsorption of L- and D-amino acids on calcite: Implications for biochemical homochirality. *Proc. Nat. Acad. Sci. USA*, **98**, 5487–5490.

Inoue Y. (1992). Asymmetric photochemical reactions in solution. *Chem. Rev.*, **92**, 741–770.

Jackson T.A. (1971). Preferential polymerization and adsorption of L-optical isomers of amino acids relative to D-optical isomers on kaolinite templates. *Chem. Geol.* **7**, 295–306.

Jacques J., Collet A., Wilen S.H. (1981). *Enantiomers, Racemates and Resolution*, J. Wiley and Sons, New York.

Jorissen A., Cerf C. (2002). Photoreactions as the Origin of Biomolecular Homochirality: A critical review. *Origins Life Evol. Biosphere*, **32**, 129–142.

Julg A., Favier A., Ozias, Y. (1989). A theoretical study of the difference in the behavior of L- and D-alanine toward the two inverse forms of kaolinite. *Struc. Chem.*, **1**, 137–141.

Julg A. (1989). Origin of the L-homochirality of amino-acids in the proteins of living organisms in Molecules in Physics, *Chemistry and Biology*, vol. IV, Ed. J. Maruani, p. 33–52, Kluwer Academic Pub., Netherlands.

Kagan H., Moradpour A., Nicoud J.F., Balavoine G., Martin R.H., Cosyn J.P. (1971). Photochemistry with circularly polarised light. Asymmetric synthesis of octa- and nonahelicene. *Tetrahedron Lett.*, **22**, 2479–2482.

Kagan H.P., Fiaud J.C. (1988). *Kinetic Resolution in Topics in Stereochemistry*, vol 18. eds. Eliel, E.L. and Willen, S.H., p. 249–330, John Wiley, New York.

Kelvin W.T. (1904). *Baltimore Lectures on Molecular Dynamics and the Wave Theory of Light*, p. 602–642, C.J. Clay, London.

Kondepudi D.K., Nelson G.W. (1984). Chiral symmetry-breaking and its sensitivity in non-equilibrium chemical systems. *Physica*, **125A**, 465–496.

Kondepudi D.K., Nelson G.W. (1985). Weak neutral currents and the origin of biomolecular chirality. *Nature*, **314**, 438–441.

Kondepudi D.K. (1996). Selection of handedness in prebiotic chemical processes, in *Physical Origin of Homochirality in Life*, Ed. Cline D.B., p. 65–72, American Institute of Physics, AIP Conference 379, Woodbury, New York.

Kondepudi D.K., Laudadio J., Asakura K. (1999). Chiral symmetry breaking in stirred crystallization of 1,1′-binaphthyl melt. *J. Am. Chem. Soc.*, **121**, 1448–1451.

Kuhn W., Braun E. (1929). Photochemische erzeugung optisch aktiver stoffe. *Naturwiss.*, **17**, 227–228.

Kvenvolden K., Lawless J., Pering K., Peterson E., Flores J., Ponnamperuma C., Kaplan I.R., Moore C. (1970). Evidence for extraterrestrial aminoacids and hydrocarbons in the Murchison meteorite. *Nature*, **228**, 923–926.

Larder D.F. (1967). Historical aspects of the tetrahedron chemistry. *J. Chem. Ed.*, **44**, 661–666.

Lee T.D., Yang C.N. (1956). Question of parity conservation in weak interactions. *Phys. Rev.*, **104**, 254–258.

Mc Bride J.M., Carter R.L. (1991) Spontaneous resolution by stirred crystallization. *Angew. Chem. Int. Ed.*, Engl. **30**, 293–295.

Mac Dermott A.J. (1996). The weak force and SETH: The search for extra-terrestrial homochirality, in *Physical Origin of Homochirality on Earth*, Ed.D.B. Cline, p. 241–254, American Institute of Physics, Proc. 379, Woodbury, New York.

Mac Dermott A.J. (2002). The origin of biomolecular chirality, p. 23–52 in *Chirality in Natural and Applied Science*, Eds. W.J. Lough and I.W. Wainer, Blackwell Science Ltd. CRC Press, USA and Canada.

Mason S.F. (1982). *Molecular Optical Activity and the Chiral Discriminations*. Cambridge University Press, Cambridge.

Mason S.F. (1997). Extraterrestrial handedness. *Nature*, **389**, 804.

Mason S.F. (2002). Pasteur on molecular handedness – and the sequel, p. 1–19 in *Chirality in Natural and Applied Science*, Eds. W.J. Lough and I.W. Wainer, Blackwell Science Ltd. CRC Press, USA and Canada.

Matsuura K., Inoue S., Tsuruta T. (1965) Asymmetric selection in the copolymerization of N-carboxy-L- and D-alanine. *Makromol. Chem.*, **85**, 284–290.

Meierhenrich U., Thiemann W.H-P., Rosenbauer, H. (1999). Molecular parity violation via comets. *Chirality*, **11**, 575–582.

Meinschein W.G., Frondel C., Laur P., Mislow K. (1966). Meteorites: Optical activity in organic matter. *Science*, **154**, 377–380.

Mileikovwsky C., Cucinotta F., Wilson F., Gladman B., Hornek G., Lindegren, L., Melosh J., Rickman H., Valtonen M., Zheng, J.Q. (2000). Natural transfer of viable microbes in space, Part 1: From Mars to Earth and from Earth to Mars. *Icarus*, **145**, 391–427.

Miller S.L. (1997). Peptide nucleic acids and prebiotic chemistry. *Nature Struct. Biol.*, **4**, 167–169.

Mills W.H. (1932). Some aspects of stereochemistry. *Chem. and Ind.*, 750–759.

Milton R.C. deL, Milton S.F.C., Kent, S.B.H. (1992). Total chemical synthesis of a D-enzyme. The enantiomers of HIV-1 protease show demonstration of remplaçable chiral-substrate-specificity. *Science*, **256**, 1445–1448.

Mislow K. (1965). *Introduction to Stereochemistry*, W.A. Benjamin, New York.

Mislow K. (1996). A commentary on the topological chirality and achirality of molecules. *Croat. Chim. Acta*, **69**, 485–511.

Mislow K. (1997). Fuzzy restrictions and inherent uncertainties in chirality studies, in *Fuzzy Logic in Chemistry*, ed. D.H. Rouvray, p. 65–88, Academic Press, San Diego (USA).

Monod J. (1970) *Le Hasard et la Nécessité; Essai sur la Philosophie Naturelle de la Biologie Moderne*, Editions du Seuil, Paris.

Mullie F., Reisse J. (1987). Organic matter in carbonaceous chondrites. *Topics in Current Chemistry (Spinger Verlag)*, **139**, 83–117.

Nielsen P.E. (1993). Peptide-Nucleic Acid (PNA)–A model structure for the primordial genetic code. *Origins Life Evol. Biosphere*, **23**, 323–327.

Nielsen P.E. (1996). Peptide Nucleic Acid (PNA). Implications for the origin of the genetic material and the homochirality of life, pages in Physical Origin of Homochirality in Life, *AIP Conference Proceedings* 379. Ed. D.B. Cline, p. 55–61,Woodbury, New York.

Nishino H., Kosaka A, Hembury G.A., Shitomi H., Onuki H., Inoue I. (2001). Mechanism of pH-dependent photolysis of aliphatic aminoacids and enantiomeric enrichment of racemic leucine by circularly polarized light. *Org. Letters*, **3**, 921–924.

Nordén B. (1977). Was photoresolution of amino acids the origin of optical activity in life. *Nature*, **266**, 567–568.

Nordén B. (1978). The Asymmetry of Life. *J. Mol. Evol* **11**, 313–332.

Penzias A.A. (1980). Nuclear processing and isotopes in the Galaxy. *Science*, **208**, 663–669.

Pizzarello S., Krishnamurty R.V., Epstein S., Cronin J.R. (1991). Isotopic analyses of amino acids from the Murchison meteorite. *Geochim. Cosmochim. Acta*, **55**, 905–910.

Pizzarello S., Cronin J.R. (1998). Alanine enantiomers in the Murchison meteorite. *Nature*, **394**, 236.

Pizzarello S., Cronin J.R. (2000). Non-racemic amino acids in the Murchison and Murray meteorites. *Geochim. Cosmochim. Acta*, **64**, 329–338.

Pizzarello S., Zolensky M., Turk K.A. (2003). Nonracemic isovaline in the Murchison meteorite: Chiral distribution and mineral association. *Geochim. Cosmochim. Acta*, **67**, 1589–1595.

Quack M. (2002). How important is parity violation for molecular and biomolecular chirality? *Angew. Chem. Int. Ed.*, **41**, 4618–4630.

Prigogine I., Kondepudi D.K. (1999). *Thermodynamique. Des Moteurs Thermiques aux Structures Dissipatives*, ed. Odile Jacob, Paris.

Rassat A, Fowler P.V. (2003). Any scalene triangle is the most chiral triangle. *Helvetic Chimica Acta*, **86**, 1728–1740.

Rau H. (1983). Asymmetric Photochemistry in Solution. *Chem. Rev.*, **83**, 355–547.

Rawn J.D. (1989). *Biochemistry*, Neil Patterson Pub. Carolina Biological Supply Company, Burlington, North Carolina.

Rikken G.L., Raupach E. (2000). Enantioselective Magnetochiral Photochemistry. *Nature*, **405**, 932–935.

Reisse J., Mullie F. (1993). On the origins of organic matter in carbonaceous chondrites. *Pure and Applied Chemistry*, **65**, 1281–1292.

Reisse J. (2001). in *L'environnement de la Terre Primitive*, eds. Gargaud M., Despois D., Parisot J.P., p. 323–342, Presses Universitaires de Bordeaux, Bordeaux, France.

Roberts J.A. (1984). Supernovae and life. *Nature*, **308**, 318.

Robinson R. (1974). Preface of the Van't Hoff-Le Bel Centenary Volume. *Tetrahedron*, **30**, 1477–1486.

Rubenstein E., Bonner W.A., Noyes H.P., Brown G.S. (1983). Supernovae and life. *Nature*, **306**, 118–120.

Salam A. (1993). The Origin of Chirality, the Role of Phase Transition and their Induction in Amino acids in *Chem. Evol. and Origin of Life*, eds. Ponnamperuma C, Chela-Flores J., Deepak Pub., Hampton, Virginia, USA.

Shibata T., Yamamoto J., Matsumoto N., Yonekubo S., Osanai S., Soai K. (1998). Amplification of a Slight Enantiomeric Imbalance in Molecules Based on Asymmetric Autocatalysis: The First Correlation between High Enantiomeric Enrichment in a Chiral Molecule and Circularly Polarized Light. *J. Am. Chem. Soc.*, **120**, 12157–12158.

Siegel J.S. (1998). Homochiral imperative of molecular evolution. *Chirality*, **10**, 24–27.

Siegel J.S. (2002). Shattered Mirrors. *Nature*, **419**, 346–347.

Singleton D.A., Vo L.K. (2002). Enantioselective synthesis without discrete optically active additives. *J. Am. Chem. Soc.*, **124**, 10010–10011.

Soai K., Shibata T., Morioka H., Choji K. (1995). Asymmetric autocatalysis and amplification of enantiomeric excess of a chiral molecule. *Nature*, **378**, 767–768.

Soai K., Osanai S., Kadowaki K., Yonebuko S., Shibata T., Sato I. (1999). d- and l-Quartz-promoted highly enantioselective synthesis of a chiral organic compound. *J. Am. Chem. Soc.*, **121**, 1235–1236.

Soai K., Shibata T., Sato I. (2000). Enantioselective automultiplication of chiral molecules by asymmetric autocatalysis. *Acc. Chem. Res.*, **33**, 382–390.

Spach P.G. (1974). Polymérizsation des énantiomères d'un acide α-aminé. Stéréosélection and amplification de l'asymétrie. *Chimia*, **28**, 500–503.

Szabo-Nagy A., Keszthelyi L. (1999). Demonstration of the Parity-Violating Energy Difference between Enantiomers. *Proc. Natl. Acad. Sci. USA*, **96**, 4225–4255.

Takats Z, Nanita S.C., Cooks R.G. (2003). Serine octamer reactions: indicators of prebiotic relevance. *Angew. Chem. Int. Ed.*, **42**, 3521–3523.

Tranter G.E. (1985). Parity-violating energy differences of chiral minerals and the origin of biomolecular homochirality. *Nature*, **318**, 172–173.

Triggle D.J. (1997). Stereoselectivity of drug action. *Drug Discovery Today*, **2**, 138–147.

Valéry-Radot P. (1968). *Pages Illustres de Pasteur*. Hachette (Paris).

Vester F. (1974). The (hi)story of the induction of molecular asymmetry by the intrinsic asymmetry in β-decay. *J. Mol. Evol.*, **4**, 1–13.

Wagnière G., Meier A. (1983). Difference in the absorption coefficient of arbitrarily polarized light in a magnetic field. *Experientia*, **39**, 1090–1091.

Wald G. (1957). The origin of optical activity. *Ann. N.Y. Acad. Sci.*, **69**, 353–358.

Wang W., Yi F., Ni Y., Jin Z., Tang Y. (2000). Parity violation of electroweak force in phase transition of single crystals of D- and L- alanine and valine. *J. of Biological Physics*, **26**, 51–65.

Wannier P.G.A. (1980). Nuclear abundances and evolution of the interstellar medium. *Ann. Rev. Astron. Astrophys.*, **18**, 399–437.

Weissbuch I., Addadi L., Berkovitch-Yellin Z., Gatti E., Lahav M., Leiserowitz L. (1984). Spontaneous generation and amplification of optical activity in alpha aminoacids by enantioselective occlusion in centrosymmetric crystals of glycine. *Nature*, **310**, 161–164.

Wolstencroft R.D. (1985). Astronomical sources of circularly polarized light and their role in determining chirality on Earth. in IAU Symp. 112. *The Search for Extraterrestrial Life*, p. 171–175, D. Reidel, Dordrecht.

Wu C.S., Ambler E., Hayward R.W., Hoppes D.D., Hudson R.P. (1957). Experimental test of parity conservation in beta-decay. *Phys. Rev.*, **105**, 1413–1415

Yamagata Y. (1966). A hypothesis for the asymmetric appearance of biomolecules on Earth. *J. Theor. Biol.*, **11**, 495–498.

Zanasi R., Lazeretti P., Ligabue A. and Soncini A. (1999) in *Advances in BioChirality* eds. Palyi, G. et al., ch. 29, pp 377–385, Elsevier, Amsterdam.

Zee A. (1999). *Fearful Symmetry: The Search for Beauty in Modern Physics*, p 224, Princeton Science Library.

Zepik H., Shavit E., Tang M., Jensen T.R., Kjaer K., Bolbach G., Leiserowitz L., Weissbuch I., Lahav M. (2002). Amplification of oligopeptides in two-dimensional crystalline self-assemblies on water. *Science*, **295**, 1266–1269.

4 Peptide Emergence, Evolution and Selection on the Primitive Earth

I. Convergent Formation of *N*-Carbamoyl Amino Acids Rather than Free α-Amino Acids?

Auguste Commeyras, Laurent Boiteau, Odile Vandenabeele-Trambouze, Franck Selsis

Abstract After summarising current knowledge about the origins of primitive Earth organic matter, we focus our attention solely on α-amino acids and their derivatives. We then analyze the mechanism for the formation of these compounds, under both extraterrestrial and primitive-Earth conditions, and show that a "multicomponent system" consisting of prebiotic molecules (hydrogen cyanide, several carbonyl compounds, ammonia, alkyl amines, carbonic anhydride, sodium bicarbonate, borate, cyanic acid) may have been the precursors of these essential compounds. We show that this multicomponent system leads reversibly to several intermediate nitriles, which irreversibly evolve, first to α-amino acids and *N*-carbamoyl amino acids via selective catalytic processes, and then to *N*-carbamoyl amino acids alone.

4.1 Introduction

The Earth is 4.6 billion years old (4.6 Gyr). Two million years after it was formed, it had attained 80% of its present-day mass, the continents and oceans were already present, the plate-tectonic process had already begun (Martin 1986, 1987; Maas and McCulloch 1991; Carlson 1996; Vervoort et al. 1996; Mojzsis et al. 2001; Wilde et al. 2001), and there was also an atmosphere, though admittedly little is known about it (Selsis 2000). The last 20 per cent of the Earth's mass $(1.2 \times 10^{21}$ tons) was added by meteorites, micrometeorites and comets between -4.4 and -4.1 Gyr (Maurette 2001). The emergence of life may possibly have taken place in this period (acceptedly before -3.5 Gyr); as a consequence its substrate, namely organic matter, was obviously of both exogenous and endogenous origins. Qualitatively and quantitatively assessing each of these two different origins is a question of current importance that we examine, limiting ourselves to amino acids.

Investigation of the chemical mechanisms involved in the formation of both exogenous and endogenous amino acids should eventually provide information on the prebiotic environment. Such research needs first to acquire knowledge on meteoritic amino acids and related compounds (Sects. 4.2 and 4.3) and on the chemical reactions responsible for their synthesis (Sects. 4.4 and 4.5), in order to find out their potential prebiotic relevance (Sect. 4.6) as well as physical constraints on parent bodies (exogenous synthesis).

We conclude by studying a series of reactions that could have led to the formation of peptides, under prebiotic conditions.

4.2 Organic Molecules on the Primitive Earth

An exhaustive inventory of the origins of all the organic molecules occurring on the primitive Earth 4 Gyr ago was carried out by Chyba and Sagan (1992). They distinguished three probable sources:

- Exogenous contribution from meteorites, micrometeorites, interplanetary dust particles (IDP), comets, etc.
- Endogenous synthesis associated with impact of such bodies
- Endogenous production associated with other available energy sources (solar UV, lightning, radioactivity, etc.)

The total contribution of these three sources would have been heavily dependent on the composition of the primitive atmosphere. It would, in fact, have ranged from $\approx 10^{11}\,\mathrm{kg\,y^{-1}}$ for a reductive atmosphere to $\approx 10^{8}\,\mathrm{kg\,y^{-1}}$ for a slightly reductive atmosphere (atmospheric $[H_2]/[CO_2]$ ratio ≈ 0.1) or less for even lower hydrogen levels. The presence of methane (CH_4) in the primitive atmosphere would have oriented endogenous syntheses towards a considerable production of nitriles, especially the simplest of these, hydrocyanic acid HCN, which amount could have reached ca. $10^{10}\,\mathrm{kg\,y^{-1}}$. Assuming the volume of the primitive oceans to have been identical to that of the present day, considering only the soluble part of the organic matter ($\approx 40\%$ of the total determined by Chyba and Sagan (1992), see Cronin and Chang 1993), and supposing this organic matter to be rather stable under its supposed conditions, the primitive oceans could have contained (at a steady state) between ca. $0.4 \times 10^{-3}\,\mathrm{g\,L^{-1}}$ of total organic matter in the case of a neutral atmosphere and $0.4\,\mathrm{g\,L^{-1}}$ in the case of a reductive atmosphere. Given the imprecision of current knowledge about the composition of the primitive atmosphere, three different eventualities can be considered:

- A highly reductive atmosphere in which syntheses due to impacts would have predominated over endogenous production due to electrical discharges and solar UV
- A slightly reductive atmosphere in which, by contrast, endogenous syntheses (solar UV, lightning) would have predominated, augmented by the contribution of organic matter from IDPs
- A very slightly reductive, or nonreductive atmosphere in which the contribution of organic matter from IDPs would have predominated

We shall now examine the qualitative aspects linked to these eventualities.

4.3 Exogenous Amino Acids and Related Compounds

Analysis of the organic matter contained in meteorites, especially carbonaceous chondrites, regularly collected on Earth's surface (Cronin and Chang 1993; Maurette 1998), has provided an approach to extraterrestrial organic chemistry (the study of dense clouds is another possible approach, doubtless more direct, but restricted mostly to volatile materials). This organic matter could correspond to that formed in the early Solar System, or even in the pre-Solar System (Cronin and Pizzarello 1983; Cronin and Chang 1993; Cooper and Cronin 1995; Cooper et al. 2001). It is mainly (> 60%) macromolecular, insoluble and difficult to characterise. The soluble part (< 40%) is a complex mixture of molecular organic compounds in which the following have been characterized: acids (carboxylic, dicarboxylic, hydroxyl, sulfonic, phosphonic); amines; amides; nitrogenous heterocycles including purines and pyrimidines; alcohols; sugars; carbonyl compounds; aliphatic and aromatic hydrocarbons; amino acids and hydantoins. Structures of compounds dealt with in this chapter are shown in Fig. 4.1.

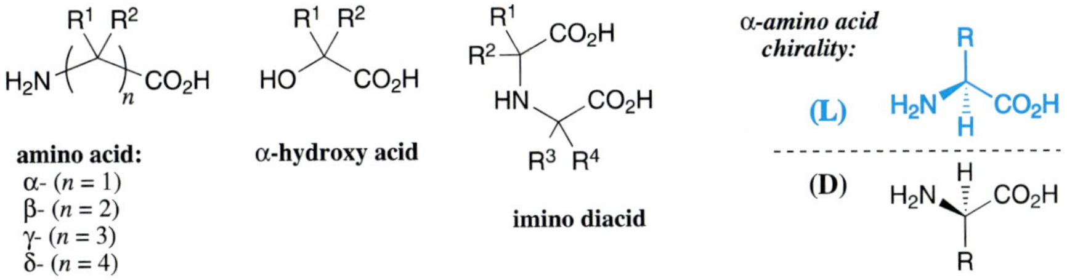

Fig. 4.1. General structure of amino acids and related compounds. All proteinic α-amino acids have the L configuration (*blue*)

4.3.1 Exhaustive Survey of Exogenous Amino Acids

More than 70 different amino acids have been characterized, principally in the Murchison and Murray meteorites (both being carbonaceous chondrites) (Tables 4.1–4.4), representing a total abundance of 75 µg per g of meteorite. Over a set of 7 other carbonaceous chondrites, the average mass of extracted amino acids was only 0.6 µg per g. It should, nevertheless, be noted that the mass of organic molecules present in chondrites decreases as a function of the time the materials have spent on the Earth (Cronin and Pizzarello 1983), thus probably leading to an underassessment of the amount of exogenous organic matter. Thus, even if the value obtained for the Murchison meteorite is representative of chondrites and if this value is doubled to take into account the existence of amino acid precursors in the total carbonaceous matter (Cronin and Pizzarello 1983; Cooper and Cronin 1995), the abundance of meteoritic amino acids remains quantitatively low, even considering that chondritic micrometeorites could have

Table 4.1. The 20 protein α-amino acids: names, 3-letter and 1-letter codes, and numbers of C atoms in their alkyl chains. *blue*: α-amino acids detected in carbonaceous chondrites

Name	3-lett code	1-lett. code	# C atoms	Name	3-lett code	1-lett. code	# C atoms
Alanine	Ala	A	3	Leucine	Leu	L	6
Arginine	Arg	R	5+1	Lysine	Lys	K	6
Aspartic Acid	Asp	D	4	Methionine	Met	M	4+1
Asparagine	Asn	N	4	Phenyl-alanine	Phe	F	3+6
Cysteine	Cys	C	3	Proline	Pro	P	5
Glutamic Acid	Glu	E	5	Serine*	Ser	S	3
Glutamine	Gln	Q	5	Threonine	Thr	T	4
Glycine	Gly	G	2	Tryptophane	Trp	W	3+8
Histidine	His	H	3+3	Tyrosine	Tyr	Y	3+6
Isoleucine	Ile	I	6	Valine	Val	V	5

* though very often detected, Ser is widely considered as a terrestrial contaminant.

delivered larger amounts of organic matter than meteorites themselves. This low rate of exogenous carbon delivery may nonetheless have been of great *qualitative* importance (see Sect. 4.3.4) for prebiotic chemistry processes on the primitive Earth.

4.3.2 Formation Mechanisms of Exogenous Amino Acids

Out of more than 70 identified extraterrestrial amino acids (Tables 4.2–4.4), 46 (including 8 diacids) have their amine function in the α-position, 13 in the β-position, 10 in the γ-position and 1 in the δ-position. The formation rate and/or the stability of α-isomers therefore appears favoured in extraterrestrial environments.

4.3.2.1 α-Amino Acids

The formation mechanism of α-amino acids has been the subject of a great deal of attention. In the historic experiment of Miller (1953) α-amino acids and α-hydroxy acids were associated with imino diacids $HN(CR_2CO_2H)_2$, these three compound classes altogether being considered as *the* signature of Urey–Miller experiments. Miller (1957) explained the formation of the α-amino acids α-hydroxy acids by the Strecker reaction (4.1), without being able to suggest a mechanism

Table 4.2. Names and structures of α-amino acids detected in carbonaceous chondrites

2–5 carbon atoms	6,7 carbon atoms	7 carbon atoms (contd)	
glycine	leucine	2-amino-4-methyl caproic acid*	
alanine	*isoleucine**	2-amino-4,4-dimethyl valeric acid	
sarcosine	norleucine	2-amino-2-ethyl-valeric acid	
2-amino-butyric acid	**2-methyl-norvaline**	2-amino-3-ethylvaleric acid	
2-amino isobutyric acid	**2-amino-2,3-dimethyl butyric acid**	2-amino-5-methyl caproic acid	
N-methyl-alanine	2-amino-2-ethylbutyric acid	*2-amino-3-methyl caproic acid**	
N-ethyl-glycine	pseudoleucine	*2-amino-3,4-dimethyl-valeric acid**	
valine	cycloleucine	2-amino-2,3,3-trimethyl butyric acid	
proline	2-amino-heptanoic acid	2-amino-2-ethyl-3-methylbutyric acid	
isovaline	*2-amino-2,3-dimethyl valeric acid**	2-amino-3,3-dimethyl valeric acid	
norvaline	2-amino-2,4-dimethyl valeric acid	**2-amino-2-methyl caproic acid**	

In blue: protein amino acids (only 8; Ser although very often detected, is widely considered as a terrestrial contaminant). * *(names in italic)*: diastereomer pairs (counted as two compounds; most amino acids are enantiomer pairs and counted as a single compound). Boxed *(names in bold)*: amino acids detected as non racemic in meteorites (1–9% ee of L configuration; the methyl group in place of the α-hydrogen makes them resistant to racemization).

Table 4.3. Names and structures of α-amino diacids detected in carbonaceous chondrites

Succinic type	Glutaric type	Others
aspartic acid	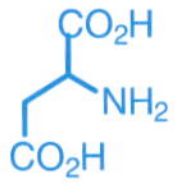 glutamic acid	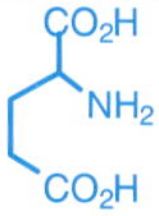 2-amino pimelic acid 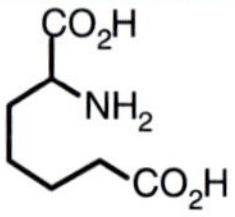
2-amino-2-methyl succinic acid	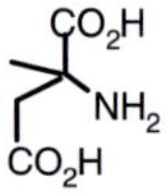 2-amino-2-methyl glutaric acid	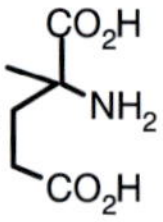 2-amino adipic acid 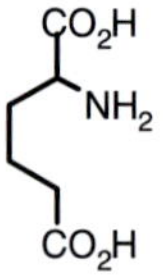
2-amino-3-methyl succinic acid* 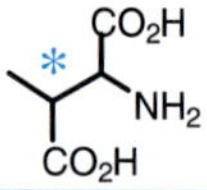		

(see notes on Table 4.2)

Table 4.4. Names and structures of non-α-amino diacids detected in carbonaceous chondrites

β-amino acids	β- (cont.), γ-amino acids	γ- (cont.), δ-amino acids
β-alanine	 3-amino-2-ethyl butyric acid*	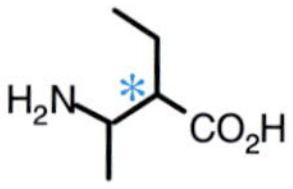 4-amino-2-methyl valeric acid* 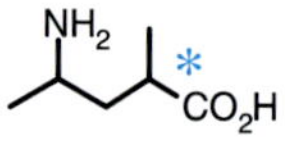
3-amino-butyric acid	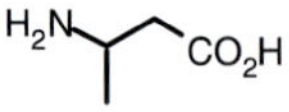 3-amino-3-methyl valeric acid	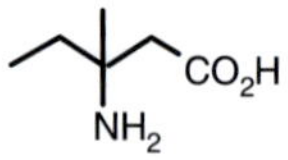 uncertain structure (either of these 2):
3-amino-2-methyl-propionic acid	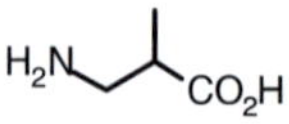 3-amino-2,3-dimethyl butyric acid	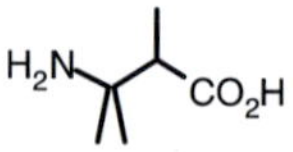 4-amino-caproic acid 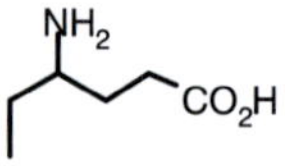
3-amino-2-methyl butyric acid	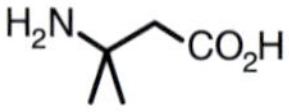 4-amino-butyric acid	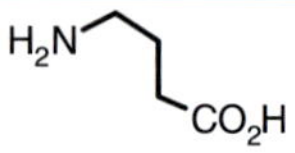 4-amino-3-ethyl butyric acid 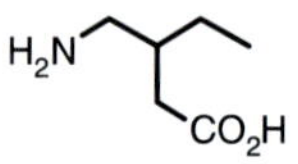
3-amino-3-methylbutyric acid*	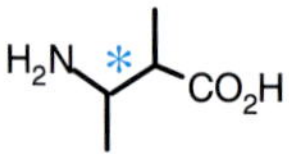 4-amino-2-methylbutyric acid	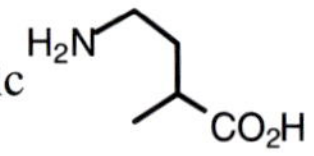 uncertain structure (either of these 2):
3-amino-pivalic acid	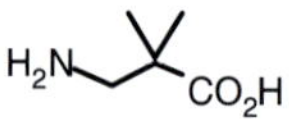 4-amino-3-methylbutyric acid	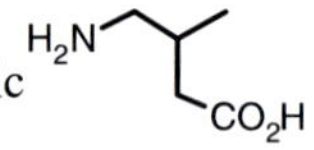 4-amino-4-methylvaleric acid 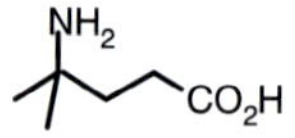
3-amino-valeric acid	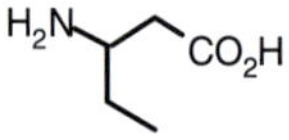 4-amino-valeric acid	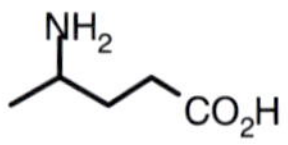 4-amino-3,3-dimethyl butyric acid 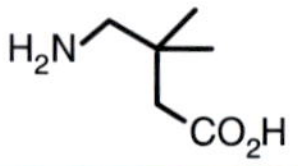
3-amino-2-ethyl-propionic acid	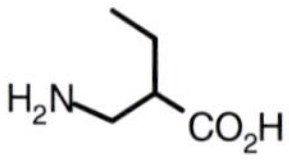 4-amino-3-methylvaleric acid*	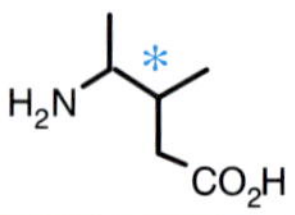 5-amino valeric acid 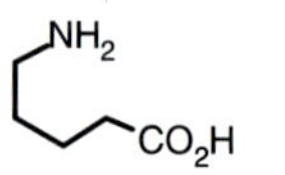

(see notes on Table 4.2)

for the imino-diacid formation. He nevertheless intuitively considered these three classes altogether to be also the signature of the Strecker reaction.

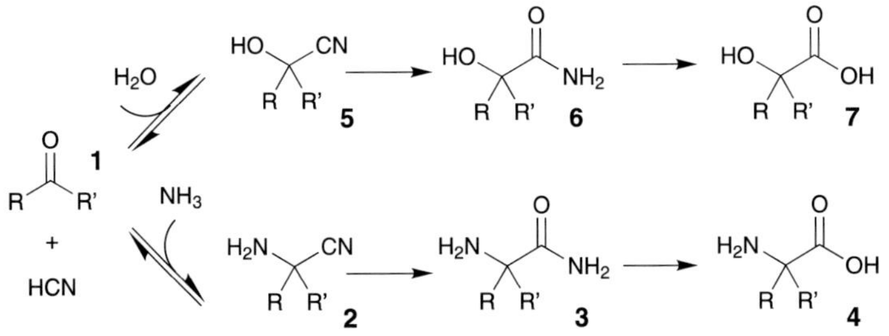

$$\begin{array}{c} \text{(scheme 4.1)} \end{array}$$
(4.1)

The early (and still accepted) hypothesis for meteoritic α-amino acid synthesis considered the Strecker reaction as the most plausible although only two (α-hydroxy- and α-amino acids) out of its three-component signature had been identified in meteorites (Peltzer et al. 1984; Cronin and Chang 1993; Botta et al. 2002). The recent discovery of imino diacids in meteorites (Pizzarello and Cooper 2001) completed this signature and confirmed Miller's hypothesis of the Strecker reaction for meteoritic α-amino acid synthesis. A better understanding of chemical mechanisms (see Sect. 4.5.2) explains the formation of imino diacids through Strecker processes.

If meteoritic α-amino acids actually originated through Strecker reactions, then their precursor carbonyl compounds are easily deduced by applying retrosynthesis to the set of meteoritic amino acids (Tables 4.2 and 4.2). This provides the 35 carbonyl derivatives listed in Table 4.5, where all possible isomers are present. The higher their molecular mass, the lower their abundance (relatively to that of α-amino acids).

The presence of all the possible isomers of these carbonyl derivatives, some of which being still present in meteorites (Jungclaus et al. 1976), clearly shows that they have previously been synthesized by nonselective processes, possibly photochemical. In the long term, this whole set of carbonyl derivatives is to be taken into consideration in a global kinetic model to explain the presence of exogenous and endogenous α-amino acids on the primitive Earth.

4.3.2.2 Non α-Amino Acids

The proposed mechanism (Cronin and Chang 1993) (4.2) for the formation of γ-amino acids involves the following initial molecules: α, β unsaturated aldehyde **8** variously substituted, hydrocyanic acid and ammonia. After addition of two HCN molecules, cyanohydrins (α-hydroxy nitriles) **9** are obtained. After hydration and hydrolysis (Strecker reaction), α-amino nitriles **10** lead to **11** and **12** (when R^1, R^2, and R^3 are hydrogens, **12** is glutamic acid). The following reaction sequence: cyclization of **12**, decarboxylation of **13** and hydrolysis of **14** produces γ-amino acids **15**, which might therefore appear as degradation products of α-amino 1,3-diacids **12**. If so, this would increase the count of exogenous α-amino acids from 46 to about 54, out of 70.

$$(4.2)$$

β-amino acids could have been formed according to (4.3) by addition of ammonia on acrylonitrile **16** ($R^1 = R^2 = R^3 = H$) or its substituted derivatives ($R^1, R^2, R^3 \neq H$), followed by hydrolysis of the nitrile function (Cronin and Chang 1993).

$$(4.3)$$

The proposed mechanism (Cronin and Chang 1993) (4.4) for the formation of the unique δ-amino acid **21** detected in the meteorites, is similar to that in (4.3), but starting with cyanobutadiene **19** as the initial substrate. This process involves, together with the above addition reaction, a reduction step on **20**.

$$(4.4)$$

The formation of proline **25** can be explained either by the mechanism described by (4.5) (detection of pyrrolidine (saturated **22**) in interstellar clouds is consistent with this explanation) (Cronin and Chang 1993), or by cyclization of $H_2N-CH_2-CH_2CH_2CHO$.[1]

$$(4.5)$$

---

[1] Another possible pathway involves cyanohydrins $H_2N-(CH_2)_n-CH(CN)-OH$, the "Strecker precursors" of ornithine ($n = 3$) and lysine ($n = 4$): if present, their intramolecular substitution would lead to proline and pipecolic acid, respectively (both compounds being identified in Murchinson and Murray meteorites, but neither lysine nor ornithine).

4 Peptide Emergence, Evolution and Selection on the Primitive Earth. I.   125

Table 4.5. The 35 carbonyl derivatives precursor of meteoritic α-amino acids and α-amino diacids. In *blue:* compounds bearing a nitrile group

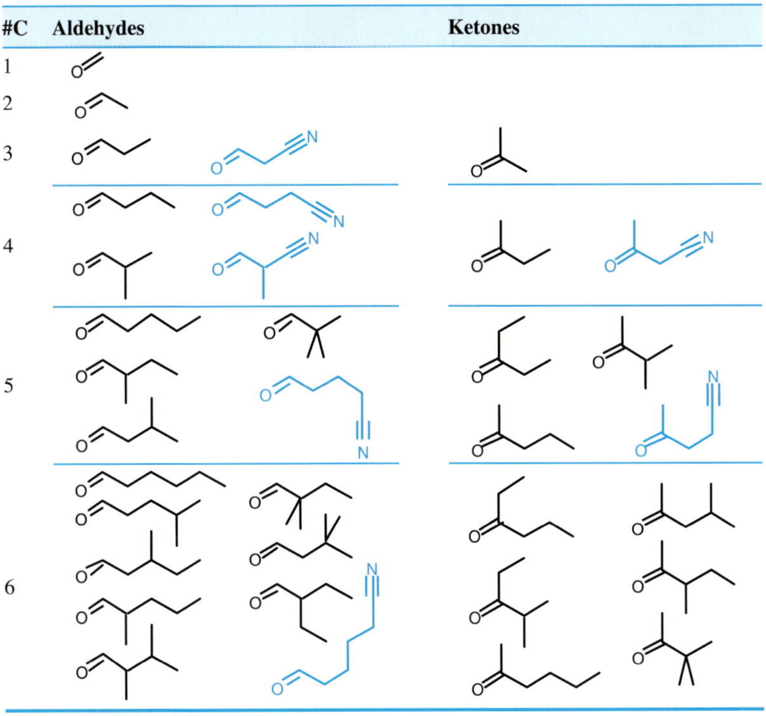

4.3.3 Other Meteoritic Compounds Closely Related to Amino Acids

Cooper and Cronin (1995) and subsequently Shimoyama and Ogasawara (2002) identified the compounds shown in Fig. 4.2 in the Murchison and Yamoto-791198 meteorites.

Lactames **13** and **14** are intermediates in γ-amino acid formation from α-amino acids (4.2). Compounds **26** to **30** are amide or imide derivatives. Compounds **31** are hydantoins (imidazolidine-2,4-diones) which can result from two pathways (Fig. 4.3):

– Either the reaction of cyanic acid on α-amino acids (formed through the Strecker reaction), followed by cyclization (Cooper and Cronin 1995), in which case hydantoins both substituted and unsubstituted on nitrogen 1 can be formed.
– Or the Bücherer–Bergs reaction (Taillades et al. 1998) via the cyclization of isocyanates **35** (see Sect. 4.5.2). In this case only 1-unsubstituted hydantoins are accessible.

Actually only 1-unsubstituted hydantoins have been identified in meteorites, which rather supports the hypothesis of a Bücherer–Bergs pathway.

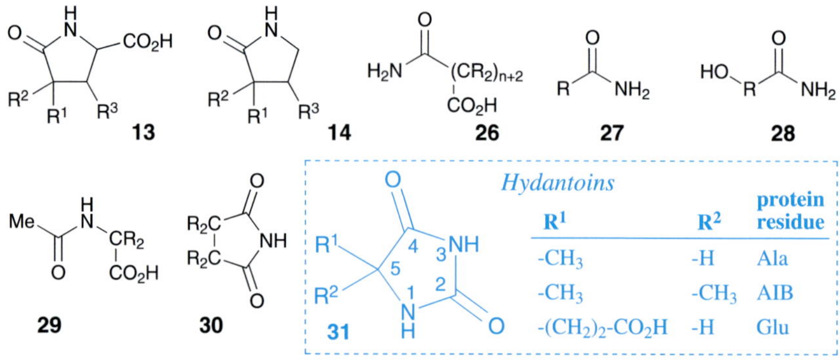

Fig. 4.2. Amides, imides and hydantoins discovered in the Murchison and Yamoto-791198 meteorites

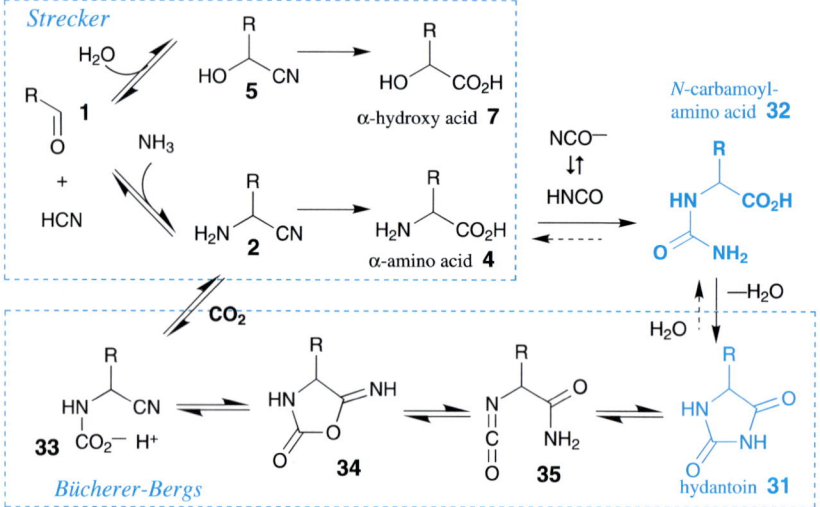

Fig. 4.3. Pathways of the Strecker and Bücherer–Bergs reactions and N-carbamoylation of α-amino acid

4.3.4 Non-Racemic Exogenous α-Amino Acids

(see also Cronin and Reisse Chap. 3, Part II)

Six extraterrestrial α-amino acids (framed structures in Tables 4.2–4.4) extracted from the Murchison and Murray meteorites were found as nonracemic (Cronin and Pizzarello 1997; Pizzarello and Cronin 2000). Their enantiomeric excesses (L) are between 1.0 and 9.2%. The presence of a CH_3 group in the α position instead of an H, argues for their nonbiological origin, which is confirmed by their D/H and $^{13}C/^{12}C$ isotopic ratios. This enantiomeric excess could be the result, not of chemoselectivity, but rather of enantioselective photodegradation of the initial racemic mixture in the presolar cloud. This photodegradation could have caused by circularly polarized UV light. Circularly polarized light, although infrared, has been observed in the Orion cloud, meanwhile the origin of such radiation is the focus of numerous questions (Jorissen and Cerf 2002). This hypothesis of partial photodegradation of racemic mixtures is nonetheless consistent with laboratory results obtained as early as 1977 (Flores et al. 1977). The mechanism of this reaction was recently developed further (Nishino et al. 2001).

The existence of these nonracemic compounds opens up a wide research area in the field of the homochiral synthesis of peptides. They could have induced a symmetry breaking during the production of peptides by a "molecular engine" described in the next chapter of this book.

4.3.5 Exogenous Peptides

It is only recently that the first exogenous peptides were looked for and found in the Murchison and Yamoto-791198 meteorites. The authors of these works (Shimoyama and Ogasawara 2002) showed, however, that the only peptides present are glycylglycine (Gly-Gly) and the corresponding diketopiperazine. The abundance of these compounds, (of the order of ca. $20\,\mathrm{pmol\,g^{-1}}$), and, above all, the absence of both Ala-Gly and Gly-Ala, despite the presence of Ala and Gly in meteorites, led Shimoyama to conclude that these compounds were probably formed by classical dehydration processes, rather than via the copolymerization of NCA, which should have produced sequential macromolecules (see our next chapter in this book). The potential exogenous imports thus appear to be of minor importance.

4.3.6 Conclusion

α-Amino acids are therefore major compounds of the meteorite amino acid pool, both in structural diversity (54/70) and abundance, including 8 of the 20 protein amino acids[2]. Their formation through Strecker reactions is consistent with the presence in meteorites of the three compound classes characteristic of this kind of reaction: α-hydroxy- and α-amino acids, imino diacids. The – yet unexplained – formation of imino diacids is detailed in Sect. 4.5 devoted to chemical studies. Since β- and δ amino acids are rather marginal, we shall not go into detail of their synthesis in that section, focusing rather on Strecker and Bücherer–Bergs processes.

---

[2] 10 if we include asparagine and glutamine (ω-amide forms of aspartic and glutamic acids, respectively). Though never detected in meteorites, these amino acids were very probably formed, but would have been readily and completely hydrolysed (into aspartic and glutamic acids respectively) during either meteorite ageing or extraction process.

4.4 Endogenous Organic Matter

The endogenous production of organic matter is unknown due to the absence of any fossil record. It must have greatly depended on the environment, especially on how reducing or oxidizing this environment was.

Knowledge of the different sources of available energy and their respective contributions to the endogenous production of organic matter, has evolved over time. While solar UV has always been considered to have played an important quantitative role (Chyba and Sagan 1992) the origin of organic nitrogen chemistry nonetheless still needs to be explained, given the short wavelengths ($\lambda < 100$ nm) required for the photolysis of N_2 (Selsis et al. 1996; Selsis 2000). In this area, the (somewhat underestimated) importance of lightning and meteoritic impacts was recently re-evaluated (Navarro-Gonzalez et al. 1998, 2001), together with its consequences on the production of nitric oxide, NO. This aspect will be presented in Commeyras et al., Chap. 5, Part II, for which it is of high importance: a part of so-produced NO, remained in the atmosphere, might have played a fundamental role in the further synthesis of peptides.

Meanwhile, the other part of this NO, dissolved in the primitive ocean, could have been reduced to NH_3 by Fe^{2+} in the ocean, even in the presence of high levels of cyanide (Summers and Chang 1993; Summers and Lerner 1998). The ammonia thus produced could have made a major contribution to primitive organic synthesis and especially to that of α-amino acids (see Sect. 4.6.2). In addition to its role as a major source for early-Earth nitrogen chemistry, NO may also have served as an energy source for carbon–oxygen chemistry, which, however, could also be sustained by solar UV.

4.4.1 Endogenous α-Amino Acids

It was the historic experiment of Miller (Miller 1953) that showed that α-amino acids could have been formed on the primitive Earth from a reductive atmosphere. In the absence of reliable information on the composition of this atmosphere, the experiment was repeated on numerous occasions (Miller 1998) under different conditions, intended to simulate many of the different possible compositions $\{CH_4, NH_3, H_2, H_2O\}$, $\{CH_4, NH_3, H_2O\}$, $\{CH_4, H_2, N_2, NH_3, H_2O\}$, $\{CH_4, H_2, N_2, H_2O\}$, $\{CO, H_2, N_2, H_2O\}$, $\{CO_2, H_2, N_2, H_2O\}$, under various H_2/CH_4, H_2/CO or H_2/CO_2 ratios. The formation of amino acids was observed in all cases, even unfavourable ones (CO_2, H_2, N_2, H_2O), provided the H_2/CO_2 ratio was greater than 2, thus involving large amounts of hydrogen in the atmosphere[3]. If we complement these atmospheric models by the sulfydryl function (–SH), phenylacetylene and indole (obtained by flash pyrolysis simulating the electrical discharges of gaseous CH_4, NH_3 mixtures, see Miller 1998), we obtain

---

[3] Recent observations show that the gases issuing from black smokers (undersea hot springs) contain considerable hydrogen amounts (Brack, lecture in Montpellier, 2002).

17 protein α-amino acids, plus a collection of nonprotein amino acids analogous to those discovered in meteorites, together with some imino diacids.

As early as 1955, Stanley Miller showed that carbonyl derivatives, hydrocyanic acid and ammonia were the raw materials that produced these α-amino acids. He proposed the Strecker reaction as the formation mechanism of these compounds (Miller 1957, 1998). This proposition was used once again to explain the formation of some of the exogenous α-amino acids (see Sect. 4.3.2). It is remarkable to see how Miller's proposition remains of current importance.

In addition to these synthetic pathways, others have been explored. We can cite:

- The one starting from HCN (Oró 1961) leading to low percentages of glycine and traces of alanine and aspartic acid.
- The one starting from 2-amino acrylonitrile $H_2C=C(NH_2)CN$ (Ksander et al. 1987), which leads, in the absence of oxygen and water, to numerous organic compounds such as glutamic acid, and a precursor of the corrinoid cycle of vitamin B12 obtained with a very high yield (50%).

This last synthesis could have been prebiotic. The ease of synthesis of an essential building block of vitamin B12 interested Eschenmoser et al. (Ksander et al. 1987), who wondered whether the complexity of vitamin B12 was only apparent, and whether the selectivities of such syntheses starting from simple reagents might not be more common in the field of prebiotic chemistry than we had imagined. The authors suggested that research efforts should be developed in that direction.

We show in the next section of this chapter that the synthesis of N-carbamoyl amino acids could be a second example of selective prebiotic synthesis. To present these results, we first summarize the knowledge acquired through studying a multicomponent reaction system made up of the following molecules $\{H_2O, HCN, R^1R^2CO, NH_3, R^3NH_2, R^3R^4NH, CO_2, B(OH)_4^-\}$. Secondly, we use this knowledge to try to understand the formation dynamics of α-amino acids and N-carbamoyl amino acids on the primitive Earth.

4.5 Formation Mechanisms of α-Amino Acids and N-Carbamoyl Amino Acids Via Strecker and Bücherer–Bergs reactions

The aim of this section is to present the reaction mechanisms of Strecker and Bücherer–Bergs reactions, involving 4 initial components $\{H_2O, R^1R^2CO, HCN, NH_3\}$ and 5 initial components $\{H_2O, R^1R^2CO, HCN, NH_3, CO_2\}$, respectively.

We shall also examine extensions of these multicomponent systems, including borate $^4B(OH)_4^-$ (see Sect. 4.5.2), primary or secondary amines (R^3NH_2 or R^3R^4NH, respectively, see Sect. 4.5.3), the fate of both Strecker and Bücherer–Bergs reactions being different for each instance. Considering then the global system made up of all 8 above components, the combination of Strecker and Bücherer–Bergs reactions together with those of borate[4] will lead to a set of α-amino acids, N-carbamoyl amino acids, α-hydroxy acids and traces of imino diacids. The great structural diversity of several reactants: R^1R^2CO, R^3NH_2, R^3R^4NH (see Tables 4.1–4.5), will logically result in highly complex mixtures.

If such complexity is examined through combinatorial chemistry techniques only, we can get information on the set of final products only, losing the dynamic and mechanistic aspects such as equilibria and kinetics. This latter information, which is nonetheless indispensable for the understanding and modelling of dynamic processes, needs to be sought by means of more detailed, less global experimental approaches. That is the price to be paid for the quantitative study of chemical-evolution processes.

Our essential results are summarized in Commeyras et al., 2003. On this basis, and for the sake of clarity, the behaviour of the above 8-component system can be schematically presented as a first group of reversible steps producing a wide variety of chemical intermediates (Fig. 4.4), followed by a second group of irreversible steps resulting into a limited number of more stable species

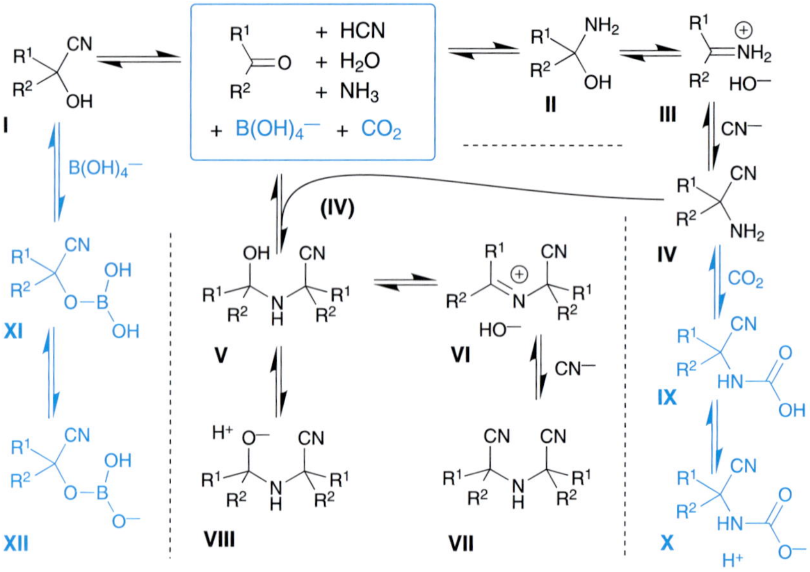

Fig. 4.4. Reversible reactions and products (at equilibrium) of the 6-component system $\{H_2O, HCN, R^1R^2CO, NH_3, CO_2, B(OH)_4^-\}$

---

[4] Boron is a non-negligible component of the Earth's crust (0.03%), with borates involved in the composition of glass (13%). Therefore it may be worthwhile analyzing the potential participation of borates in organic molecules formation under both prebiotic and laboratory conditions, because of their catalytic role in some reactions.

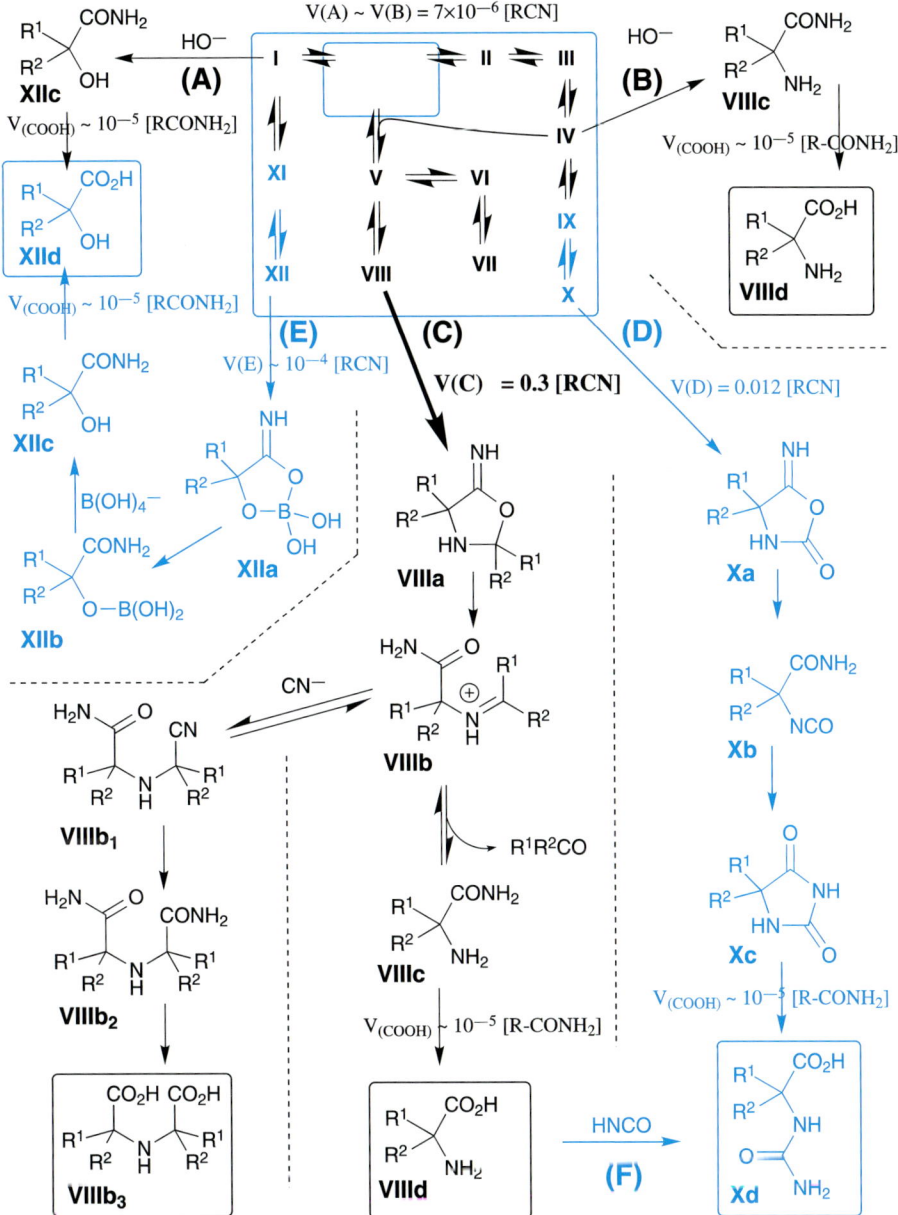

Fig. 4.5. The set of (competing) irreversible reactions in the reaction network from the 6-component system {H_2O, HCN, R^1R^2CO, NH_3, CO_2, $B(OH)_4^-$}, showing pathways (A)–(E)

(Fig. 4.5).[5] The special case of formaldehyde in the presence of high amounts of ammonia will not be examined however, since we considered it as too far from reasonable prebiotic conditions (see Sect. 4.6.2).

4.5.1 The Set of Reversible Reactions

We shall first examine the set of reversible reactions of the initial 6-component system $\{H_2O, HCN, R^1R^2CO, NH_3, CO_2, B(OH)_4^-\}$ (additional constraints introduced by primary and secondary amines R^3NH_2 and R^3R^4NH will be examined later). This set of reactions[6] (Fig. 4.4) is the starting point of Strecker and Bücherer–Bergs reactions that involve further, irreversible steps.

The reactions able to take place in this initial system (box on top of Fig. 4.4) mostly depend on the reactivity of the carbonyl derivatives R^1R^2CO. They can react:

- with water to give hydrates (not shown)
- with hydrocyanic acid to give cyanohydrins **I** (fast)[7]
- with ammonia to give α-amino alcohols **II**

Amino alcohols **II** (slowly) lose an HO^- anion to produce iminium cations **III**, which then react with the cyanide anion to give α-amino nitriles **IV**, the amine function of which reacts with carbonyl derivatives to produce α-amino alcohols **V**. **V** are subject to two competitive reactions: either they lose an HO^-, yielding iminium cations **VI**, which further react with cyanide to produce α-amino dinitriles **VII**, or **V** lose a proton to give alcoholate ions **VIII**.

CO_2 specifically reacts with α-amino nitriles **IV** to give carbamic acids **IX** that ionize into carbamates **X**. Boric acid $B(OH)_3$ (in equilibrium with borate $B(OH)_4^-$) specifically reacts with the cyanohydrins **I** to produce boric esters **XI**, which ionize into borates **XII**.

Under certain conditions (experimental but probably not prebiotic), the equilibria corresponding to these reversible reactions can be reached, thus constituting a reservoir of various nitrile families (cyanohydrins, amino nitriles, amino dinitriles, etc.). Their relative abundance depends on reactant structures (the nature of R^1, R^2), initial stoichiometry and concentration, and on the experimental conditions, especially pH, temperature and CO_2 pressure (Commeyras et al. 2003).

While such information is easily accessible experimentally, the main concern is then to determine whether this product distribution at equilibrium is representative or not of the final product distribution when further irreversible steps are

---

[5] This separation into two sets is made for the sake of clarity. Under prebiotic conditions, the equilibrium corresponding to the first set is very unlikely to be reached due to both reactant (e.g. ammonia) fluxes and competitive reactions from the second set.

[6] Equilibria of both hydration of carbonyl derivatives R^1R^2CO and carbonation of ammonia NH_3 (which are also part of this set) have been omitted to make Fig. 4.4 simpler.

[7] Molecular structures involved in Figs. 4.4–4.7 are labelled with bold Roman figures.

involved. In other words whether the final product distribution is under thermodynamic or kinetic control. This requires determination of kinetic data on both reversible and irreversible steps.

4.5.2 The Set of Irreversible Reactions

In these reaction media, irreversibility appears when nitriles (–CN) are transformed into amides (–CONH$_2$). Five different, competitive pathways (A–E, Fig. 4.5) can bring about this transformation[8]. Their starting points are the species **I**, **IV**, **VIII**, **X**, **XII** respectively, as shown in Fig. 4.4.

The first two pathways (A) and (B) are well known as the *Strecker reaction*[9] (Strecker 1850; Miller 1957; see Sect. 4.3.2). They correspond to the intermolecular attack of the hydroxide anion (HO$^-$) on the nitrile group of the cyanohydrins **I** and α-amino nitriles **IV**. They can explain the formation of both α-hydroxy amides **XIIc** and α-amino amides **VIIIc**, but not the formation of imino diacids **VIIIb$_3$**.

The other three reaction pathways (C), (D), (E) evidenced by us, result in the hydration of the nitrile group of **I** and **IV** in a more efficient way than both pathways (A) and (B). In addition, the reaction pathway (C) can explain the formation of imino diacids **VIIIb$_3$**.

The starting point of the pathway (C) is the alcoholate anions **VIII**. These anions react *intra*molecularly with their nitrile group, to give the α-amino amides **VIIIc** via two intermediates **VIIIa** and **VIIIb**. This pathway (C) leads to the same products as (B), but approximately 10^5 times faster, through catalysis by carbonyl derivatives.

Concerning imino diacids **VIIIb$_3$**, Garrel (personal communication) recently showed that they originate from **VIIIb**. When the initial reagent concentrations are such as [R^{1}R^{2}CO] > [CN$^-$] ≫ [NH$_3$], the addition of cyanide to intermediates **VIIIb** occurs, that forms new nitriles **VIIIb$_1$** as minor products, which then evolve into **VIIIb$_3$**. Therefore we consider that this reaction pathway (C) and its side-products must be completely included into the Strecker reaction, which should not be limited to reaction pathways (A) and (B).

The starting point of the pathway (D) is the carbamate anions **X**, which react *intra*molecularly with their nitrile group to give the hydantoins **Xc** via two intermediates **Xa** and **Xb**. This reaction is almost as fast as (C), and is known as the Bücherer–Bergs reaction.

---

[8] An additional pathway to amino acids and involving hydrogen peroxide, was not taken into consideration because poorly compatible with prebiotic condition (Rossi et al. 1996).

[9] The name "Strecker reaction" was initially given to the synthesis of α-amino acids by *acidic hydrolysis* (quenching of equilibrated aqueous alkaline) mixtures of carbonyl compounds, hydrocyanic acid and ammonia, referring to his original experiments. Later this name covered any experimental conditions using reaction pathways (A)+(B)+(C).

The starting point of the pathway (E) is the borate anions **XII**, which react *intra*molecularly with their nitrile group to give the α-hydroxy amides **XIIc**, regenerating, en route, the boric acid, which thus behaves as a catalyst in this reaction.

Kinetic studies enabled us to determine the kinetic laws and constants (4.6–4.10) of these reactions (Taillades et al. 1998; Commeyras et al. 2003), the following equations being valid in the pH 5–9 range, with values given at 25°C:

$$v_{(A)} = k_{(A)}[\text{HO}^-][\textbf{I}] \quad k_{(A)} = 3.8 \times 10^{-4}\,\text{L}\,\text{mol}^{-1}\,\text{s}^{-1} \tag{4.6}$$

$$v_{(B)} = k_{(B)}[\text{HO}^-][\textbf{IV}] \quad k_{(B)} = 1.4 \times 10^{-3}\,\text{L}\,\text{mol}^{-1}\,\text{s}^{-1} \tag{4.7}$$

$$v_{(C)} = k_{(C)}[\text{HO}^-][\text{R}^1\text{R}^2\text{CO}][\textbf{IV}] \quad k_{(C)} = 83\,\text{L}^2\,\text{mol}^{-2}\,\text{s}^{-1} \tag{4.8}$$

$$v_{(D)} = \frac{k_{(D)}}{1 + \frac{K_w}{K_1}[\text{HO}^-]}[\text{CO}_2]^t\,[\textbf{IV}] \quad k_{(D)} = 1.2 \times 10^{-4}\,\text{L}\,\text{mol}^{-1}\,\text{s}^{-1} \tag{4.9}$$

$$v_{(E)} = k_{(E)}[\text{B(OH)}_3]^t\,[\text{HO}^-]\,[\textbf{I}] \quad k_{(E)} = 180\,\text{L}^2\,\text{mol}^{-2}\,\text{s}^{-1}, \tag{4.10}$$

with K_w and K_1 in (4.9) being the dissociation constants of water and of carbonic acid respectively (first dissociation), $[\text{CO}_2]^t$ in (4.9) and $[\text{B(OH)}_3]^t$ in (4.10) being the total concentration (acid + base forms) of carbonate + CO_2 and borate, respectively. It is noteworthy that above pH 7.5 the rate law (4.9) is independent of the pH. Further studies will be necessary to determine the temperature dependence of these rate constants.

In a first approach, however, at any pH between 5 and 9, and provided the concentration of either carbonyl compounds, carbonate or borate are above the millimolar range, the reaction pathways (A) and (B) (the only ones dealt with in previous prebiotic studies, see Sect. 4.3.2) should play a minor role comparatively to much faster pathways (C), (D), (E) using the same starting nitriles **I**, **IV**, **VIII**. The latter pathways would thus be responsible for the formation of α-amino amides **VIIIc**, hydantoins **Xc**, and α-hydroxyamides **XIIc**. Nevertheless, such a conclusion must be adapted to prebiotic conditions, where the concentrations of the various nitrile families **I**, **IV**, **VIII** should be far from identical.

Studying the subsequent evolution of these α-amino amides **VIIIc**, hydantoins **Xc**, and α-hydroxyamides **XIIc** into α-amino acids **VIIId**, N-carbamoyl amino acids **Xd** and α-hydroxy acids **XIId**, respectively, showed that these three hydrolytic reactions have low but equivalent rates ($v_{(\text{CO}_2\text{H})} \approx 10^{-5}[\text{R} - \text{CONH}_2]$, see Commeyras et al. 2003). From this we can deduce that these reactions are incapable of modifying the selectivities induced by reactions (C), (D) and (E).

If the multicomponent system under consideration (Fig. 4.4) is completed with cyanate (the actual reacting species being cyanic acid HNCO), the system is driven to evolve towards N-carbamoyl amino acids (CAA) **Xd** through pathway (F). Kinetic studies of this reaction (Taillades et al. 2001) showed that the operating conditions required for carrying out this reaction are fully compatible with those of the other reactions described in Fig. 4.5. In addition, cyanic acid also reacts with amino amides **VIIIc** to give N-carbamoylamino *amides*.[10] We observed (unpublished results) that the nitrosation of such compounds by gaseous NO_x (see next chapter of this book) quantitatively yields hydantoins, the hydrolysis of which produces N-carbamoylamino *acids*.

Globally, we can thus conclude that:

The "multicomponent" system represented in Figs. 4.4 and 4.5 is kinetically controlled by irreversible reactions proceeding through reaction pathways (C), (D), (E).

This "multicomponent" system convergently evolves towards two families of compounds: α-hydroxy acids (minor) and N-carbamoyl amino acids (major).

4.5.3 Fate of Primary and Secondary Amines (R^3NH_2, R^3R^4NH)

Primary amines R^3NH_2 follow the same reaction pathways as ammonia, with some limitations. Indeed while they can take pathways (B) and (C) to form α-NR^3-amino acids **VIIId**, they are blocked on pathway (D) by the impossibility of forming isocyanates **Xb**, so that hydantoins **Xc** substituted on nitrogen-1 are inaccessible. A roundabout way to obtain such substituted hydantoins is to react cyanate with α-amino acids-NR^3 according to pathway (F), then to dehydrate the NR^3-carbamoylamino acids **Xd** thus obtained. However, no such product has been identified in meteorites, which leads to the conclusion that these cyanates only played a minor role in the parent bodies, doubtlessly for kinetic reasons (see Sect. 4.7). This obviously does not exclude their having played a role on the primitive Earth. In the parent bodies, hydantoins could have been formed via the Bücherer–Bergs reaction.

The reactivity of secondary amines (R^3R^4NH) is even more severely limited. They can only lead to α-NR^3R^4-aminonitriles **IV** and then α-NR^3R^4-amino acids **VIIId** through pathway (B).

Taking these limitations into consideration provides interesting information in the analysis of prebiotic scenarios (see Sect. 4.6).

4.5.4 Conclusion

On the basis of the above chemical data, it is possible to explain the formation of imino diacids through Strecker processes, which therefore include the three pathways A, B, C allowing the initial multicomponent mixture {H_2O,

---

[10] Cyanic acid reacts with amines in general; its reaction with ammonia produces urea (this simple reaction is not included in Fig. 4.5).

HCN, R^1R^2CO, R^3R^4NH, CO_2, $B(OH)_4^-$} to be driven out of equilibrium. The Bücherer–Bergs reaction is a fourth pathway D allowing this equilibrium to be broken, and leading to CAA. The kinetics of pathways C, D, E that predominate over A and B determine the relative hydroxy acid/amino acid/CAA formation ratio. The presence of cyanic acid may have convergently driven AA evolution towards CAA, which would then become the most abundant products of this chemical process.

In the next section we examine possible prebiotic implications of such chemistry.

4.6 Prebiotic Formation of α-Amino Amides and Hydantoins Through Strecker and Bücherer–Bergs Reactions

Dynamic study of complex systems generally provides information about the preferential directions taken by the system under the pressure of a given environment. In the case of chemical evolution, these studies enable us to imagine how the molecular complexification operated, and what were the most probable mechanisms for this complexification.

4.6.1 Formation of Exogenous α-Amino Amides and Hydantoins

4.6.1.1 Environment and Reagents

Both Strecker and Bücherer–Bergs reactions require the presence of (rather liquid) water, which puts constraints on the places where they might have occurred. Exogenous organic matter delivered by meteorites is considered to have been formed in parent bodies, the history of which indicates that they could have met temperature ranges of ca. 263–298 K (Clayton and Mayeda 1984), with the possible presence of (interstitial or permafrost) liquid water.[11]

The analysis of meteorite extracts led Peltzer et al. (1984) to propose the Strecker reaction to explain the formation of α-amino acids in parent bodies. The presence of α-hydroxy acids and (recently discovered) of imino diacids (Pizzarello and Cooper, 2001) strongly confirm this hypothesis.

Besides, investigation by Lerner (1997) on the Strecker reaction in aqueous solution simulating the putative parent-body composition ([NH_3] = 2 mM, [CN] = 5 mM, [aldehydes] = 7 mM, [ketones] = 0.75 mM) at various temperatures (263 and 295 K), shows the formation of the 3 components of the Miller–Strecker signature (α-hydroxy- and α-amino acids, imino diacids) after a 4-month reaction. When the reaction was performed at 295 K, only the compounds

---

[11] The synthesis of amino acids and related compounds in interstellar molecular clouds is also probable; however, the preservation of such organics from presolar origin during accretion of the (hot) protoplanetary disc is very doubtful.

derivating from aldehydes were observed, whereas at 263 K derivatives from both aldehydes and ketones were observed.[12] The formation of these imino diacids appeared not to be influenced by the presence of meteoritic minerals.

From these data we can reasonably consider that meteoritic amino acids were synthesised in parent bodies through the Strecker reaction. We can add that the temperature was probably lower than 263 K (considering the abundance of α-alkyl, α-amino acids in meteorites), and that the ammonia concentration was quite low compared to that of other reagents such as HCN or carbonyl compounds.

We must, however, take care on requirements put on the prebiotic environments to fit the Strecker reaction with meteorite amino acids, since their analysis through extraction processes gives merely an estimation of meteorite content that can easily be modified by the extraction process itself.

4.6.1.2 Suitability of Reactions to Prebiotic Conditions

We have seen that reaction pathway (C) is catalyzed by carbonyl derivatives. When the catalyst used is formaldehyde, the activation energy of this reaction is zero (Pascal et al. 1980; Commeyras et al. 2003). Such low activation energy values are generally the consequence of favoured intramolecular reactions, as it is the case of the hydration of nitriles by alcoholate anions **VIII** in pathway (C). Noteworthy that the rate constant for this reaction ($k_{H_2CO} = 3.4 \times 10^7 \, L^2 \, mol^{-2} \, min^{-1}$) is of the same magnitude order as enzymatic reaction rate constants. Current research into the emergence of the catalytic activities of enzymes makes reference to such reactions (Pascal 2003).

A similar cyclization of carbamates **X** occurs in pathway (D), for which the (yet unknown) activation energy is probably quite low, thus allowing the reaction to occur at low temperatures. However, this step is probably not the rate-determining one for hydantoin **Xc** formation, which might rather be the opening of **Xa** with formation of isocyanate **Xb** (requiring base catalysis). Further investigation is necessary to confirm this hypothesis.

Conversely to (D) and similarly to (C), the (fast) cyclisation of **XII** in pathway (E) is probably the rate-determining step in borate-catalysed formation of α-hydroxyamides **XIIc**. However, there is no information about the concentration of borate in prebiotic media.

The above examples of intramolecular catalysis are certainly not unique in prebiotic chemistry; they could justify continuing research in this direction, especially since they have already led to the development of energy-economical, waste-free industrial processes (Commeyras et al. 1976; Taillades et al. 1986).

---

[12] We explain this by the higher thermolability of tertiary α-amino nitriles (ketone derivatives) compared to that of secondary α-amino nitriles (aldehyde derivatives), the former thus being unable to react with carbonyl compounds (pathway C on Fig. 4.5) at higher temperature to form the α-alkyl amino amides and acids (Commeyras et al. 2003).

4.6.2 Endogenous Formation of α-Amino Amides and Hydantoins

To try to explain the formation of endogenous α-amino acids it is obviously desirable to have information about the nature and concentrations of the species present on the primitive Earth, and about the pH and temperature of the oceans. This knowledge is still imprecise and can thus only lead to hypotheses.

4.6.2.1 Availability of Starting Compounds

The evaluation range for the partial pressure of CO_2 is very wide, going from 10^{-4} bar (Pinto et al. 1980) to more likely estimates, ranging between 0.1 and 10 bar (Owen et al. 1979; Kasting and Ackerman 1986; Mojzsis et al. 1999).

While the formation of carbonyl derivatives and cyanide does not seem to be problematic, evaluation of their relative and especially absolute concentrations remains highly uncertain. Values of 2×10^{-2} mol L^{-1} for these different constituents (Pinto et al. 1980) have been proposed. To our knowledge, these estimations have not been updated.

Concerning ammonia, its origins and concentration have been the subject of numerous investigations. The initial proposition, that it was formed by reduction of nitrogen by atmospheric hydrogen ($N_2 + 3 H_2 \rightarrow 2 NH_3$) (Bada and Miller 1968), has been contested. Two other reactions have been proposed. The first is based on hydrolysis of cyanic acid ($HNCO + H_2O \rightarrow NH_3 + CO_2$), itself formed by electrical discharges in a hypothetical primitive atmosphere made up of N_2, CO_2 and H_2, (Yamagata and Mohri 1982). The other involves reduction of NO by Fe^{2+}, which was common in the primitive ocean (Summers and Chang 1993; Summers 1999). The formation of NO in considerable quantities (Prinn and Fegley 1987; Navarro-Gonzalez et al. 1998, 2001) could thus have led to regular and sufficient production of NH_3, a requirement for the production of amino acids. If, however, we take into account the fact that NH_3 readily photolyzes, it is likely that the ammonia concentration in the primitive atmosphere was still very low. Summers (1999) estimated the (pH dependent) steady-state concentration of ammonia (both NH_3 and NH_4^+) in primitive Earth oceans to have remained within the 2×10^{-7} mol L^{-1} (pH 8.5) to 3×10^{-5} mol L^{-1} (pH 5) limits. These very low concentrations led these authors to doubt the efficiency of the Strecker process, although afterwards, being unable to find an alternative, they accepted it as probable.

These facts must therefore be taken into consideration when examining the endogenous synthesis of α-amino acids.

4.6.2.2 Thermodynamic Aspects

From a thermodynamic point of view, taking into account these low reactant concentrations and the equilibrium constants K_{cya} and K_{an} associated to the reactions forming cyanohydrins **I** and α-amino nitriles **IV** respectively, it is easy to estimate their relative concentration. In the case of formaldehyde, $K_{cya} = 4.76 \times 10^5 \, \text{L mol}^{-1}$ and $K_{an} = 4 \times 10^7 \, \text{L}^2 \, \text{mol}^{-2}$, therefore at equilibrium the concentration ratio of **I** to **IV** would be of the order of 1000/1. Nevertheless, the situation is different due to the fact that the relative concentrations of these two products is determined by their formation rates (kinetic control) and not by their relative stability (thermodynamic control).

4.6.2.3 Kinetic Aspects

From a kinetic point of view, cyanohydrins are formed within a few minutes ($t_{1/2} = 1$ min at any pH) whereas α-amino nitriles are formed much more slowly ($t_{1/2} = 6$ h, maximum rate at pH 9).[13] As soon as formed, α-amino nitriles **IV** react, partly decomposing back to initial reactants (with a pH-dependent rate, maximum above pH 5–6) and partly hydrating into α-amino amides **VIIIc** via the irreversible reaction (C). We have shown that this hydration reaction (C) is 10 times faster than the decomposition reaction (Pascal et al. 1978; Commeyras et al. 2003). The consequence (Fig. 4.6) is that the formation equilibria of the α-amino nitriles **IV** should never be attained under prebiotic conditions, because these compounds **IV** are immediately and irreversibly transformed into α-amino amides. We can add that the Bücherer–Bergs reaction (D), leading to the hydantoins **Xc**, should have amplified this process of withdrawing α-amino nitriles from the reversible reaction system.

Therefore even for ammonia concentrations of 10^{-6} mol L^{-1}, equilibrium shifts towards amino amides and hydantoins may have allowed the major formation of these compounds to the detriment of hydroxy amides. Then the summary of thermodynamic and kinetic data seems in agreement with: (1) a similar abundance of amino acids compared to hydroxy acids in meteorites, and (2) the fact that compounds originating from irreversible reactions of α-amino nitriles α-amino amides, hydantoins) might have formed more abundantly on the primitive Earth in presence of carbonyl derivatives and ammonia (in lower concentration), if the pH of primitive oceans was around 5–6.

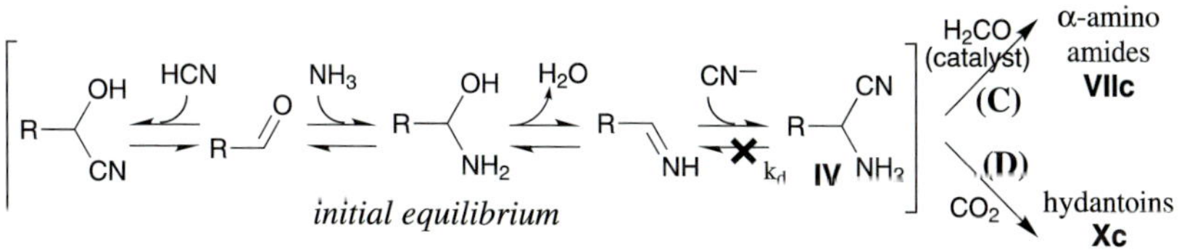

Fig. 4.6. Rapid and irreversible reaction of α-amino nitriles through pathways (C) and (D) does not enable the formation equilibrium of α-amino nitriles to be established

---

[13] Cyanohydrins (α-hydroxy nitriles) **I** are stable in aqueous solution, unlike carbonyl derivatives or hydrocyanic acid. We can thus state that they protected both carbonyl derivatives and hydrocyanic acid against degradation, enabling these compounds to wait for the gradual arrival of ammonia.

In summary, it is reasonable to conclude that the formation rates of α-amino amides and hydantoins on the primitive Earth, via the Strecker and Bücherer–Bergs reactions, could have been controlled by the formation rate of ammonia.

4.7 Convergent Evolution Towards *N*-Carbamoyl Amino Acids under Prebiotic Conditions

On the primitive Earth, even when controlled by the ammonia formation rate, a "multicomponent" system represented by Figs. 4.4 and 4.5 could only have spontaneously evolved (Fig. 4.7) into the three families of compounds analyzed above (see Sect. 4.5.2), i.e. a minority of α-hydroxy amides **XIIc** (omitted in Fig. 4.7), and a majority of α-amino amides **VIIIc** and hydantoins **Xc**. The subsequent evolution of these compounds into α-hydroxy acids **XIId**, α-amino acids **VIIId** and *N*-carbamoyl amino acids **Xd** with equivalent reaction rates ($3.6 \times 10^{-5} h^{-1}$ and $2.2 \times 10^{-5} h^{-1}$ measured at pH 8 and 25°C, see Commeyras et al. 2003) cannot have influenced the selectivity generated by the fast steps (C) (D) (E) investigated above.

Concerning the *N*-carbamoylation reaction, it most certainly played an important prebiotic role[14], but it can only have been involved in the slow step

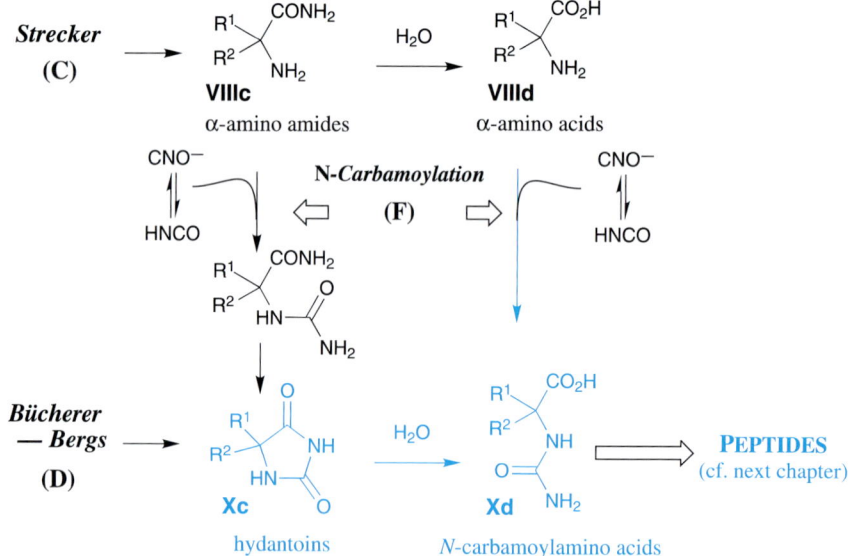

Fig. 4.7. Convergent evolution of reaction pathways (C), (D) and (F) towards *N*-carbamoyl amino acids and thence to peptides

---

[14] The estimated pH of the primitive ocean (5–7) (Mojzsis et al. 1999) is favorable to carbamoylation reactions. Cyanic acid HNCO is formed in model gas mixtures of

($k_{(F)} \approx 1.7 \times 10^{-3}\,\mathrm{L\,mol^{-1}\,s^{-1}}$ at 50°C and pH 6.5) (Taillades et al. 2001; Commeyras et al. 2003), considerably later than the preceding fast steps which controlled selectivity.[15]

It is therefore reasonable to consider cyanate (in equilibrium with cyanic acid) in the primitive ocean at pH 5–7, to have slowly transformed α-amino amides **VIIIc** and α-amino acids **VIIId** into N-carbamoyl amino acids **Xd** by two different pathways: either via N-carbamoyl amino amides and hydantoins, or directly from α-amino acids (Fig. 4.7).

The case of meteorites, where amino acids would be more abundant than both hydantoins and (yet unobserved) N-carbamoyl amino acids, appears contradictory with a convergent N-carbamoyl amino acid synthesis. Although the measured ratio may be biased from the original meteorite composition (by e.g. impact or extraction processes), this is probably rather due to quite different conditions having occurred in parent bodies comparatively to the primitive Earth, especially concerning CO_2 and cyanic acid abundance. Further analytic and chemical investigations should be able to provide useful information on these parent-body conditions.

In summary, on the primitive Earth, the set of reaction processes Fig. 4.7 could have converged not towards free α-amino acids **VIIId**, but rather towards N-carbamoyl amino acids **Xd**. In the next chapter we shall examine how these N-carbamoyl amino acids could have led to peptides.

4.8 Conclusions

In this chapter, we have shown that the multicomponent system {H_2O, HCN, R^1R^2CO, NH_3, R^3NH_2, R^3R^4NH, CO_2, $B(OH)_4^-$} could have been the origin of the α-amino acids and derivatives discovered in meteorites and formed on the primitive Earth.

The formation mechanisms of these derivatives have been exhaustively examined, and the Strecker and Bücherer–Bergs reactions can be said to have played a major, even unique, role in the formation of these compounds. It was these reactions that led the above multi-component system, made up of essential prebiotic molecules, to evolve quickly, majoritarily and selectively into α-amino acids and N-carbamoyl amino acids.

Under prebiotic conditions, the formation rates of these compounds could have been controlled by the ammonia formation rate.

The involvement of cyanic acid, an essential prebiotic molecule, in addition to the above reactions, in a subsequent slower step, could have led to N-carbamoyl amino acids alone.

---

the primitive atmosphere (Yamagata and Mohri 1982) and is detected as its trimer (cyanuric acid) in meteorites (Hayatsu et al. 1975).

[15] This additional example confirms (were it still necessary) the absolute necessity of studying the dynamics of potentially prebiotic systems in order to understand their chemical evolution.

The knowledge acquired about these reactions enables us to envisage kinetic modelling of the whole system, the only possible way to understand the respective contributions of all the forces involved.

Acknowledgement

For their fruitful collaboration to this work, we are indebted to our colleagues from Montpellier: Hélène Collet, Jacques Taillades, Laurence Garrel, Robert Pascal, Alain Rousset, Jean-Christophe Rossi, Hervé Cottet, Jean-Philippe Biron, Olivier Lagrille, Raphaël Plasson, and Grégoire Danger, as well as to Hervé Martin from the laboratory Magmas & Volcans at the University of Clermont-Ferrand (France), and to Michel Dobrijevic from the Observatoire Aquitain des Sciences de l'Univers (Bordeaux, France).

We are also grateful to:

The Centre National de la Recherche Scientifique (CNRS), especially its Chemical Science Department and the Institut National des Sciences de l'Univers (INSU); the Exobiology Research Group of the Centre National d'Etudes Spatiales (CNES); and the European Community through the COST D27 action (prebiotic chemistry), for their support.

The companies Aventis Animal Nutrition (France) and Degussa (Germany) for financial supports.

Professor Dr. Jacques Reisse from the Université Libre de Bruxelles (Belgium), and Dr. Muriel Gargaud from the Observatoire Aquitain des Sciences de l'Univers (Bordeaux, France) for many helpful discussions.

References

Bada J, Miller SL (1968). Ammonium ion concentration in the primitive ocean. *Science* **159**: 423–425.
Botta O, Glavin DP, Kminek G, Bada J (2002). Relative amino acid concentrations as a signature for parent body processes of carbonaceous chondrites. *Orig. Life Evol. Biosphere* **32**: 143–163.
Carlson RW (1996). Where has all the old crust gone? *Nature* **379**: 581–582.
Chyba CF, Sagan C (1992). Endogenous production, exogenous delivery and impact-shock synthesis of organic molecules: an inventory for the origins of life. *Nature* **355**: 125–132.
Clayton RN, Mayeda TK (1984). The oxygen isotope record in Murchison and other carbonaceous chondrites. *Earth Planet. Sci. Lett.* **67**: 151–161.
Commeyras A, Taillades J, Mion L, Pascal R, Lasperas M, Rousset A (1976). Procédé d'hydrolyse catalytique chimique d'alpha-aminonitrile ou de leurs sels. French Patent nr 76 365 20.
Commeyras A, Taillades J, Collet H, Boiteau L, Pascal R, Vandenabeele-Trambouze O, Pascal R, Rousset A, Garrel L, Rossi JC, Cottet H, Biron JP, Lagrille O, Plasson R, Souaid E, Selsis F, Dobrijevic M (2003). Approche dynamique de la synthèse des peptides et de leurs précurseurs sur la Terre primitive. In: *Les Traces du Vivant*.

Gargaud M, Despois D, Parisot JP (eds.), Presses Universitaires de Bordeaux, (Chap. 5) pp 115–162 (see especially the appendix).

Cooper GW, Cronin JR (1995). Linear and cyclic aliphatic carboxamides of the Murchison meteorite: hydrolyzable derivatives of amino acids and other carboxylic acids. *Geochim. Cosmochim. Acta* **59**: 1003–1015.

Cooper GW, Kimmich N, Beslisle W, Sarinana J, Brabham K, Garrel L (2001). Carbonaceous meteorites as a source of sugar-related organic compounds for the early Earth. *Nature* **414**: 879–883.

Cronin JR, Pizzarello S (1983). Amino acids in meteorite. *Adv. Space Res.* **3**: 5–18.

Cronin JR, Chang S (1993). Organic Matter In Meteorites: Molecular and Isotopic Analyses of the Murchison Meteorite. In: *The Chemistry of Life's Origins,* Greenberg JM et al. (eds.), Kluwer Academic Publishers, Dordrecht, The Netherlands, pp 209–258.

Cronin JR, Pizzarello S (1997). Enantiomeric excesses in meteoritic amino acids. *Science* **14**, 275(5302): 951–955.

Flores JJ, Bonner WA, Massey GA (1977). Asymmetric photolysis of (RS)-leucine with circularly polarized ultraviolet light. *J. Am. Chem. Soc.* **99**: 3622–3625.

Hayatsu R, Studier MH, Moore LP, Anders E (1975). Purines and trieazines in the Murchison meteorite. *Geochim. Cosmochim. Acta* **39**: 471–488.

Jorissen A, Cerf C (2002). Asymmetric photoreactions as the origin of biomolecular homochirality: a critical review. *Orig. Life Evol. Biosphere* **32**: 129–142.

Jungclaus GA, Yuen GU, Moore CB (1976). Evidence for the presence of low molecular weight alcools and carbonyl compounds in the Murchison météorite. *Meteoritics* **11**: 231–237.

Kasting JF, Ackerman TP (1986). Climatic consequences of very high carbon dioxide levels in the Earth's early atmosphere. *Science* **234**: 1383–1985.

Ksander G, Bold G, Lattmann R, Lehmann C, Früh T, Xiang Y, Inomata K, Buser H, Schreiber J, Zass E, Eschenmoser A (1987). Chemie der α-aminonitrile. *Helv. Chim. Acta* **70**: 1115–1172.

Lerner NR (1997). Influence of Allende minerals on deuterium retention of products of the Strecker synthesis. *Geochim. Cosmochim. Acta* **61**: 4885–4893.

Maas R, McCulloch MT (1991). The provenance of Archean clastic metasediments in the Narryer Gneiss complex, western Australia: trace elements geochemistry, Nd isotopes, and U-Pb ages for detrital zircons. *Geochim. Cosmochim. Acta* **55**: 1915–1932.

Martin H (1986). Effect of steeper Archean geothermal gradient on geochemistry of subduction-zone magmas. *Geology* **14**: 753–756.

Martin H (1987). Petrogenesis of Archaean trondhjemites, tonalites and granodiorites from eastern Finland: major and trace element geochemistry. *J. Petrology* **28**: 921–953.

Maurette M (1995). Were micrometeorites a source of prebiotic molecules on the early Earth? *Adv. Space Res.* **15**: 113–126.

Maurette M (1998). Micrometeorites on the early Earth. In: *The Molecular Origin of Life: Assembling Pieces of the Puzzle.* Brack A (ed.), Cambridge University Press, Cambridge, UK, pp 147–186.

Maurette M (2001). La matière extraterrestre primitive et les mystères de nos origines. In: *L'environnement de la Terre primitive.* Gargaud M, Despois D, Parisot JP (eds.), Presses Universitaires de Bordeaux, Bordeaux, Fr, pp 99–127.

Miller SL (1953). Production of aminoacids under possible primitive Earth conditions. *Science* **117**: 528–529.

Miller SL (1957). The mecanism of synthesis of amino acids by electric discharges. *Biochim. Biophys. Acta* **23**: 480–489.

Miller SL (1998). The endogenous synthesis of organic compounds. In: *The Molecular Origins of Life*. Brack A (ed.). Cambridge University Press, Cambridge, UK, pp 59–85.

Mojzsis SJ, Krishnamurthy R, Arrhenius G (1999). Before RNA and after: Geophysical and geochemical constraints on molecular evolution. In: *The RNA Word (second edition)*. Gesteland RF, Cech TR, Athins JF (eds.), Cold Spring Harbor University Press, NY, USA, pp 1–47.

Mojzsis SJ, Harrison MT, Pidgeon RT (2001). Oxygen-isotope evidence from ancient zircons for liquid water at the Earth's surface 4,300 Myr ago. *Nature* **409**: 178–181.

Nagata Y (1999). D-amino acids in nature. In: *Advances in BioChirality*. Pályi G, Zucchi C, Caglioti L (eds.), Elsevier, Amsterdam , The Netherlands, pp 271–283.

Navarro-Gonzalez R, Molina MJ, Molina LT (1998). Nitrogen fixation by volcanic lightning in the early Earth. *Geophys. Res. Lett.* **25**: 3123–3126.

Navarro-Gonzalez R, McKay CP, Nna Mvondo D (2001). A possible nitrogen crisis for Archaean life due to reduced nitrogen fixation by lightning. *Nature* **412**: 61–64.

Nishino H, Kosaka A, Hembury G, Shitomi H, Onuky H, Inoue Y (2001). Mechanism of pH-Dependent Photolysis of Aliphatic Amino Acids and Enantiomeric Enrichment of Racemic Leucine by Circularly Polarized Light. *Organic Lett.* **3**: 921–924.

Oró J (1961). Amino-acid synthesis from hydrogen cyanide under possible primitive Earth conditions. *Nature* **190**: 389–390.

Owen T, Cess RD, Ramanathan V (1979). Enhanced carbon dioxide greenhouse to compensate for reduced solar luminosity on early Earth. *Nature* **277**: 640–642.

Pascal R (2003). Catalysis by Induced Intramolecularity: What Can Be Learned by Mimicking Enzymes with Carbonyl Compounds that Covalently Bind Substrates? *Eur. J. Org. Chem*, pp 1813–1824.

Pascal R, Taillades J, Commeyras A (1978). Systèmes de Strecker et Apparentés. X. Décomposition et hydratation en milieu aqueux basique des α-aminonitriles secondaires. Processus d'hydratation autocatalytique et catalyse par l'acétone. *Tetrahedron* **34**: 2275–2281.

Pascal R, Taillades J, Commeyras A (1980). Systèmes de Strecker et Apparentés. XII. Catalyse par les aldéhydes de l'hydratation intramoléculaire des α-aminonitriles. *Tetrahedron* **36**: 2999–3008.

Peltzer ET, Bada JL, Schlesinger G, Miller SL (1984). The chemical conditions on the parent body of the Murchison meteorite: some conclusions based on amino, hydroxy, and dicarboxylic acids. *Adv. Space Res.* **4**: 69–74.

Pinto JP, Gladstone GR, Yung YL (1980). Photochemical production of formaldehyde in Earth's primitive atmosphere. *Science* **210**: 183–185.

Pizzarello S, Cronin JR (2000). Non-racemic amino acids in the Murray and Murchison meteorites. *Geochim. Cosmochim. Acta* **64**: 329–338.

Pizzarello S, Cooper GW (2001). Molecular and chiral analyses of some protein amino acid derivatives in the Murchison and Murray meteorite. *Meteorit. Planet. Sci.* **36**: 897–909.

Prinn RG, Fegley B (1987). Bolide impacts, acid rain, and biospheric traumas at the Cretaceous-Tertiary boundary. *Earth Planet. Sci. Lett.* **83**: 1–4.

Rossi JC, Garrel L, Taillades J, Commeyras A (1996). Hydrolyse et oxydation d'α-aminonitriles en présence de solution aqueuse basique de H_2O_2. *CR Acad. Sci. Paris* **322**: 767–773.

Selsis F (2000). *Modèle d'évolution physico-chimique des atmosphères de planètes telluriques. Application à l'atmosphère primitive terrestre et aux planètes extrasolaires.* PhD thesis, Université Bordeaux 1 (France).

Selsis F, Parisot JP, Dobrijevic M, Toublanc D (1996). *Photochemical modeling of the primitive atmosphere of telluric planets.* 11th International Conference on the Origin of Life, Orléans (France).

Shimoyama A, Ogasawara R (2002). Dipeptides and Diketopiperazines in the Yamato-791198 and Murchison Carbonaceous Chondrites. *Orig. Life Evol. Biosphere* **32**: 165–179.

Strecker A (1850). *Liebigs Ann. Chem.* **75**: 27–51.

Summers DP (1999). Sources and sinks for ammonia and nitrite on the early Earth and the reaction of nitrite with ammonia. *Origins of Life and Evolution of the Biosphere* **29**: 33–46.

Summers DP, Chang S (1993). Prebiotic ammonia from reduction of nitrite by iron(II) on the early Earth. *Nature* **365**: 630–633.

Summers DP, Lerner NR (1998). Ammonia from iron (II) reduction of nitrite and the Strecker synthesis: do iron (II) and cyanide interfere with each other? *Origin of Life Evolution of the Biosphere* **28**: 1–11.

Taillades J, Beuzelin I, Garrel L, Tabacik V, Commeyras A (1998). N-carbamoyl-α-aminoacids rather than free α-aminoacids formation in the primitive hydrosphere: a novel proposal for the emergence of prebiotic peptides. *Orig. Life Evol. Biosphere* **28**: 61–77.

Taillades J, Brugidou J, Pascal R, Sola R, Mion L, Commeyras A (1986). Nouvelles voies de synthèse d'acides α-aminés. *L'Actualité Chimique*, pp 13–20.

Taillades J, Boiteau L, Beuzelin I, Lagrille O, Biron JP, Vayaboury W, Vandenabeele-Trambouze O, Giani O, Commeyras A (2001). A pH-dependent cyanate reactivity model: application to preparative N-carbamoylation of amino acids. *Perkin Trans. 2*, pp 1247–1253.

Vervoort JD, Patchett PJ, Gehrels GE, Nutman AJ (1996). Constraints on early Earth differentiation from hafnium and neodymium isotopes. *Nature* **379**: 624–627.

Wilde SA, Valley JW, Peck WH, Graham CM (2001). Evidence from detrital zircons for the existence of continental crust and oceans on the Earth 4.4 Ga ago. *Nature* **409**: 175–178.

Yamagata Y, Mohri T (1982). Formation of cyanate and carbamyl phosphate by electric discharges of model primitive gas. *Orig. Life* **12**: 41–44.

5 Peptide Emergence, Evolution and Selection on the Primitive Earth
II. The *Primary Pump* Scenario

Auguste Commeyras, Laurent Boiteau, Odile Vandenabeele-Trambouze, Franck Selsis

Abstract We propose a dynamic scenario for the emergence and evolution of peptides on the primitive Earth, through a molecular engine (the primary pump), which works at ambient temperature and continuously generates, elongates and complexifies sequential peptides. This new scenario is based on a cyclic chemical reaction sequence that could have taken place on tidal beaches; it requires a buffered ocean, emerged land and a nitrosating atmosphere. We show that the primitive Earth during the Hadean may have satisfied all of these requirements.

This scenario is not necessarily what actually happened, but it represents a global approach of peptide prebiotic synthesis, and most of its parts are accessible to experiment. As it develops, it may open up a gateway to the emergence of homochirality and the catalytic activities of peptides.

5.1 From N-carbarmoyl Amino Acids (CAA) to Peptides

5.1.1 Introduction

The hypothesis of the prebiotic RNA world (Gilbert 1986; Gesteland et al. 1999) is hampered by the infeasibility of producing these macromolecules without the aid of enzymatic catalysts or modern organic synthetic tools (Joyce and Orgel 1999). In the absence of any satisfactory scenario to explain the origin of peptides, the coemergence of RNA and peptides as suggested by Kauffman (1993) or De Duve (1998), remains itself a fragile hypothesis. The prebiotic emergence of peptides meets two major obstacles: the activation of amino acids, then the possibility of forming long peptide chains under aqueous conditions, namely the only reasonable prebiotic solvent until now. Three different approaches have been developed in this direction.

The first approach focuses on peptide synthesis under extraterrestrial conditions (e.g. interstellar ice grains), through the analysis of meteoritic content as the only available record. This approach remains disappointing: only the Gly-Gly dipeptide has actually been detected in the Murchison and Yamoto-791198 meteorites (Shimoyama and Ogasawara 2002). Experimental models of interstellar ice do not show evidence of peptide formation.[1]

---

[1] In our opinion, the formation of peptides in an extraterrestrial ice analogue claimed by Munoz Caro et al. (2002) results from a hasty interpretation of predictable results.

The second consists in studying the behaviour of amino acids at high temperature. At 100–180 °C racemic compounds are obtained with insufficient evidence for the presence of peptide bonds (Rohlfing 1976; Fox and Dose 1977). Wet/dry cycles at 25–100 °C, in the presence or absence of clays or silicates, lead to dipeptides and traces of racemic trimers to pentamers (Lahav et al. 1978; Rode et al. 1997). At 100 °C in the presence of NiS/FeS and thiols to simulate volcanic zones or undersea vents, only dipeptides and traces of racemic tripeptides are detected (Huber and Wächtershäuser 1998). This approach is therefore highly contested (Bengston and Edstrom 1999).

The third approach investigates activated α-amino acids (esters, thioesters, N-carboxy anhydrides) or activating agents (carbonyl diimidazoles or carbodiimides) with amino acids. It is more encouraging since it effectively allows for oligocondensation under smooth, nonracemising conditions, whether in homogenous aqueous solution, on the surface of rocks, (Paecht-Horowitz and Eirich 1988; Orgel 1989, 1998; Bohler et al. 1996; Weber 1998), or in double-layer membranes (Luisi et al. 2000); however, it has not yet led to the expected *sequential* peptides[2] (Luisi 2000). For this approach to be credible we need to understand how activated molecules, which are highly reactive and sensitive to hydrolysis, could have been continuously formed on the primitive Earth. The relevance to prebiotic chemistry of activating agents such as carbonyl diimidazole or dicyclohexyl carbodiimide, or of amino acid N-carboxy anhydrides (NCA) remains contested (Liu and Orgel 1998; Blair and Bonner 1980, 1981).

Therefore the question of the prebiotic formation of long peptides remains open. We shall address this question by investigating a cyclic reaction scheme occurring in an open system out of equilibrium. The dynamics of such system (which is an important feature of living organisms) is very likely to result in self-organisation, as highlighted by, e.g. Nicolis and Prigogine (1977). It has been the object of numerous, mostly theoretical investigations in many fields – including the origins of life. Meanwhile, very few experimental studies have been published on the behaviour of such open chemical systems, despite their fundamental interest. Investigations (including experimental) on the prebiotic emergence and (self-) organisation of complex chemical reaction schemes, including their dynamical aspects, seems therefore important.

In the previous chapter we showed that the conditions that existed on the primitive Earth could have selectively led to N-carbamoyl amino acids (CAA), which have, however, been considered for long as an evolutionary cul-de-sac under prebiotic conditions. In this chapter we shall examine a possible activation pathway from CAA to sequential peptides through amino-acid N-carboxy anhy-

---

In such experiments amino acids identified after acidic hydrolysis (6N HCl, 120 °C for 6 h) probably do not actually originate from the hydrolysis of peptides, but rather of α-amino nitriles, namely the Strecker precursors of observed α-amino acids.

[2] In this chapter, sequential peptides are referred to as peptides made of different amino-acid residues, (of optionally known sequence), oppositely to homopeptides in which all residues are identical.

drides (NCA), based on a dynamic process, that might also have contributed to the emergence of homochirality and of catalytic activity. Our attention will be focused on the relevance of the required conditions with respect to the primitive Earth environment.

5.1.2 The Primary Pump

Our approach of the prebiotic emergence of peptides is based on a cycle of chemical reactions forming the basis of a molecular engine, called by us the primary pump and shown in Fig. 5.1 (Commeyras et al. 2001, 2002).

- **First cycle.**

The first step **(1)** is the formation of N-carbamoyl amino acids (CAA) either by reaction of free amino acids with cyanate, or in other ways (see previous chapter). The second step **(2)** is the concentration of these CAA by water evaporation. This could have taken place on the shores of Hadean continents, at

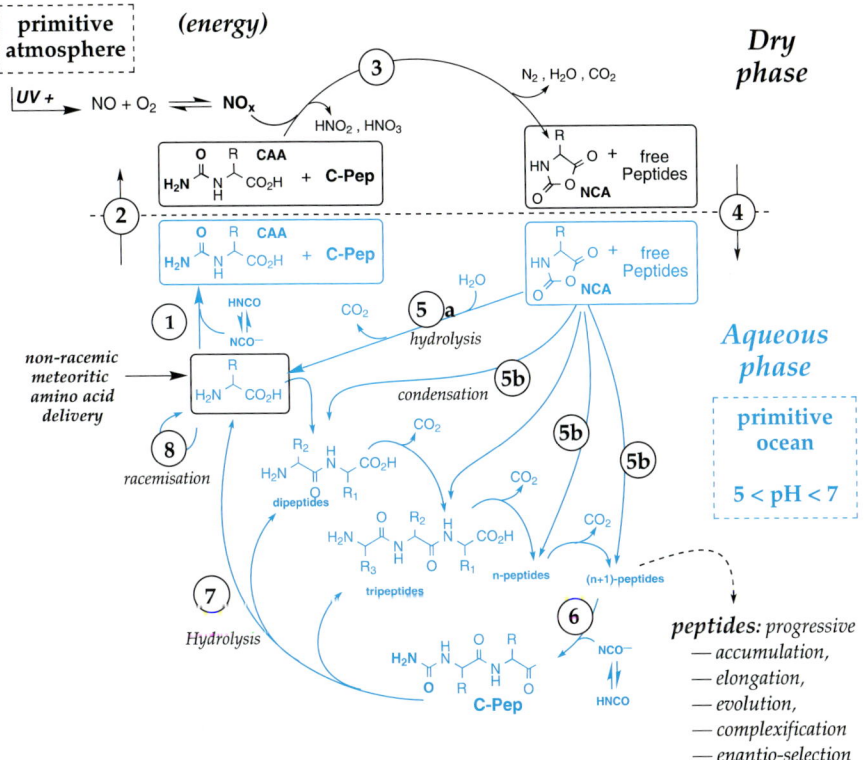

Fig. 5.1. The primary pump: a molecular engine leading to the production of evolutionary sequential peptides, starting from CAA and from NCO^-, NO, O_2, HCO_3Na, H_2O

low tide. In the dry phase, NO_x[3] from the atmosphere (see Sect. 5.2.3 and previous chapter) react with the CAA (nitrosation), leading to NCA through the cyclisation of unstable nitroso urea intermediates, step **(3)**. This reaction is quantitative at ambient temperature, producing stoichiometric amounts of nitrogen, water and nitrous/nitric acid together with NCA. In such acidic medium NCA degradation is relatively slow; within a few hours they undergo hydrolysis ($t_{1/2} \approx 1$ h) into α-amino acids, thus being recycled. At step **(4)** the rising tide quickly puts NCA into contact with ocean water (this is also conceivable for tidal pools or lagoons), the pH of which (5–7) allows peptide formation from NCA in competition with their hydrolysis, step **(5)**. This step is complete within a few minutes. Afterwards, the so-produced peptides can react with cyanate to give N-carbamoyl peptides, step **(6)**: this reaction occurs as easily as amino-acid N-carbamoylation. Over longer time in the hydrosphere, peptides and N-carbamoyl peptides undergo slow hydrolysis, step **(7)**: this leads to shorter peptides, eventually down to α-amino acids. Meanwhile both undergo N-carbamoylation.

- **Next cycles.**

The reaction of NCA with N-carbamoyl peptides is impossible, thus theoretically preventing peptide elongation in subsequent drying/wetting cycles. However, when N-carbamoyl peptides arrive on the shore and are dried, they are immediately unprotected by reaction with NO_x (Collet et al. 1999; Lagrille et al. 2002), step **(3)**. This reaction occurs under the same conditions as NCA formation from CAA. Therefore at the next high tide, unprotected peptides are elongated by reaction with newly formed NCA.

Such successive elongation/partial hydrolysis cycles may have allowed peptide chemical evolution. The intrinsic rate of peptide formation is higher than that of peptide hydrolysis, which should be favourable to peptide accumulation; however, actual rates are highly dependent on the reactant fluxes and, moreover, on the dry/wet alternation period.

When considering peptides, it is difficult not to associate homochirality with emergence, as hypothesized by George Wald (1957). It is well known that the extraterrestrial input of α-amino acids was not totally racemic, see for instance Pizzarello and Cronin (2000). If these nonracemic α-amino acids played a role in the synthesis of peptides, then their (low) enantiomeric excesses may have been amplified. Such amplification phenomena have been demonstrated by Blair and Bonner (1980), however, in organic solvents. If in addition, amino acids or CAA racemise faster than peptide-chain residues[4], then peptide prebiotic synthesis

---

[3] In this chapter NO_x is referred to as the result of mixing NO and O_2 in any ratio, which can equivalently be obtained in other ways, e.g. mixing NO and NO_2.

[4] In aqueous solution it is likely that free amino acids epimerise slower than peptide residues. However, the latter may epimerise slower than some amino-acid derivatives such as CAA (recent, unpublished results of our group are consistent with this possibility), or even slower than free amino acids/peptide end-residues if epimerisation catalysts are involved.

may have progressed towards homochirality. A great deal of work still needs to be done to test this hypothesis.

5.2 Environmental Requirements

Each step of the so-called primary pump has been studied carefully and can be considered as well known. Nevertheless, for the primary pump to produce peptides, all steps must be well-controlled. Under prebiotic conditions, is it conceivable that, spontaneously, the correct sequence of steps took place and gave peptides? What are the minimal requirements to fulfill such a goal? Without any doubt, oceans and emerged land are required and they must be in contact with an atmosphere containing nitrogen oxides. The time window for this scenario lies between -4.4 Gyr and -3.8 Gyr. The following discussion will consider the probability of these environmental requirements, with special emphasis on the pH of the aqueous phases, which is critical in our scenario.

5.2.1 Primitive Earth

The oldest terrestrial rocks recognized so far are the Acasta gneisses in Canada, dated at -4.030 ± 0.003 Gyr (Bowring and Williams 1999), which consist of a tonalite, trondhjemite and granodiorite suite (TTG) that is typical of the primitive Archaean continental crust. The oldest known terrestrial materials are zircon crystals, extracted very recently from Jack Hill metaquartzites (Australia) and dated at -4.404 ± 0.008 Gyr (Wilde et al. 2001). Their chemical characteristics demonstrate that they crystallized in TTG-like magmas, thus establishing the existence of a stable continental crust as early as 4.4 Gyr ago. In addition, hafnium and neodymium isotopic studies in both old zircons and early Archaean TTG (McCulloch and Bennet 1993; Wilde et al. 2001) show that, when these materials were generated, significant volumes of continental crust had already been extracted from the terrestrial mantle.[5] Based on these isotopic data, McCulloch and Bennet (1993) calculated that between 8 and 15% of the volume of the present-day continental crust had already been extracted prior to -4.0 Gyr.

Petrogenetic studies performed on the Archaean TTG continental crust have shown that it was generated by partial melting of hydrous basalts (Martin 1986, 1987). Available water lowers the temperature of basalt solidus by several hundred degrees, and is therefore absolutely necessary for melting basalt and generating TTG. Basalts, generated at great depth by anhydrous melting of mantle peridotite, contain only very small amounts of water. Today they incorporate water through alteration by seawater, for instance in midocean-ridge systems. As an essential consequence, the existence of a Hadean continental crust implies

---

[5] Isotopic deviations ε_{Hft} (for Hf) and ε_{Ndt} (for Nd) were both found to be positive: ε_{Hft} (resp., ε_{Ndt}) represents the difference between the isotopic ratio $^{176}Hf/^{177}Hf$ (resp., $^{143}Nd/^{144}Nd$) of a rock and that of chondrites, taken as the reference.

that huge volumes of liquid water must have been available on the surface of the primitive Earth. Mojzsis et al. (2001) and Wilde et al. (2001) have analyzed the oxygen isotopic composition of the 4.4- and 4.3-Gyr zircons: the $\delta^{18}O$ (5.4–15.0‰) clearly indicates that the source of the magma in which these zircons crystallized had already strongly reacted with a hydrosphere. This demonstrates that oceans existed as early as 4.4 Gyr ago on the Earth's surface; this conclusion is also supported by the fact that the Isua gneisses in Greenland (dated at 3.865 Gyr) consist of sediments accumulated in an aqueous environment.

The existence of Hadean beaches acted upon by ocean tides (the presence of the Moon at that time is now widely accepted) is therefore shown to be a consistent hypothesis.

5.2.2 Primitive Atmosphere

The abundance of atmospheric nitrogen oxides (NO_x) is a limiting factor for the efficiency of nitrosation and thus for the production of peptides through the primary pump. Unfortunately, the nature (pressure, temperature, composition) of the prebiotic atmosphere remains obscure in the absence of geological archives, preventing us from addressing quantitatively this point. Still, we can discuss the availability of NO_x considering the few constraints on the primitive terrestrial environment.

The formation of NO_x is considered to occur through two successive steps: (1) formation of nitric oxide NO through activation of N_2; (2) formation of higher NO_x species by reaction of NO with atomic or molecular oxygen. The major oxygen source is photodissociation of CO_2, with a mixing ratio highly dependent on CO_2 pressure (Rosenqvist and Chassefière 1995; Selsis et al. 2002). In a 1-bar CO_2 atmosphere, O_2 pressures up to 20 mbar can be maintained at a steady state in this way, though this would depend upon there being no O_2 consumption at the Earth's surface. Such an uptake would almost certainly have occurred, through the oxidation of rocks and reducing volcanic gases (Selsis et al. 2002), the amount of which is unknown, however. While it has been suggested that the early upper mantle was more reducing than it is today (Pavlov et al. 2000), its present oxidation state could have been reached as early as 4.3 Gyr ago (Tolstikhin and Marty 1998). We performed computer simulation of atmosphere models under photochemical activation, using the computer code PHOEBE (photochemistry for exobiology and exoplanets, Selsis et al. 2002), starting from $N_2 + CO_2$ background atmospheres including water vapour at thermodynamic equilibrium. For initial mixtures of 0.8 bar of N_2 plus 0.2 bar (A) or 3.2 bar (B) of CO_2, O_2 mixing ratios of 10^{-4} and 10^{-3} were found for configurations A and B, respectively, if no O_2 uptake is taken in account. Supposing this O_2 uptake to be at the same level as today, the calculated mixing ratio drops to 10^{-6} and 5×10^{-5} for configurations A and B, respectively. An example of such simulation (configuration A) taking in account other NO sources (see infra) is shown in Fig. 5.2.

Concerning NO formation, the partial pressure of N_2 is thought to have reached a level close to its present value (0.8 bar) earlier than 4.3 Gyr (Tolstikhin

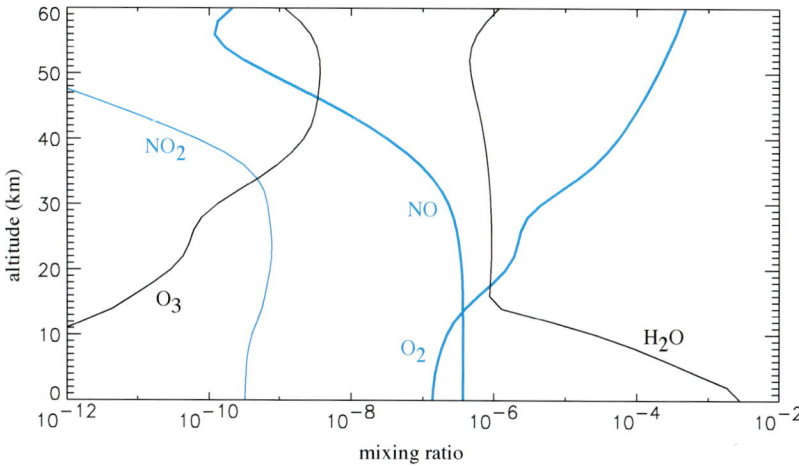

Fig. 5.2. Simulation of the primitive Earth atmosphere composition with the numerical code PHOEBE. The 1-bar background atmosphere of N_2 (80%) + CO_2 (20%) includes NO production by volcanic lightning (Navarro-Gonzalez et al. 1998)

and Marty 1998). A high CO_2 partial pressure (p_{CO_2}) is usually assumed, although a recent study (Sleep and Zahnle 2001) claims that the formation of carbonates was so efficient on the Early Earth that CO_2-dominated atmospheres have occurred only for geologically brief periods following giant impacts. In our case, we assume p_{CO_2} to be at least high enough to insure a mean surface temperature above $0\,°C$. Depending on the period considered (from 4.3 to 3.6 Gyr ago), this implies a minimum p_{CO_2} being in the range 100–400 mbar (Caldeira and Kasting 1992). If p_{CO_2} was lower, two cases can be considered:

1. No other greenhouse gas warmed the surface and the Earth was globally frozen: surface liquid water was then restricted to the vicinity of active volcanic zones, or at the surface of ice near the equator and during the warmest day hours. The efficiency of the primary pump in such conditions has not been addressed yet.
2. The surface is warmed above $0\,°C$ by CH_4 produced in hydrothermal systems (Pavlov et al. 2000): In that situation the presence of abundant reduced species in the atmosphere considerably lowers the photochemical lifetime of the NO_x and inhibits the nitrosation.

Thus, a prebiotic atmosphere for which nitrosation cannot be a priori ruled out contains at least 100 mbar of CO_2 but no abiotic methane. This picture is compatible with our knowledge of the primitive Earth as recently reviewed by Kasting and Catling (2003).

In the abiotic early atmosphere, NO_x have to be produced from N_2 (and not from the biogenic N_2O as it is now). The N–N bond energy is particularly high and consequently, short wavelengths are required for the photolysis of N_2 ($\lambda <$

100 nm). Under such conditions the photochemistry initiated by solar UV does not produce significant levels of NO_x (Selsis 2000). Such a conclusion is valid for the present-day situation, but what about the Hadean era? Indeed, observations of young Sun-like stars (Guinan and Ribas 2000) leads to the conclusion that the soft X-ray and hard UV (XUV: 1–100 nm) emission of the young Sun was very probably much higher than its present value. The luminosity in the XUV range (L_X) varies as $L_X(t)/L_{X,Sun} = 6.16 \times t^{-1.19}$ where t is the age of the star in Gyr, and $L_{X,Sun}$ the present solar XUV luminosity. The photochemistry initiated by this intense XUV flux on the early Earth still has to be explored but it surely led to a more efficient production of NO_x at high and middle altitudes. However, even under such irradiation conditions and in the presence of an atmosphere without reduced species such as CH_4 and H_2, the surface mixing ratio of cumulated NO_x species do not reach 10^{-9}.

Another, more efficient way to produce NO_x consists in shocking up to more than 1500 K a gas mixture containing N_2 and oxygen-bearing species (O_2, CO_2, H_2O) and cooling it as rapidly as possible. In this case, the oxygen molecules, or atoms generated at high temperature, react with N_2 producing NO. Such chemistry occurs during lightning, meteors and impacts.

The efficiency of these processes in our present O_2-rich atmosphere is higher than in an anoxic atmosphere where the reaction $N + O_2 \rightarrow NO + O$ does not occur. In the absence of O_2, the efficiency depends on the ratio CO_2/N_2. When the ratio $[N_2]/[CO_2]$ is above 5, the formation of NO (through the reaction $N + CO_2 \rightarrow NO + CO$) becomes inefficient because of the fast recombination of N atoms into N_2 through the three-body reaction $2N + N_2 \rightarrow 2N_2$. But when the CO_2 and N_2 abundances are comparable, the energy yield of the NO production is only 10 times lower than in an O_2-rich atmosphere (Navarro-Gonzalez et al. 2001).

Navarro-Gonzalez et al. (1998, 2001) studied both experimentally and theoretically the production of NO by lightning in CO_2-rich atmospheres. Considering the early volcanic activity and the number of lightning strikes associated to volcanic plumes, these authors estimated the continuous production of NO on Early Earth and found 3×10^9–3×10^{10} mol y^{-1} for an atmosphere with $[CO_2]/[N_2] > 0.3$. As an upper limit, if all of these radicals could be used by the primary pump, this rate would allow a NCA production of 10^{-9} mol cm^{-2} s^{-1}. Of course, a fraction of the produced NO is reduced by volcanic gases, or reacts to form HNO_3 that is rained out through acid rains. This continuous production occurring in the lower atmosphere could maintain an average abundance of about 10^{-6}, 100 times higher than in the present-day atmosphere.

In the case of meteors, few quantitative studies have been done (see Jenniskens et al. 2000, for the meteor-induced chemistry in a CO_2-rich atmosphere). NO is produced in the plasma of the meteor trails in which air temperature is briefly raised to about 5000 K: NO and NO_2 lines have thus been observed in Leonid meteor spectra (see, for instance, Carbary et al. 2003). The NO_x produced by these small-sized bodies is generated in the upper atmosphere, above

50 km and, even in the absence of detailed modelling, this source does not seem to contribute to the delivery of NO_x to the lower atmosphere.

More studies are available for larger impacts because of the environmental consequences they may induce. Prinn and Fegley (1987) modelled the K/T impact[6] of a 5×10^{14}-kg asteroid, or a 1.25×10^{16}-kg comet. They suggested that the explosion could have produced from 5×10^{14} to 1.2×10^{17} moles of NO, explaining by the same way the acidic content of the worldwide K/T geological layer. However, Zahnle et al. (1988) showed in a short note that Prinn and Fegley overestimated these numbers by considering an instantaneous cooling. Later, Turco et al. (1982) estimated that the Tungunska bolid (probably a comet) that disrupted in the atmosphere in 1908 generated about 3×10^{12} moles of NO between 10 and 50 km of altitude. In an early, CO_2-dominated atmosphere, this would produce in a single event 3×10^{11} moles of NO, the same amount as the whole production of NO by volcanic lightning. Again, most of the NO is generated in the middle atmosphere and rained out in the form of nitric acid. Further modeling would be required to know how much NO reaches the surface. Indeed, this source may be very efficient and nearly continuous as Tunguska-like events may have occurred 0.1–1 times a year on the prebiotic Earth. In a review paper addressing the environmental consequence of large impacts, Toon et al. (1994) inferred a power law giving the mass production of NO as a function of the impact energy.

We summarized these results in Fig. 5.3, using Navarro-Gonzalez et al. (2001) for scaling the result to a CO_2-dominated atmosphere. We applied the standard evolution of the impact rate: an exponential decrease with a timescale of 144 Myr (Chyba 1990) until –3.5 Gyr ago. The impact flux 3.5 Gyr ago, equal to the present-day one, is taken from Morrison et al. (1994). We assume that the impactors are asteroids with a typical density of $3\,g\,cm^{-3}$, a mass distribution that is constant in time, and an average impact velocity of $20\,km\,s^{-1}$.

For impact energies above 10^{21} J (equivalent to an impactor diameter of 3 km), the induced NO is well mixed down to the surface. For smaller bodies, 40 m–3 km, the abundance of NO peaks up to 10^{-3} in the middle atmosphere and diffuses to the surface where it can reach 10^{-6} as an upper limit. Impactors with diameter below 40 m do not deliver efficiently NO to the surface. For the largest impacts (similar to those that formed the main Lunar Basins around 3.9 Gyr ago) the entire atmosphere is heated to $T > 1500\,K$ and may contain 0.1% NO, (10^{19} moles). Because such hot conditions obviously frustrate organic chemistry, we did not consider impacts with energy above 10^{25} J (equivalent to a 50-km impactor) as a source for "useful NO". Converted in $mol\,yr^{-1}$, the production of NO by impactors in the 3–50 km range is quite low ($< 1.3 \times 10^7\,mol\,yr^{-1}$, see Fig. 5.3), but every 10^4–10^5 years, the abundance of NO reaches 10^{-3} and

---

[6] Of a ca. 10-km diameter object, having occurred 65 Myr ago at the Cretaceous/Tertiary era (K/T) boundary, and plausibly responsible for the extinction of the dinosaurs. Such impact energy is estimated between 10^{22} and 10^{24} J (equivalent to 0.25 to 25 billion Hiroshima bombs).

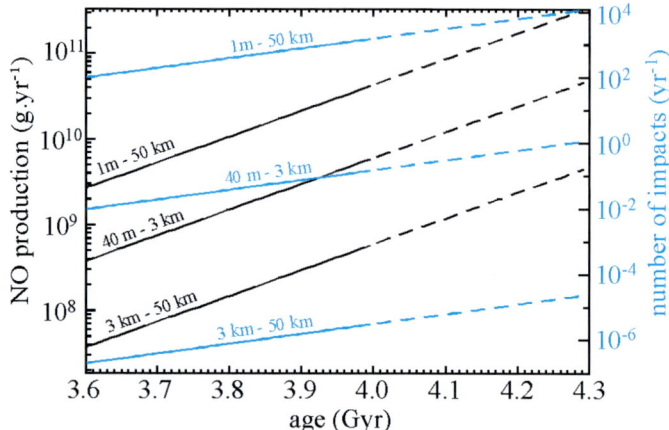

Fig. 5.3. Nitric-oxide production by impacts. *Black lines*: impact frequency as a function of age, for three diameter ranges of impactors. *Blue lines*: corresponding production of nitric oxide. *Dashed lines*: extrapolation beyond 4.0 Gyr (age of oldest available Lunar cratering records)

decreases with a typical timescale of 1 yr (much shorter than the typical interval between two impacts of this size).

In summary, the continuous source of NO on the early Earth was due to volcanic lightning. In the most favourable conditions (a negligible abiotic production of methane and a moderate release of reduced volcanic species) it could have maintained an average abundance of NO around 10^{-6}, this level being higher or lower, by at least one order of magnitude, depending on the proximity (in space and time) to volcanic plumes. Violent and stochastic impacts resulted in higher abundance (up to 10^{-3}) decreasing slowly for months or years. From the present knowledge, we can thus conclude that, if the primary pump did contribute to the production of peptides on the prebiotic Earth, it may have been fed with NO by single and violent events (impacts or volcanic eruptions) rather than continuously. Complementary experimentation and modelling will be necessary anyway in the future to substantiate this very hypothetical scenario.

5.2.3 About the pH of Primitive Oceans

5.2.3.1 Requirements from the Primary Pump

The pH of the aqueous phase (the primitive ocean) is a critical parameter of our scenario, especially for steps (**1**) and (**5**). Concerning step (**1**), the N-carbamoylation of amino acids occurs at any pH comprised between 5 and 12, with a maximum rate at pH 7–8. CAA are stable over the whole pH 5–12 range. Concerning step (**5**), the polycondensation of NCA in water has been studied from the middle 1950s. (Bartlett and Dittmer 1957; Bartlett and Jones 1957). Below pH 4, only NCA hydrolysis occurs. Above pH 4, there is a competition between NCA hydrolysis and NCA condensation on amino acids or peptides, with increasing condensation yields while increasing the pH. In the pH range 4–6.5 this competition creates the most favourable conditions for an efficient kinetic selectivity in peptide formation, thus allowing the chemical evolution of the growing peptide pool. This is important especially for the selection of homochiral peptides.

5.2.3.2 Environmental Data.

With respect to these requirements, it is important to have an idea of the pH of the primitive ocean. Due to the presence of large amounts of CO_2 in the primitive Earth atmosphere, (Kasting 1993; Zahnle 1998), the early ocean was probably initially quite acidic (its pH being possibly as low as 3.5). The pH should then have gradually increased to its current value (8.2) through progressive extraction of alkaline materials (Na^+, K^+, Ca^{2+}, etc.) from reducing rocks.

Therefore the primary pump process could have started as soon as the pH of the ocean reached 4, and could have worked until it reached ca. 6.5. This pH window is estimated to have taken place during the Hadean. Moreover, the pH of the ocean may have remained around 6.3 (pK_A of carbonic acid) for a long time due to the buffer effect. More precise knowledge of this primitive-ocean pH chronology is necessary since it would allow the corresponding time windows to be refined.

5.2.3.3 pH Change During Water Evaporation (Step 2).

A possible drawback in the activation of CAA can be the acidification of the medium during evaporation: CAA are well known to convert into hydantoins under acidic conditions (usually pH < 2.5), what we did not observe in our experiments simulating prebiotic conditions. In fact the evaporation of an aqueous solution of pH 6.5 containing (bi)carbonate salts as the major components does not result into its acidification since carbonic acid will evaporate as well as CO_2 during the concentration process (in fact we even observed an increase in pH; ultimately an alkaline salt of the CAA is obtained). Therefore the possibility of hydantoin formation during step (**2**) appears very limited. Conversely, a part of the NO_x from the atmosphere will be necessary to neutralise alkali before the nitrosation takes place.

5.3 Investigation of the Primary Pump

In the above section we have seen that α-amino acids N-carboxy anhydrides (NCA), may have been synthesized in a fairly continuous way on the primitive

Earth. This lends credence to the argument for peptide synthesis through activated α-amino acids (see introduction) as it argues in favour of the prebiotic relevance of NCA, which are also the major monomers for the preparation of synthetic polypeptides from laboratory to industrial scale.

At the same time our approach acquires a dynamic dimension. It gives prebiotic significance to Blair and Bonner's results concerning the amplification of low enantiomeric excesses during polymerization of NCA, even though their results were obtained in nonaqueous solvents (Blair and Bonner 1980, 1981). It provides arguments for carrying out the necessary research to evaluate the potentialities of the primary pump. We investigated (see next Section) the different steps of the primary pump, first separately, focusing on steps **(1)**, **(3)** and **(5)**, then in a more integrated way.

5.3.1 Step-By-Step Experimental Investigation

5.3.1.1 Step (1): *N*-carbamoylation

Cyanate (NCO^-) irreversibly reacts with α-amino acids to produce CAA acids. Study of this reaction shows that it can occur in aqueous media over a very wide pH range: 5 to 13 (Taillades et al. 2001), with the rate law (5.1) involving the hydrolysis of cyanate (k_0), the catalysis by carbonates (k_1), the formation of urea (k_2) and the *N*-carbamoylation (k_3) of amino acids (AA) (t = total: acid + base forms).

$$v = d[NCO]^t/dt = [NCO]^t \times (k_0 + k_1 \times [CO_3]^t + k_2 \times [NH_3]^t + k_3 \times [AA]^t). \quad (5.1)$$

At 50°C and pH 6.5, the *N*-carbamoylation rate constants k_3 were measured: 1.7×10^{-3} L mol^{-1} s^{-1} for glycine, 1.3×10^{-3} L mol^{-1} s^{-1} for valine, and 3.5×10^{-3} L mol^{-1} s^{-1} for threonine. This reaction is possible only above pH 5, with almost quantitative yields at pH 7 for cyanate to α-amino acid ratios above 1.5.

In addition to its interest for the quantification of a future computer kinetic model, this reaction might allow refinement of the moment the primary pump start-up, namely when the pH of primitive oceans reached 5 (see Sect. 5.2.3).

5.3.1.2 Step (3): CAA Nitrosation

The gaseous mixture O_2/NO reacts on CAA even in the absence of water, as shown in Fig. 5.4. (Collet et al. 1996; Taillades et al. 1999). Photo (a) shows crystals of anhydrous *N*-carbamoyl valine. In the presence of an excess of anhydrous gaseous mixture O_2 + NO *N*-carbamoyl valine is quantitatively (according to NMR analysis) converted into valine-NCA at 20°C within half an hour (photo c) (Collet et al. 1996). The middle photo (b) shows the surface of the crystal during the reaction. The bubbles that are forming are of nitrogen. The liquid consists of the water and nitrous acid produced by the reaction (step **(3)**). Even

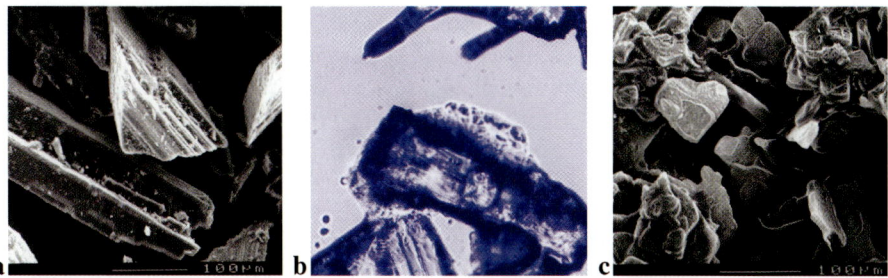

Fig. 5.4. (From left to right) (**a**) crystals of *N*-carbamoyl valine (SEM); (**b**) action of $NO + O_2$ (optical microscopy, false colours) (**c**): valine-NCA (SEM). Approximately same magnification for **a–c**

under initially anhydrous conditions, the solid CAA is completely converted into solid NCA (Taillades et al. 1999).

The reaction efficiency depends on the O_2/NO ratio. In the absence of O_2 there is no reaction. When the O_2/NO ratio is equal to, or greater than, 1 the reactivity is below its maximum, which is observed for an O_2/NO ratio of 0.25 (this ratio probably remained quite low during the Hadean). The NO_x gaseous mixture actually contains the species NO, O_2, N_2O_4 and N_2O_3 in equilibrium, the ratio of which depends on the O_2/NO mixing ratio. We evaluated the rate constants of CAA nitrosation in the solid-gas phase by N_2O_4, N_2O_3 (nitrosating agents present in the gaseous NO_x mixture) and also by nitrous acid HNO_2 (a side-product of both CAA nitrosation and N_2O_3/N_2O_4 hydrolysis). The values obtained at $25\,°C$ ($k_{N_2O_4} = 1.7\,s^{-1}, k_{N_2O_3} = 10^8\,s^{-1}$ and $k_{HNO_2} = 0.1\,s^{-1}$, respectively) show the very high reactivity of N_2O_3 (Lagrille 2001): even traces of this species in the primitive atmosphere might have been sufficient for activating CAA into NCA.

Further studies will be necessary to evaluate the working limits of the primary pump, in terms of, e.g. minimum NO_x/CAA ratio, or NO_x concentration in the atmosphere.

5.3.1.3 Step (5): Insights in Peptide Synthesis

According to George Wald's suggestions (Wald 1957) the α-helical secondary structure of a growing homochiral polypeptide chain might ensure the selection of α-amino acids of the same chirality as those already present in the chain during its lengthening, thus preserving and increasing the homochirality of the growing chain.

These suggestions were proved by Idelson and Blout (1958) and Lundberg and Doty (1957), by studying the polymerization (in dioxane) of the NCA. Stereoselectivity was negligible at the simple peptide level and much higher for the α-helix at higher degrees of polymerization. In the same paper Wald said "If one could grow such polymers in a reversible system in which synthesis was partly balanced by hydrolysis, the opportunity for selection would be greatly improved." Obviously, Wald was underlining the importance of water. Our results giving prebiotic status to NCA led us to carry out these studies again using water as solvent.

Oligocondensation of NCA in water: stereoselectivity.

NCA in aqueous solutions undergo two competitive reactions during step **(5)**, hydrolysis and
condensation (aminolysis), with different kinetic laws:

$$v_{\text{hyd}} = k_{\text{hyd}} \times [\text{HO}^-] \times [\text{NCA}], \tag{5.2}$$

and

$$v_{\text{cond}} = k_{\text{cond}} \times [\text{R}-\text{NH}_2] \times [\text{NCA}], \tag{5.3}$$

where $\text{R}-\text{NH}_2$ is the amine form of amino acid or peptide.

We reconsidered the work carried out by Bartlett (Bartlett and Dittmer 1957; Bartlett and Jones 1957), using capillary electrophoresis. With this technique, we directly measured the rate of NCA hydrolysis and of peptide formation (Plasson et al. 2002). Partial results for valine-NCA give $k_{\text{hyd}} = 4.40 \times 10^{-4}\,\text{s}^{-1}$ for hydrolysis, and $k_{\text{LL}} = 1.65 \times 10^{-2}\,\text{L}\,\text{mol}^{-1}\,\text{s}^{-1}$, and $k_{\text{LD}} = 1.14 \times 10^{-2}\,\text{L}\,\text{mol}^{-1}\,\text{s}^{-1}$ for the formation of Val$_2$ from L-Val-NCA reacting on L-Val and D-Val, respectively.

So, in water, the hydrolysis rate of the NCA is low compared to their condensation rate. The condensation rate constant k_{LL} is much greater than k_{LD}, and for the formation of dipeptide, the $k_{\text{LL}}/k_{\text{LD}}$ ratio is already much greater than 1. This ratio must be higher for the synthesis of larger peptides (octapeptides and above).

Fast Measurement of Peptide pK_A.

In water, the (pH-dependent) formation rate of peptides depends on their pK_A, which are, however, not very well known. In our group, Cottet et al. (personal communication, to be published) have developed a fast method for measuring the pK_A of oligopeptides using capillary electrophoresis (CE), with no need to isolate them from the mixture. Figure 5.5 illustrates the pk_A measurement of oligoglycines.

From this information, we deduce that, in aqueous solution, longer peptides (lower pK_A) react faster with NCA than the shorter ones (higher pK_A), assuming their intrinsic rate constants of coupling with NCA to be identical (Fig. 5.6). This last point needs to be confirmed.

On the basis of these experimental data, we are running computer simulation to investigate possible enantiomeric amplification of the peptide pool by the primary pump.

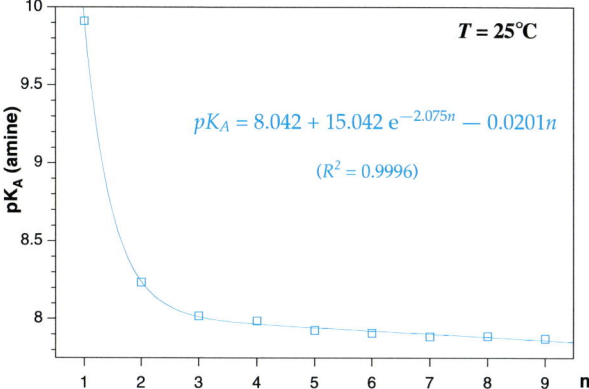

Fig. 5.5. pK$_A$ of oligoglycines Gly$_n$ (1–9 residues) as a function of n (Cottet, to be published)

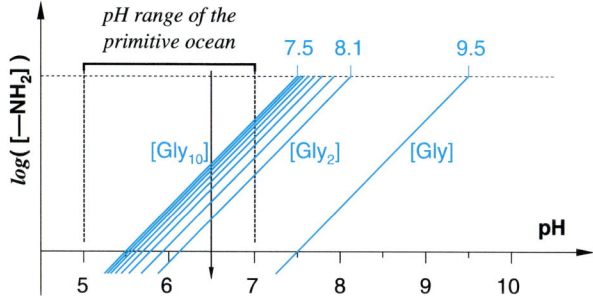

Fig. 5.6. In aqueous solution, for equal concentration of peptides, the concentration of the nonprotonated amino group of the longer peptides (lower pK$_A$) is higher than the concentration of the nonprotonated amino group of shorter peptides

5.3.2 Integrated Experimental Approach: Chemoselectivity

This physicochemical approach needs to be checked by synthesis techniques because, in water, other parameters are involved in the lengthening of peptide chains, and these parameters need to be understood.

We investigated the chemoselectivity of peptide synthesis under the conditions of the primary pump, by experimentally cycling its essential steps (**1–5**) several times (NaHCO$_3$ buffer 100 mM, pH 6.7) containing amino acids or their N-carbamoyl derivatives (Commeyras et al. 2002). Results using two amino acids valine (V) and glutamic acid (E) are shown in Fig. 5.7. Many peptides are formed but not all of them. Figure 5.8 shows the distribution of peptides formed in these experiments and shows also a very effective chemoselectivity during the elongation process.

With four amino acids (glycine, alanine, valine, glutamic acid) as starting materials, MALDI-TOF analysis shows that the length of the peptides increases with the cycle number, but insoluble peptides are formed. The insoluble fraction (separated after 7 cycles) is made up of peptides whose molecular weigh is distributed around 800 daltons, whereas the mass of the soluble compounds is higher than 2000, thus evidencing the effect of additional selectivity factors, physical rather than chemical.

The stimulating aspect of this point is that a stress (here possibly due to hydrophobicity of peptide chains) separates the insoluble peptides. The only

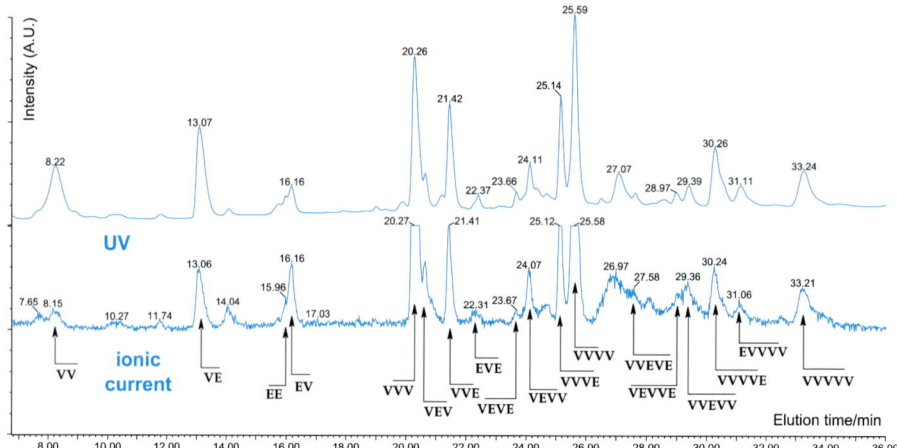

Fig. 5.7. Peptide elongation and selection after 5 cycles under primary pump conditions, running with 2 amino acids valine (V) and glutamic acid (E). Analysis by HPLC/MS (ESI/Q-TOF)

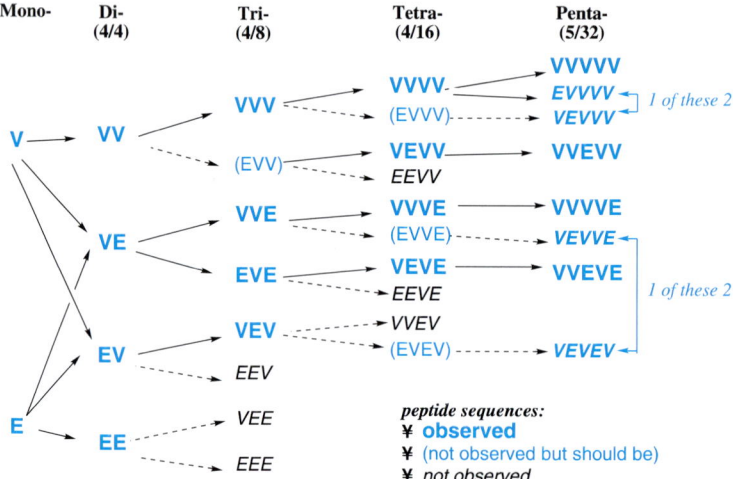

Fig. 5.8. Observed chemoselectivity during the formation of peptides under primary pump conditions, from 2 amino acids (V, E) as starting materials

peptides capable of evolving in the short term are the soluble ones, while the others – inappropriately adapted – are "frozen" as formed, and probably slowly hydrolyzed.

In conclusion we can consider that in aqueous solution, the above-described selectivities could have considerably decreased the number of possible peptides formed ($2n^x$ with $n > 20$ and $x > 100$), and have thus made this synthesis realistic. Detailed analysis of these various selectivities will require additional research.

5.4 Energy

Chemical evolution involves the interconversion of energy under its various forms, from *physical*: heat, electromagnetic (radioactivity, UV light, etc.), mechanic (tides, impacts, etc.) to *chemical*, through the formation of activated molecules, which could prefigure the increasing complexity of metabolic intermediates in the living world.

The primary-pump scenario is one example of this first step of converting energy from physical to chemical, in a permanent, continuous way: primary energy sources (heat, UV light, etc.) promote the formation of small activated molecules such as cyanic acid and NO_x, then followed by anhydrides such as NCA, responsible for the polymerisation of amino acids. As an energy carrier, NCA may also have been involved in other processes. Furthermore, as chemical structures increased in complexity, other activated/activating species should have emerged. Why not therefore consider NCA as possible ancestors (another could be thioesters, see De Duve 1998) of adenosine triphosphate (ATP), the "modern", somewhat ultimate energy carrier in today's biological world?

The kinetic model of the *primary pump* shown in Fig. 5.1 could, in a future study, be broadened to include how it was fed with matter and energy as shown in Fig. 5.9.

The energy aspect, with photochemical reactions leading to the essential prebiotic molecules, for which there are already rich sources of information and methods, could be developed through the utilization of, dedicated software such as PHOEBE (photo-chemistry for exobiology and exoplanets: see Selsis et al. 2002).

Concerning the feeding of matter, the selective synthesis process of CAA is described in the previous chapter. We showed that about 50 carbonyl derivatives, combined with ammonia, methylamine, ethylamine, hydrocyanic acid, and carbonic anhydride, are sufficient to model the formation of all meteoritic α-amino acids and terrestrial peptides. The mechanisms of the reactions involved in these syntheses are known, along with most of their kinetic constants (Commeyras et al. 2003).

Although complex, only kinetic modelling of the whole system can provide understanding or insight into the evolution of such systems.

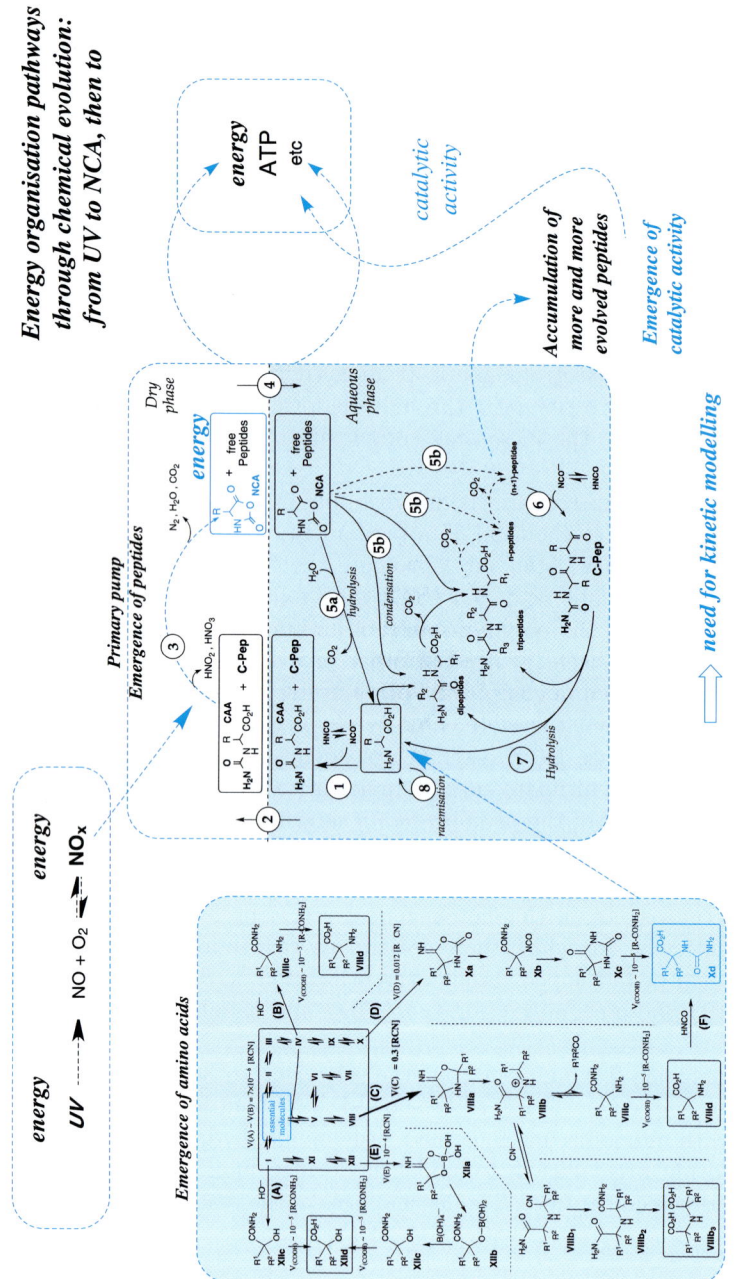

Fig. 5.9. Global view of the primary pump with matter and energy feeding

5.5 Conclusions and Perspectives

Our primary-pump scenario appears to be the first that could have been capable of supplying *sequential* peptides under rather realistic prebiotic conditions. Maintained out of thermodynamic equilibrium by continuous/sufficient inputs of matter (amino acids) and energy (through activated molecules such as cyanate and NO_x), this dissipative system has the ability to recycle its reagents and to increase the complexity of the reaction products (peptides). The minimum requirements for the prebiotic relevance of the model, namely the presence of oceans, emerged land and nitrogen oxides in the atmosphere, seem to have been fulfilled on the primitive Earth and the primary pump could work as soon as the pH of the oceans rose to 4–5. Our results underline the importance of such a buffering effect and suggest the most suitable locations for this process were beaches, rather than closed lakes or lagoons without any contact with the ocean. The emerging (eventually homochiral) peptides should progressively have begun to act as catalysts (Pascal 2003), which may have helped the emergence of self-replicating systems. In addition, the role of NO in the metabolism of currently living organisms, being the remnant of a past, much more important contribution (e.g. as a source of either organic nitrogen or energy), raises questions about a possible key role of this compound in prebiotic chemistry.

Although several questions remain unanswered, such as the effective abundance of NO_x in the atmosphere, or of cyanate in the hydrosphere in regard to the requirements of the primary pump, this model presents the unique advantage to allow experimental testing. This scenario should work as well with NCA produced by other prebiotic pathways. To our opinion, NCA should be considered as a major prebiotic intermediate.

Acknowledgement

For their fruitful collaboration to this work, we are indebted to our colleagues from Montpellier: Hélène Collet, Jacques Taillades, Laurence Garrel, Robert Pascal, Jean-Christophe Rossi, Hervé Cottet, Jean-Philippe Biron, Olivier Lagrille, Raphaël Plasson, and Grégoire Danger, as well as to Hervé Martin from the laboratory Magmas & Volcans at the University of Clermont-Ferrand (France), and to Michel Dobrijevic from the Observatoire Aquitain des Sciences de l'Univers (Bordeaux, France).

We are also grateful to:

- The Centre National de la Recherche Scientifique (CNRS), especially its Chemical Science Department and the Institut National des Sciences de l'Univers (INSU); the Exobiology Research Group of the Centre National d'Etudes Spatiales (CNES); and the European Community through the COST D27 action (prebiotic chemistry), for their support.
- The companies Aventis Animal Nutrition (France) and Degussa (Germany) for financial supports.

- Prof. Dr. Jacques Reisse from the Université Libre de Bruxelles (Belgium), and Dr. Muriel Gargaud from the Observatoire Aquitain des Sciences de l'Univers (Bordeaux, France) for many helpful discussions.

References

Bartlett PD, Dittmer DC (1957). A kinetic study of the Leuchs anhydrides in aqueous solution. II. *J. Am. Chem. Soc.* **79**: 2159–2161.
Bartlett PD, Jones RH (1957). A kinetic study of the Leuchs anhydrides in aqueous solution. I. *J. Am. Chem. Soc.* **79**: 2153–2159.
Bengston M, Edstrom ED (1999). A new method for testing models of prebiotic peptide assembly. In: *Advances in BioChirality*, Pályi G, Zucchi C and Caglioti L (eds), Elsevier. pp 115–123.
Blair NE, Bonner WA (1980). Experiments on the amplification of optical activity. *Origin of Life and Evolution of the Biosphere* **10**: 255–263.
Blair NE, Bonner WA (1981). A model for the enantiomeric enrichment of polypeptides on the primitive Earth. *Origin of Life* **11**: 331–335.
Bohler C, Hill AR, Orgel LE (1996). Catalysis of the oligomerization of o-phosphoserine aspartic acid, or glutamic acid by cationic micelles. *Origins of Life and Evolution of the Biosphere* **26**: 2–5.
Bowring SA, Williams IS (1999). Priscoan (7,00–4,03 Ga) orthogneisses from northwestern Canada. *Contribution to Mineralogy and Petrology* **134**: 3–16.
Caldeira K, Kasting JF (1992). Susceptibility of the early Earth to irreversible glaciation caused by carbon dioxide clouds. *Nature* **359**: 226–228.
Carbary JF, Morrison D, Romick GJ, Yee JH (2003). Leonid meteor spectrum from 110 to 860 nm. *Icarus* **161**: 223–234.
Chyba CF (1990). Impact delivery and erosion of planetary oceans in the early inner solar system. *Nature* **343**: 129–133.
Collet H, Boiteau L, Taillades J, Commeyras A (1999). Solid phase decarbamoylation of N-carbamoylpeptides and monoalkylureas using gaseous NO_x: a new simple deprotection reaction with minimum waste. *Tetrahedron Letters* **40**: 3355–3358.
Collet H, Bied C, Mion L, Taillades J, Commeyras A (1996). A new simple and quantitative synthesis of α-Aminoacid-N-Carboxyanhydrides. *Tetrahedron Letters* **37**: 9043–9046.
Commeyras A, Collet H, Boiteau L, Taillades J, Vandenabeele-Trambouze O, Cottet H, Biron J-P, Plasson R, Mion L, Lagrille O, Martin H, Selsis F, Dobrijevic M (2002). Prebiotic Synthesis of Sequential Peptides on the Hadean Beach by a Molecular Engine Working with Nitrogen Oxides as Energy Sources. *Polymer International* **51**: 661–665.
Commeyras A, Taillades J, Collet H, Mion L, Boiteau L, Trambouze-Vandenabeele O, Cottet H, Biron JP, Schué F, Giani O, Lagrille O, Plasson R, Vayaboury W, Martin H, Selsis F, Dobrijevic M, Geffard M (2001). La Terre, matrice de la Vie: émergence avant-gardiste des peptides sur les plages de l'Hadéen. In: *L'environnement de la Terre Primitive*, Gargaud M, Despois D, Parisot JP (eds.), Presses Universitaires de Bordeaux, pp 361–380.
Commeyras A, Taillades J, Collet H, Boiteau L, Pascal R, Vandenabeele-Trambouze O, Pascal R, Rousset A, Garrel L, Rossi JC, Cottet H, Biron JP, Lagrille O, Plasson

R, Souaid E, Selsis F, Dobrijevic M (2003). Approche dynamique de la synthèse des peptides et de leurs précurseurs sur la Terre primitive. In: *Les Traces du Vivant*, Gargaud M, Despois D, Parisot JP (eds.). Presses Universitaires de Bordeaux, Bordeaux, France, pp 115–162.

De Duve C (1998). Clues from present-day biology: the thioester world. In: *The Molecular Origin of Life*, Brack A (ed.), Cambridge University Press. Cambridge, UK, pp 219–236.

Fox SW, Dose H (1977). *Molecular Evolution*, New York, Academic Press. New York, (USA).

Gesteland RF, Cech TR, Athins JF (1999). *The RNA Word*. 2nd edition, Cold Spring Harbor Laboratory Press. New York (USA).

Gilbert W (1986). The RNA World. *Nature* **319**: 618.

Guinan E, Ribas I (2002). Our changing Sun: the role of solar nuclear evolution and magnetic activity on Earth's atmosphere and climate. *ASP Conference Series*, **269**: 85–106.

Huber C, Wächtershäuser G (1998). Peptides by activation of amino acids with CO on (Ni,Fe)S surfaces: implications for the origin of life. *Science* **281**: 670–672.

Idelson M, Blout ER (1958). Polypeptides. XVIII. A kinetic study of the polymerisation of amino acid N-carboxyanhydrides initiated by strong bases. *J. Am. Chem. Soc.* **79**: 2387–2393.

Jenniskens P, Wilson MA, Packan D, Laux CO, Krüger CH, Boyd ID, Popova OP, Fonda M (2000). Meteors: A delivery mechanism of organic matter to the early Earth. *Earth, Moon and Planets* **82/83**: 57–70.

Joyce GF, Orgel LE (1999). Prospects for understanding the origin of the RNA world. In: *The RNA World*. 2nd edn, Cold Spring Harbor Laboratory Press (New York), pp 49–77.

Kasting JF (1993). Earth's early atmosphere. *Science* **259**: 920–926.

Kasting JK, Catling D (2003). Evolution of a habitable planet. *Annual Review of Astronomy and Astrophysics* **41**: 429–463.

Kauffman SA (1993). *The Origin of Order. Self-organisation and Selection in Evolution*, 1st edn. Oxford University Press (see especially Chap. 7).

Lagrille O (2001). *Nitrosation de N-Carbamoylaminoacides Solides par le mélange gazeux NO/O_2. Synthèse de N-Carboxyanhydrides (Anhydrides de Leuchs)*. PhD thesis, Université de Montpellier 2 (France).

Lagrille O, Taillades J, Boiteau L, Commeyras A (2002). N-Carbamoyl Derivatives and their nitrosation by gazeous NO_x–A new promising tool in stepwise peptide synthesis. *Eur. J. Org. Chem*: 1026–1032.

Lahav N, Wite DH, Chang S (1978). Peptide formation in the prebiotic Era: thermal condensation of glycine in fluctuating clay environments. *Science* **201**: 67–69.

Liu R, Orgel LE (1998). Polymerization of β-amino acids in aqueous solution. *Orig. Life Evol. Biosphere* **28**: 47–60.

Luisi PL (2000). L'assemblage des macromolécules. *La Recherche* **336**: 25–29.

Luisi PL, Walde P, Blocher M, Liu D (2000). Research on the origin of Life: membrane-assisted polycondensations of amino acids and peptides. *Chimia* **54**: 52–53.

Lundberg RD, Doty P (1957). Polypeptides. XVII. A Study of the Kinetics of the Primary Amine-initiated Polymerisation of N-carboxy-anhydrides with special reference to configurational and stereochemical effects. *J. Am. Chem. Soc.* **78**: 3961–3972.

Martin H (1986). Effect of steeper Archean geothermal gradient on geochemistry of subduction-zone magmas. *Geology* **14**: 753–756.

Martin H (1987). Petrogenesis of Archaean trondhjemites, tonalites and granodiorites from eastern Finland: major and trace element geochemistry. *Journal of Petrology* **28**: 921–953.

McCulloch MT, Bennet VC (1993). Evolution of the early Earth: constraints from ^{143}Nd-^{142}Nd isotopic systematics. *Lithos* **30**: 237–255.

Mojzsis SJ, Krishnamurthy R, Arrhenius G (1999). Before RNA and after: Geophysical and geochemical constraints on molecular evolution. In: *The RNA Word* 2nd edn. Gesteland RF, Cech TR and Athins JF (eds.), Cold Spring Harbor Laboratory Press, pp 1–47.

Mojzsis SJ, Harrison MT, Pidgeon RT (2001). Oxygen-isotope evidence from ancient zircons for liquid water at the Earth's surface 4,300 Myr ago. *Nature* **409**: 178–181.

Morrison D, Chapman CR, Slovic P (1994). The impact hazard. In: *Hazards due to Comets and Asteroids*, Gehrels T, Matthews MS and Schumann A (eds.). Univ. Arizona Press, Tucson, USA, pp 59–92.

Munoz Caro GM, Meierhenrich UJ, Schutte WA, Barbier B, Arcone Segovia A, Rosenbauer H, Thiemann WHP, Brack A, Greenberg GM (2002). Amino acids from ultraviolet irradiation of interstellar ice analogues. *Nature* **416**: 403–406.

Navarro-Gonzalez R, Molina MJ, Molina LT (1998). Nitrogen fixation by volcanic lightning in the early Earth. *Geophys. Res. Lett.* **25**: 3123–3126.

Navarro-Gonzalez R, McKay CP, Nna Mvondo D (2001). A possible nitrogen crisis for Archaean life due to reduced nitrogen fixation by lightning. *Nature* **412**: 61–64.

Nicolis G, Prigogine I (1977). *Self-organization in Nonequilibrium Systems*. Wiley & Sons. New York (USA).

Orgel LE (1989). The origin of polynucleotide-directed protein synthesis. *J. Mol. Evol.* **29**: 465–474.

Orgel LE (1998). Polymerization on the rocks: Theoretical introduction. *Origins of Life and Evolution of the Biosphere* **28**: 227–237.

Paecht-Horowitz M, Eirich FR (1988). *Origins of Life and Evolution of the Biosphere* **18**: 359.

Pascal R (2003). Catalysis by induced intramolecularity: what can be learned by mimicking enzymes with carbonyl compounds that covalently bind substrates? *Eur. J. Org. Chem.*: 1813–1824.

Pavlov AA, Kasting JF, Brown LL, Rages KA, Freedman R (2000). Greenhouse warming by CH_4 in the atmosphere of early Earth. *J. Geophys. Res.* **105**: 11981–11990.

Pizzarello S, Cronin JR (2000). Non-racemic amino acids in the Murray and Murchison meteorites. *Géochimica Cosmochemica Acta* **64**: 329–338.

Plasson R, Biron JP, Cottet H, Commeyras A, Taillades J (2002). Kinetic study of α-aminoacids N-carboxyanhydrides polymérisation in aqueous solution using capillary electrophoresis. *J. Chrom. A* **952**: 239–248.

Plasson R. (2003). *Origine moléculaire de la vie: étude de la polymérisation de aminoacides N-carboxyanhydrides dans des conditions prébiotiques, par électrophorèse capillaire*. PhD thesis, Université de Montpellier 2 (France).

Prinn RG, Fegley B (1987). Bolide impacts, acid rain, and biospheric traumas at the Cretaceous-Tertiary boundary. *Earth and Planetary Science Letters* **83**: 1–4.

Rode BM, Eder AH, Yongyai Y (1997). Amino acid sequence preferences of the salt-induced peptide formation reaction in comparison to archaic cell protein composition. *Inorganica Chemica Acta* **254**:309–314.

Rohlfing DL (1976). Thermal polyamino acids: synthesis at less than 100°C. *Science* **193**: 68–69.

Rosenqvist J, Chassefière E (1995). Inorganic chemistry of O_2 in a dense primitive atmosphere. *Planet. Space Sci.* **43**: 3–10.

Selsis F (2000). *Modèle d'évolution physico-chimique des atmosphères de planètes telluriques. Application à l'atmosphère primitive terrestre et aux planètes extrasolaires.* PhD thesis, Université Bordeaux 1, France.

Selsis F, Despois D, Parisot J-P (2002). Signature of life on exoplanets: Can Darwin produce false positive detections? *Astronomy and Astrophysics* **388**: 985–1003.

Shimoyama A, Ogasawara R (2002). Dipeptides and Diketopiperazines in the Yamato-791198 and Murchison Carbonaceous Chondrites. *Orig. Life Evol. Biosphere* **32**: 165–179.

Sleep NH, Zahnle K (2001). Carbon dioxide cycling and implications for climate on ancient Earth. *Journal of Geophysical Research* **106**: 1373–1399.

Taillades J, Collet H, Garrel L, Beuzelin I, Boiteau L, Choukroun H, Commeyras A (1999). N-Carbamoylaminoacid solid-gas nitrosation by NO/NO_x : a new route to oligopeptides via α-aminoacid N-carboxyanhydride. Prebiotic implications. *Journal of Molecular Evolution* **48**: 638–645.

Taillades J, Boiteau L, Beuzelin I, Lagrille O, Biron JP, Vayaboury W, Vandenabeele-Trambouze O, Giani O, Commeyras A (2001). A pH-dependent cyanate reactivity model: application to preparative N-carbamoylation of amino acids. *Perkin Trans. 2:* 1247–1253.

Tolstikhin IN, Marty B (1998). The evolution of terrestrial volatiles: a view from helium, neon, argon and nitrogen isotope modelling. *Chem. Geol.* **147**: 27.

Toon OB, Zahnle K, Turco RP, Covey C (1994). *Environmental perturbations caused by asteroid impacts, in Hazards due to comets and asteroids.* Univ. Arizona Press, Tucson, USA, pp 791–826.

Turco RP, Toon OB, Park C, Whitten RC, Pollack JB, Noerdlinger P (1982). An analysis of the physical, chemical, optical, and historical consequence of the 1908 Tungunska meteor fall. *Icarus* **50**: 1–52.

Wald G (1957). The origin of Optical Activity. *Ann. NY Acad. Sci.* **69**: 352–368.

Weber AL (1998). Prebiotic aminoacid thioester synthesis: Thiol-dependent amino acid synthesis from formose substrate (formaldehyde and glycolaldehyde) and ammonia. *Origins of Life and Evolution of the Biosphere* **28**: 259–270.

Wilde SA, Valley JW, Peck WH, Graham CM (2001). Evidence from detrital zircons for the existence of continental crust and oceans on the Earth 4.4 Ga ago. *Nature* **400**: 175–178.

Zahnle K (1998). Origins of atmospheres. *ASP Conf Series 148: Origins*: 364.

Zahnle K, Kasting JF, Sleep N (1988). Impact production of NO and reduced species. In: LPI Contribution 673 (abstracts of the *Topical conference on global catastrophes in Earth history: interdisciplinary conference on impacts, volcanism, and mass mortality*, Snowbird, Utah, Oct. 20–23, 1988), Lunar & Planetary Institute, Houston, pp 223–224.

6 The RNA World: Hypotheses, Facts and Experimental Results

Marie-Christine Maurel, Anne-Lise Haenni

A biochemical world that would have existed before the contemporary DNA-RNA-protein world, and named in 1986 "The RNA World" by Walter Gilbert (Gilbert, 1986), such a world had already been proposed during the preceding decades by Carl Woese, Francis Crick and Leslie Orgel (Woese, 1965; Crick, 1968; Orgel, 1968).

By demonstrating the remarkable diversity of the RNA molecule, molecular biology proved these predictions. RNA, present in all living cells, performs structural and metabolic functions many of which were unsuspected only a few years ago. A truly modern "RNA world" exists in each cell; it contains RNAs in various forms, short and long fragments, single and double-stranded, endowed with multiple roles (informational, catalytic, that can serve as templates, guides, defense, etc.), certain molecules even being capable of carrying out several of these functions.

Are the sources of this RNA world to be found in the bygone living world?

6.1 The Modern RNA World

6.1.1 Where in the Living Cell is RNA Found?

Synthesized (transcribed) in the nucleus, mature messenger RNAs (mRNAs), transfer RNAs (tRNAs) and ribosomal RNAs (rRNAs) are exported as single strands to the cytoplasm of the cell after various maturation steps. A ribonucleic acid (RNA) is formed by linking nucleotides[1], themselves composed of heterocyclic bases associated with a sugar, β-D-ribofuranose, and a phosphate molecule (phosphoric acid). The four main nucleotides contain the heterocyclic purine (adenine and guanine) or pyrimidine (cytosine and uracil) bases[2]. However, RNAs, in particular rRNAs and tRNAs contain a very large diversity of modified nucleotides, since more that a hundred modified nucleotides[3] have now been identified in these two classes of molecules (Grosjean and Benne, 1998).

---

[1] To yield a polyribonucleotide
[2] Adenine, A; guanine, G; cytosine, C; uracil, U
[3] Post-transcriptional modifications

RNAs are usually single stranded[4]. Nevertheless, these strands can base pair locally or over long stretches (intramolecular pairing). Finally, from a structural point of view, they contain a reactive hydroxyl group in the 2' position of ribose (a group that is absent in DNA). The stacking forces and pairing of bases produce "stems and helices"; defined structures bring together the helices and the regions separating them, into "motifs".

RNA helices: Through the action of the stacking forces, the skeleton of the single strand by itself tends to take the shape of a simple, right-handed and irregular helix. However, the important conformation is the double helix composed of two strands of RNA or of RNA/DNA (hybrids formed transiently during transcription) or that occurs when two distantly located complementary segments of the same RNA base pair.

The motifs identified are bulges, elbows, or loops.

Hairpins are other important structural motifs related to certain functions of RNAs. They can lead to interactions with special sequences, such as the GNRA loops[5], seven-base-long loops, etc. Large RNAs possess independent domains formed by the arrangement of a certain number of motifs. An RNA molecule can adopt several reversible conformations, depending on the presence of ions, specific surfaces or bound ligands. RNAs possess a repertory of structures reminiscent of proteins (motifs or domains) allowing them to express certain functions such as catalysis. Finally, non-Watson–Crick base pairs[6] are frequently encountered in RNAs (G-U pairs are common) and modified bases are involved, and by their strong steric hindrance with the bases, the 2' OH groups of the ribose moieties tend to prevent folding in the B helical conformation[7].

6.1.1.1 The Three Large Classes of RNA

• Messenger RNAs (mRNAs of 400 to 6000 nucleotides) are the copy of DNA genes[8]. The RNA transcripts are considerably modified in the nucleus during maturation, and during transcription of DNA into RNA, short hybrids of the A conformation appear. Their life is short in prokaryotes (a few minutes to a few dozen minutes) and can be of several hours in higher eukaryotes; mRNAs correspond to only a few per cent of the total cellular RNAs. The step-by-step decoding of the mRNA by the ribosome known as translation is regulated by specific proteins, and in some cases also by hairpin motifs and/or by pseudoknots.

---

[4] Paired *two-stranded* RNAs are exceptions found in a few rare viruses
[5] N is any nucleotide, R is a purine nucleotide
[6] See glossary. Watson–Crick pairings are the standard pairs (A-U and G-C)
[7] The bends they impose to the plane of the bases – of about 20° – on the axis results in a structure resembling the A conformation (also designated RNA 11 to stress the 11 base pairs per turn). The A form of RNA double helices is characterized by 11 base pairs per helical turn (instead of 10 for the B form), and by bending of the base pairs by 16°/helical axis (instead of 20° for DNA A)
[8] A gene is a fragment of DNA whose information is expressed via the genetic code

Pseudoknots result from base pairing between nucleotides within a loop and complementary nucleotides outside of the loop.

• Transfer RNAs (tRNAs) are small molecules whose maximum length is about 100 nucleotides. They are strongly conserved and are involved in the central metabolism of all types of cells. Their main function is to ensure the interaction between the codon presented by the mRNA and the specific amino acid (corresponding to this codon) and contained in the anticodon of the aminoacyl-tRNA. tRNAs possess two extremely specific sites: the first is the sequence CCA located at the 3' OH of the molecule; the second site is located in a loop that contains the anticodon. The cloverleaf-shaped secondary structure (Fig. 6.1) possesses several motifs. tRNAs also serve as primers during replication of certain viruses and are involved in the activity of telomerases. Synthesized as pre-tRNAs they undergo a maturation step during which RNAse P cleaves off a short fragment from the 5' end of the RNA (Guerrier-Takada et al., 1983). As already mentioned, tRNAs contain a large number of modified bases that are probably the most visible "relics" of an ancient RNA world (Cermakian and Cedergren, 1998).

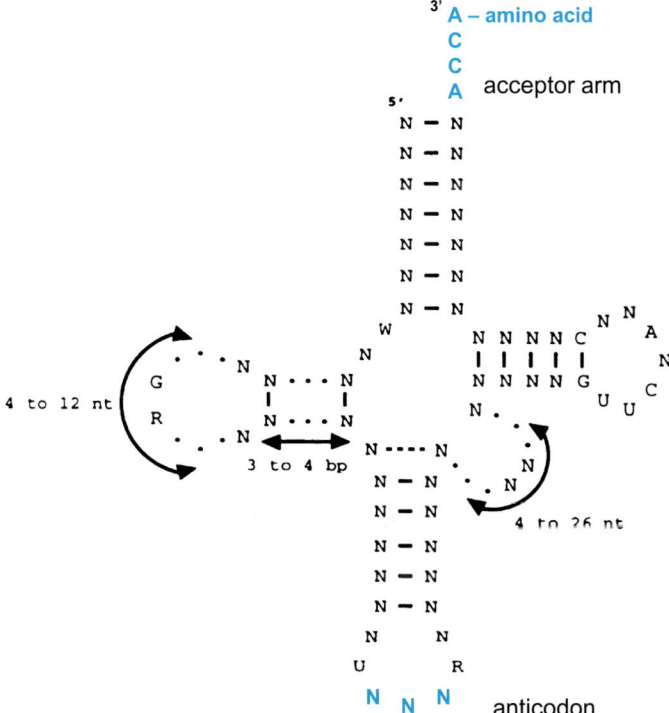

Fig. 6.1. Secondary cloverleaf structure of a tRNA. *Arrows* indicate number of nucleotides in the loop, stem and bulge

- The size of the ribosomal RNAs (rRNAs) is variable, from 120 to 4718 nucleotides. rRNAs are located in the ribosome, the site of protein synthesis. In addition to about fifty proteins, the prokaryotic ribosome contains three rRNAs and the eukaryotic ribosome four rRNAs. The rRNAs are methylated (sometimes in the 2'OH position of the ribose, protecting the polymer from hydrolysis). Their typical secondary structure is remarkably well conserved (Fig. 6.2).

They possess complex global tertiary conformations that compact the molecule into different domains, and it has now been clearly demonstrated that the rRNA catalyzes the formation of the peptide bond during protein biosynthesis (Ban et al., 2000; Nissen et al., 2000; Ysupov et al., 2001).

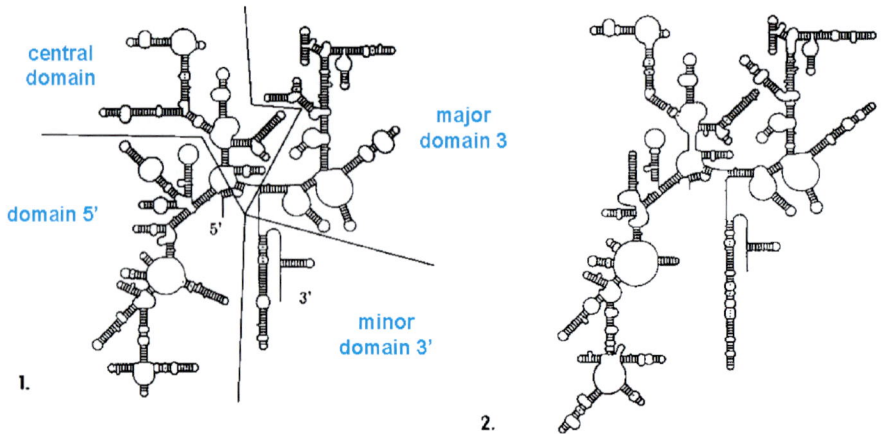

Fig. 6.2. Typical secondary structure 1) 16S rRNA of the bacterium *Escherichia coli*, 2) 18S rRNA of the yeast *Saccharomyces cerevisiae*

6.1.1.2 Noncoding RNAs (ncRNAs)

In addition to rRNAs, tRNAs and mRNAs a variety of RNA molecules have been discovered that possess very diverse functions in the living cell (Maurel, 1992; Meli et al., 2001; Zamore, 2002; Grosshans and Slack, 2002; Westhof, 2002). Before involvement of the ribosome, the RNA transcripts must undergo maturation steps. In eukaryotes, these post-transcriptional modification steps require the participation of small RNAs, the snoRNAs (small nucleolar RNAs) that together with proteins, form the snoRNP (small nucleolar ribonucleoprotein particles). Over 150 snoRNPs have been described in eukaryotes (in different lineages). They form a snoRNP complex, the snorposome, that participates in RNA maturation. The origin of the modification systems is still unknown. One of the various hypotheses put forward suggests that the snoRNAs of the RNA world would have been involved in the assembly of the protoribosomes, and more generally in the scaffolding of ribozymes (Terns and Terns, 2002).

Moreover, large snRNPs (small nuclear ribonucleoprotein particles) responsible for intron excision from pre-mRNAs have been identified. Each snRNP is composed of snRNA and about a dozen snRNP proteins. Two classes of such spliceosomes cleave different introns, whereas excision and ligation of the exons is achieved by the same biochemical mechanism (Tarn and Steitz, 1997). Spliceosomes are restricted to eukaryotes, even though containing introns bacteria have been reported.

The telomerase is an enzyme that uses a small RNA as primer during replication to elongate the linear DNA located at the end of eukaryotic chromosomes (Maizels et al., 1999).

Vault RNAs are ribonucleoprotein particles located in the cytoplasm of eukaryotes (Kong et al., 2000). They are associated with the nuclear "pore complex"; their function has not been clearly defined, but their structure suggests that they may be involved in cell transport or in the assembly of macromolecules. The history of the evolution of vault RNAs remains unknown, but these RNAs could have participated in primitive compartmentation.

Finally, an RNA-protein complex, the SrpRNA (signal recognition particle RNA) is highly conserved in the three kingdoms (Wild et al., 2002). It is involved in translation, and during secretion of proteins from the plasma membrane or from the endoplasmic reticulum.

About 15 years ago, the existence of a correcting mechanism, «editing», was demonstrated (Lamond, 1988). This co- or post-transcriptional mechanism modifies the sequence of the mRNA by the insertion or deletion of nucleotides, or by the modification of bases. Up to 55% modifications can take place with respect to the gene (in this case it is designated "cryptogene"). The sites where editing takes place are determined by the structure of the RNA, or by guide-RNAs (Stuart and Panigrahi, 2002). In kinetoplastid protozoa, guide RNAs are required to edit mitochondrial pre-mRNAs by inserting or deleting uridylate residues in precise sites (Kable et al., 1997).

Finally, the tmRNA (transfer-messenger RNA) is a stable cytoplasmic RNA found in eubacteria. tmRNAs contain a $tRNA^{Ala}$-like structure (with pairing between the 5' and 3' ends) and an internal reading frame that codes for a short peptide (peptide tag) (Fig. 6.3). It is thus at variance with the strict definition of snRNAs, since it encompasses a short reading frame. It performs a new type of recently discovered translation, known as trans-translation, during which a peptide is synthesized starting from two distinct mRNAs. tmRNA acts as tRNA and as mRNA to "help" ribosomes that are blocked on a trunctated mRNA lacking a termination codon. tmRNA participates by adding alanine to the growing peptide chain. Thus, tmRNA plays a dual role: as $tRNA^{Ala}$ it can be aminoacylated by the corresponding alanyl-tRNA synthetase, and as mRNA its open reading frame can be translated by the ribosome (Withey and Friedman, 2002; Valle et al., 2003). Could tmRNA be a bacterial adaptation, or could it have been lost by the archae and the eukaryae?

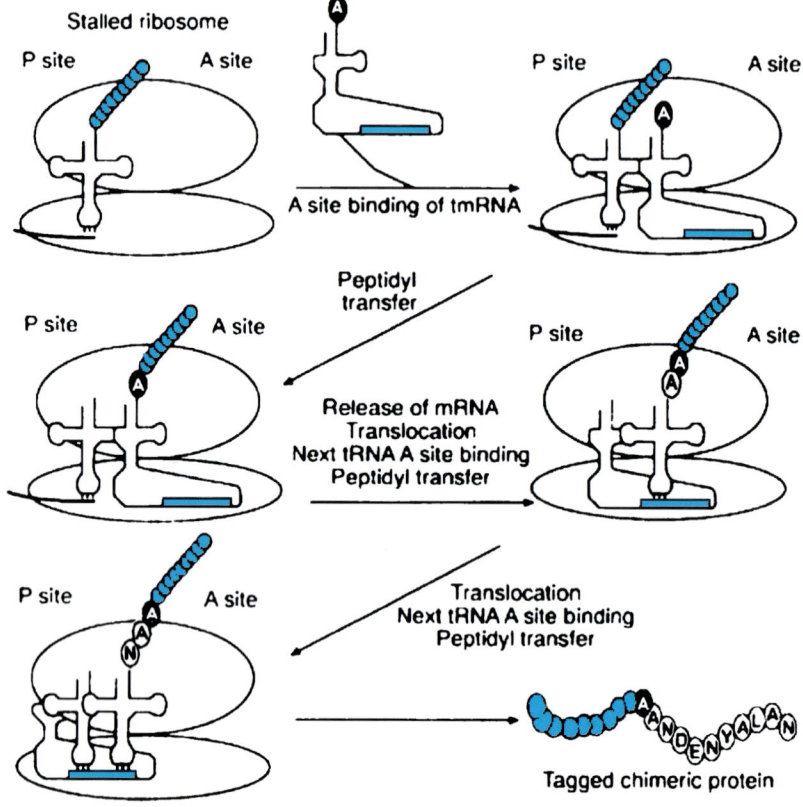

Fig. 6.3. How tmRNA functions

A eukaryotic system distantly related to tmRNA has recently been described (Barends et al., 2003) in the single-stranded Turnip yellow mosaic virus (TYMV) RNA. The 3' end of the viral genome harbors a tRNA-like structure that is indispensable for the virus viability and can be valylated. During protein biosynthesis programmed with valylated TYMV RNA, the valine residue is N-terminally incorporated into the viral polyprotein, thereby introducing a novel mechanism of initiating protein synthesis (Fig. 6.4). Here again, the viral RNA would be bifunctional, serving both as tRNA and as mRNA.

It will be interesting to determine whether other viral RNAs whose 3' end bears an aminoacylatable tRNA-like structure (Fechter et al., 2001) can also donate their amino acid for mRNA translation.

Viroids are subviral plant pathogens responsible for economically important diseases. They are small (246–401 nucleotides), single-stranded closed circular RNA molecules characterized by a highly compact secondary structure. They are devoid of coding capacity and replicate autonomously in the plant host. Two families of viroids have been characterized, the Pospiviroidae (type-member:

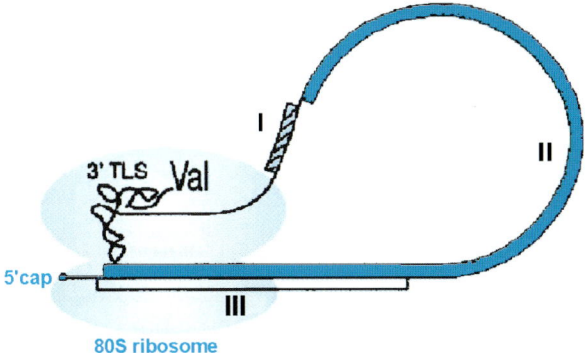

Fig. 6.4. Model of the tRNA-like structure-mediated internal initiation mechanism of TYMV RNA for polyprotein translation. I: Coat protein gene; II: Polyprotein gene; III: Movement protein gene; Adapted from Barends et al., 2003

Potato spindle tuber viroid, PSTVd) that replicates in the nucleus, and the Avsunviroidae (type-member: Avocado sun blotch viroid, ASBVd) that replicates in chloroplasts and possesses conserved hammerhead structures in the viroid and in the complementary RNA orientation. It has been suggested that the presence of hammerhead structures could reflect the early appearance of viroids in the course of evolution; they could correspond to "living fossils" of the primitive RNA world (Diener, 2001).

The few ncRNAs described here are probably only the tip of a huge iceberg (Bachellerie et al., 2002) since most of the transcriptional output of superior eukaryotes is nonprotein coding (97% for human). These ncRNAs could constitute a real RNA world in complex organisms (Eddy, 2001; Mattick, 2003). Their study may open new perspectives about the importance of RNA in primitive life. Certain RNAs that are presently being investigated are those involved in RNA interference (RNAi): the RNAs responsible for RNAi are the small interfering RNAs that target and cleave mRNAs (Nykanen et al., 2001). *Micro-RNAs*, another class of small RNAs, are involved in translation regulation (Grosshans and Slack, 2002). In eukaryotes, guide snoRNAs participate in selecting the sites on rRNAs that undergo modifications such as Ψ formation or 2'-O-methylation (Lafontaine and Tollervey, 1998).

6.2 An RNA World at the Origin of Life?

The scenario of evolution postulates that an ancestral molecular world existed originally that was common to all the present forms of life; the functional properties of nucleic acids and proteins as we see them today would have been produced by molecules of ribonucleic acids (Joyce, 1989; Orgel, 1989; Benner et al., 1989, 1993; Joyce and Orgel, 1999; Gesteland et al., 1999; Bartel and Unrau, 1999; McGinness et al., 2002; Joyce, 2002).

6.2.1 Facts

As we have seen, RNAs occupy a pivotal role in the cell metabolism of all living organisms and several biochemical observations resulting from the study of contemporary metabolism should be stressed. For instance, throughout its life cycle, the cell produces deoxyribonucleotides required for the synthesis of DNA that derive from ribonucleotides of the RNA. Thymine, a DNA specific base is obtained by transformation (methylation) of uracil a RNA specific base, and RNAs serve as obligatory primers during DNA synthesis (Fig. 6.5). Finally, the demonstration that RNAs act as catalysts is an additional argument in favour of the presence of RNAs before DNA during evolution.

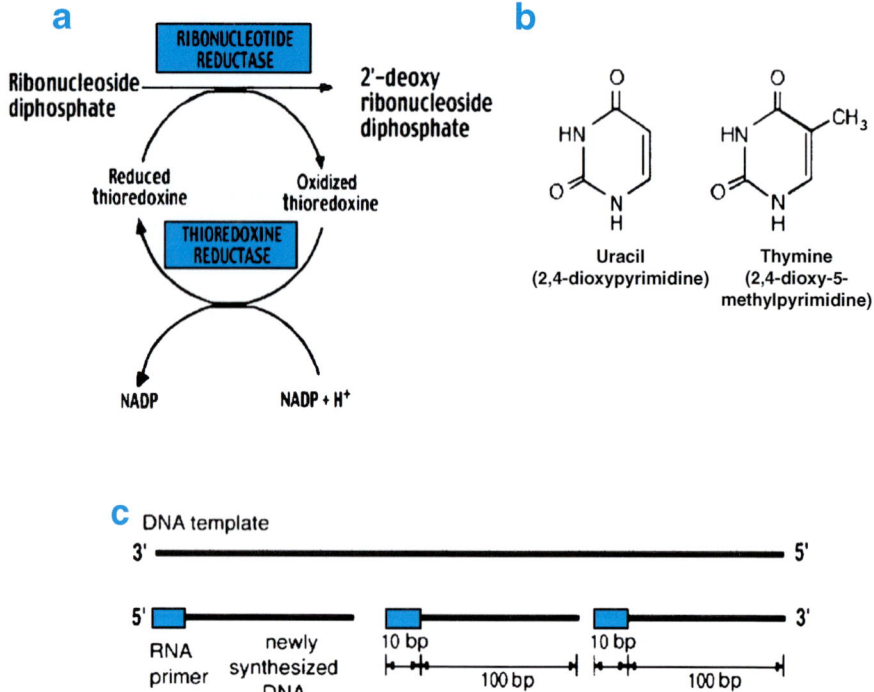

Fig. 6.5. Facts in favour of an RNA world. (**a**) Synthesis of deoxyribonucleotide; (**b**) Structure of uracil and thymine; (**c**) DNA synthesis primed by RNA

6.2.2 Hypotheses

DNA replication triggered by ribonucleotide primers can be considered as a modified transcription process during which polymerisation of RNA is "replaced" by that of DNA. In addition, DNA a double-stranded molecule lacking a hydroxyl group in 2' of the desoxyfuranose, appears more stable than RNA. Therefore it seems highly likely that RNA arose before DNA during biochemical evolution, and for this reason DNA is sometimes considered as modified RNA better suited for the conservation of genetic information. This genetic privilege would constitute a logical step in an evolutionary process during which other molecules could have preceded RNA and transmitted genetic information.

The idea of an "RNA" world rests primarily on three fundamental hypotheses, developed by Joyce and Orgel (1999):

- during a certain period in evolution, genetic continuity was assured by RNA replication,
- replication was based on Watson–Crick type base pairing,
- genetically coded proteins were not involved in catalysis.

6.2.3 But What do We Know about Primitive Replication?

Synthesis of a strand complementary of the template was studied extremely thoroughly in vitro in the group of Orgel (Inoue and Orgel, 1983; Joyce and Orgel, 1986; Orgel, 1992). During this directed synthesis, the mononucleotides (activated under the form of 5'-phosphorimidazolides) are positioned according to the Watson and Crick pairing rules along a preformed polypyrimidine template. Since these monomers are activated, they can bind together to form the complementary strand (Fig. 6.6). Orgel and his coworkers showed that starting from activated monomers, it is possible in certain conditions to copy a large number of oligonucleotide sequences containing one or two different nucleotides in the absence of enzyme (Hill et al., 1993).

Ferris and his coworkers spent some 15 years studying the assembly of RNA oligomers on the surface of montmorillonite (clay of Montmorillon in the Vienne region in France) (Ferris, 1987; Ferris and Ertem, 1992). The monomers used, nucleoside 5'-phosphorimidazolides, were probably not prebiotic molecules. Nevertheless, experimental results demonstrated that minerals that serve as adsorbing surfaces and as catalysts (Paecht-Horowitz et al., 1970; Ferris et al., 1996), can lead to accumulation of long oligonucleotides, as soon as activated monomers are available. One can thus envisage that activated mononucleotides assembled into oligomers on the montmorillonite surface or on an equivalent mineral surface. The longest strands serving as templates, direct synthesis of a complementary strand starting from monomers or short oligomers, and double-stranded RNA molecules accumulate. Finally, a double RNA helix of which one strand is endowed with RNA polymerase activity, would dissociate to copy the complementary strand to produce a second polymerase that would copy the first to produce a second complementary strand, and so forth. The RNA world would thus have emerged from a mixture of activated nucleotides. However, a mixture of activated nucleotides would need to have been available! In addition, this nucleotide chemistry is restricted in another way, since a copy of the template can be started only if the nucleotides are homochirals (Joyce et al., 1987).

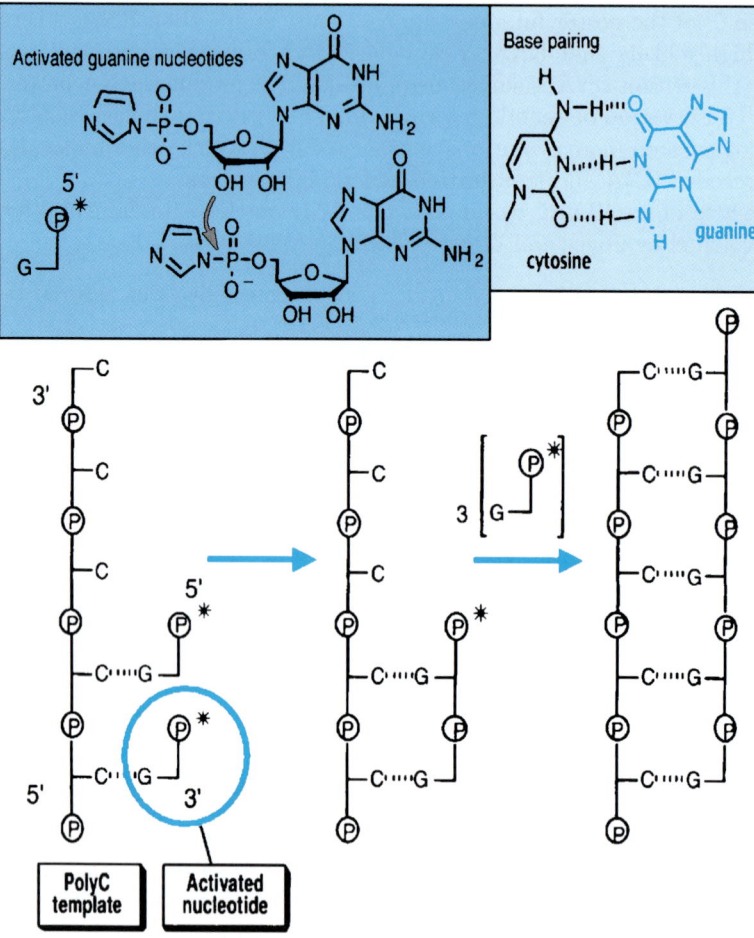

Fig. 6.6. Template-directed RNA synthesis

Finally, when either the first replicative molecule, the template or one of its elements (nucleotides) is to be synthesized from the original building blocks in particular the sugars that are constituents of nucleotides, a number of difficulties are encountered (Sutherland and Whitfield, 1997). Synthesis of the sugars from formaldehyde produces a complex mixture in which ribose is in low amounts. On the other hand, production of a nucleoside from a base and a sugar leads to numerous isomers, and no synthesis of pyrimidine nucleosides has so far been achieved in prebiotic conditions. Finally, phosphorylation of nucleosides also tends to produce complex mixtures (Ferris, 1987). Onset of nucleic acid replication is almost inconceivable if one does not envisage a simpler mechanism for the prebiotic synthesis of nucleotides. Eschenmoser succeeded in producing 2,4-diphosphate ribose during a potentially prebiotic reaction between glycol aldehyde[9] monophosphate and formaldehyde (Eschenmoser, 1999). It is thus possible that direct prebiotic nucleotide synthesis can occur by an alternative chemical pathway. Nevertheless, it is more likely that a certain organized form of chemistry preceded the RNA world, hence the notion of "genetic take-over". Since the ribose-phosphate skeleton is theoretically not indispensable for the transfer of genetic information, it is logical to propose that a simpler replication system would have appeared before the RNA molecule.

6.3 A Pre-RNA World

6.3.1 Evolutive Usurpation

During the evolutionary process, a first genetic inorganic material, would have been replaced by organic material. The hypothesis of a precursor of nucleic acid[10] (Cairns-Smith, 1966, 1982) is a relatively ancient idea, but it is only within the last few years that research has been oriented towards the study of molecules simpler than present-day RNAs, yet capable of autoreplication. Models with predictably retroactive activities can thus be tested experimentally.

6.3.2 Alternative Genetic Systems

In the peptide nucleic acids (PNA) of Nielsen and coworkers, the ribofuranose-phosphate skeleton is replaced by a polyamidic skeleton on which purine and pyrimidine bases are grafted (Fig. 6.7). PNAs form very stable double helices with an RNA or a complementary DNA (Egholm et al., 1993) and can serve as template for the synthesis of RNA, or vice versa (Schmidt et al., 1997). PNA-DNA chimeras containing two types of monomers have been produced on DNA or PNA templates (Koppitz et al., 1998). The information can be transferred from PNAs (achiral monomers) to RNA during directed synthesis; the double-helical molecule with a single complementary RNA strand is stable. Transition from a "PNA world" to an "RNA world" is hence possible. Nevertheless, the formation of oligomers from PNA monomers seems particularly difficult in pre-biotic conditions.

Eschenmoser (1994) explored the properties of nucleic acid analogues in which ribofuranose is replaced by one of its isomers, ribopyranose (Furanose, 5–membered ring; pyranose, 6–membered ring). p-RNAs (pyranosyl RNAs) (Fig. 6.7) form more stable double helices (with Watson–Crick pairings) than RNA with ribofuranose. In addition, the double helices of p-RNA wind and unwind more easily than those formed with standard nucleic acids, and this should facilitate their separation during replication. p-RNAs could therefore constitute good candidates as precursor genetic systems, but a p-RNA strand cannot pair

---

[9] Recently shown to exist in interstellar clouds and comets (Cooper et al., 2001)
[10] This is the idea of genetic take-over developed by Cairns-Smith in the 1960s

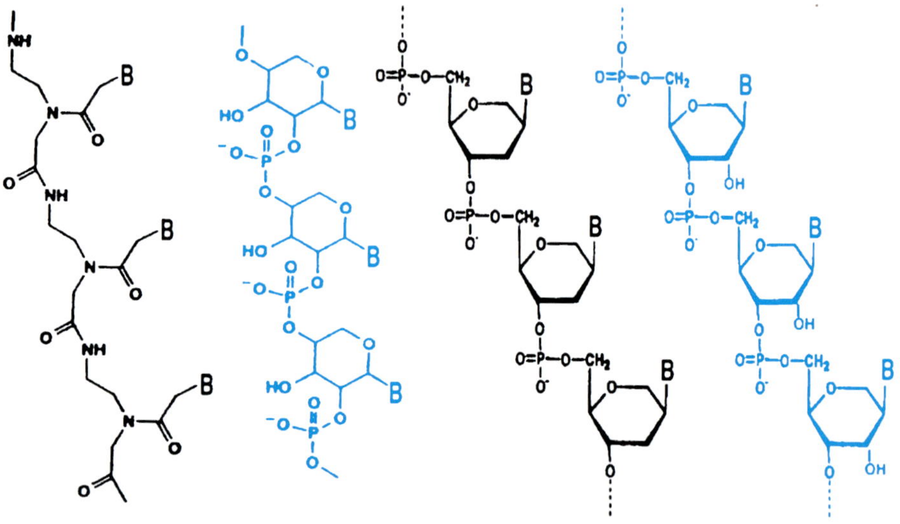

Fig. 6.7. Alternative genetic systems (B = base)

with an RNA of complementary sequence, and this makes it difficult to imagine a transition from p-RNA to RNA.

The group of Eschenmoser recently replaced the ribose moiety by a four-carbon sugar, threose, whose prebiotic synthesis seems easier. The resulting oligonucleotides designated TNAs, $(3' \rightarrow 2')$-α-L-threose nucleic acid (Fig. 6.8), can form a double helix with RNA (Schöning et al., 2000). TNA is capable of antiparallel, Watson–Crick pairing with complementary DNA, RNA and TNA oligonucleotides. Furthermore, Szostak and his collaborators have recently found that certain DNA polymerases can copy limited stretches of a TNA template, despite significant differences in the sugar-phosphate backbone, (Chaput and Szostak, 2003).

Finally, hexitol nucleic acids (HNA) (Fig. 6.7), whose skeleton is composed of 1, 5-anhydrohexitol (six-membered cyclic hexitol) and their isomers altritol nucleic acids (ANA), form stable duplexes with complementary oligonucleotides, and are very efficient templates since they favour assembly of a complementary strand during directed synthesis (Kozlov et al., 1999a, 1999b, 2000). The shape of the duplexes formed is reminiscent of that of DNA in the A form. Double-helical DNA is mainly in the B form[11], whereas the double helices of RNA in the DNA-RNA hybrids adopt the A form[12]. Kozlov et al. (1999c) have demonstrated that

---

[11] d-ribose in the 2'-endo form. The characteristics of these forms are indicated above in the text

[12] In this case, the sugar is in the 3'-endo conformation. In the A form of RNA double helices, there are 11 base pairs per helical turn (instead of 10 for the B form); the inclination of the base pairs is 16°/helical axis (20° for DNA A)

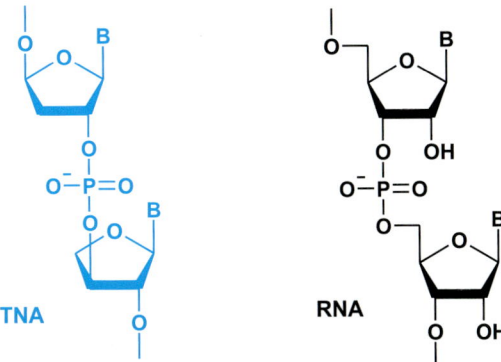

Fig. 6.8. Structure of TNA and RNA

the more the template is in the A form, the better the efficiency of directed synthesis. Based on these studies one can imagine an entire series of templates that would supply the "good" structural preorganization. Furthermore, these same authors have shown that RNA partially preorganized in the A form, is a more efficient matrix than single-stranded DNA. Finally, whatever the precursor skeleton adapted to the formation of stable duplexes may have been, the bond at the mineral surface could have imposed the necessary geometrical constraints: yet this still remains to be experimentally demonstrated.

This leads us to two major conclusions, namely that on the one hand a transition may have occurred between two different systems without loss of information, and that on the other hand the HNA and ANA nucleic acids are very efficient templates. Even if it is difficult to imagine prebiotic synthesis of these molecules, they are good model systems that show the importance of a necessary structural preorganization for directed synthesis by a template.

From the point of view of evolution, the studies described previously demonstrate that other molecules capable of transmitting hereditary information may have preceded our present day nucleic acids. This is what Cairns-Smith coined the "take-over" (Cairns-Smith, 1982), the evolutionary encroachment or genetic take-over, or to some extent what François Jacob (1970) calls genetic tinkering, in other words, making new material from the old. This also sheds light on the precision with which the various elements or processes progressively adjusted themselves, thanks to successive trials and errors.

6.4 Optimizing the Functional Capacities of Ribonucleic Acids

6.4.1 Coenzymes and Modified Nucleosides

The nucleotides that by post-transcriptional modification can today acquire the majority of functional groups present in amino acids, possess a great potential diversity that is expressed at the level of ribonucleotide coenzymes (several coenzymes derive from AMP), and of the modified bases of tRNAs (Fig. 6.9). The role of cofactors at all steps of the metabolism and their distribution within the three kingdoms suggest that a great variety of nucleotides was present in the ancestor common to all forms of life.

Several authors have underscored the possible presence of coenzymes before the appearance of the translation machinery (White, 1976). Proteins would have appeared only at a later stage, coenzymes and ribozymes being fossil traces of past catalysts. Indeed, in the living cell, only a minority of enzymes function without coenzyme; they are mostly hydrolases, and apart from this group, 70% of the enzymes require a coenzyme. If metal coenzymes involved in catalysis are considered, the number of enzymes that depend on coenzymes increases further. Present-day coenzymes, indispensable cofactors for many proteins, would be living fossils of primitive metabolism catalysts.

Most coenzymes are nucleotides (NAD, NADP, FAD, coenzyme A, ATP, etc.) or contain heterocyclic nitrogen bases that can originate from nucleotides (thiamine pyrophosphate, tetrahydrofolate, pyridoxal phosphate, etc.).

Coenzymes would be vestiges of catalytic nucleic enzymes that preceded ribosomal protein synthesis, and tRNAs can be viewed as large coenzymes participating in the transfer of amino acids. It is even possible to consider that catalytic groups that were part of nucleic enzymes were incorporated in specific amino acids rather than being "retained" as coenzymes. This could be the case of imidazole, the functional group of histidine, whose present synthesis in the cell is triggered by a nucleotide.

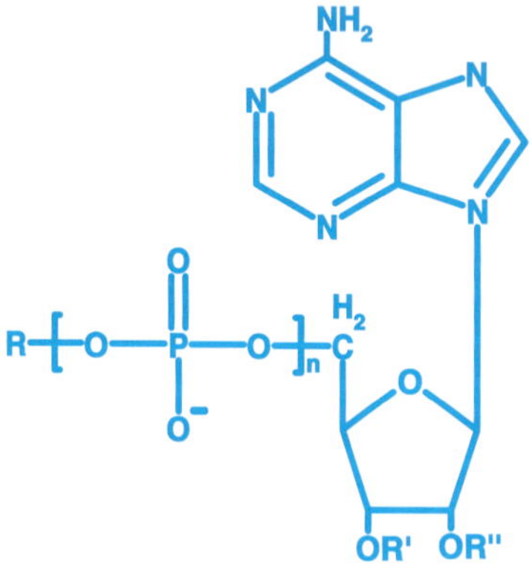

Coenzyme	R	R'	R"	n
Activated methionine	methionine	H	H	0
Amino acid adenylate	amino acid	H	H	1
Activated sulfate	SO_3^{2-}	H	PO_3^{2-}	1
Cyclic 3'-5' AMP	H	H	PO_3^{2-}	1
NAD		H	H	2
NADP		PO_3^{2-}	H	2
FAD		H	H	2
CoA-SH		H	H	2

Fig. 6.9. List of coenzymes derived from AMP

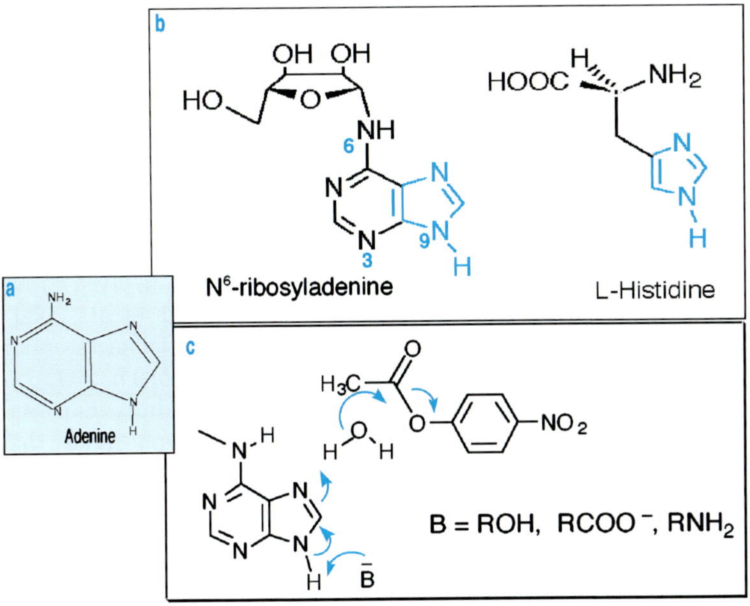

Fig. 6.10. (**a**) Adenine; (**b**) Comparison of modified adenosine and histidine; (**c**) Catalytic activity of adenine residue

The modified nucleosides present today in RNAs result from post-transcriptional modifications. Nevertheless, modified nucleosides could have been present in the primitive world and their distribution would have become established in the RNAs of the three living kingdoms (Cermakian and Cedergren, 1998).

Our working hypothesis is based on the demonstration of esterase activity in a nucleoside analogue N^6-ribosyladenine (Fuller et al., 1972; Maurel and Ninio, 1987). This activity, which is due to the presence of an imidazole group that is free and available for catalysis, is comparable to that of histidine placed in the same conditions (Fig. 6.10). We have studied the kinetic behaviour of this type of catalyst (Ricard et al., 1996) and have shown that the catalytic effect increases greatly when the catalytic element, pseudohistidine, is placed in a favourable environment within a macromolecule (Décout et al., 1995). Moreover, primitive nucleotides were not necessarily restricted to the standard nucleotides encountered today, and because of their replicative and catalytic properties, the N6 and N3 substituted derivatives of purines could have constituted essential links between the nucleic acid world and the protein world.

6.4.2 The Case of Adenine

Purine nucleotides, and in particular those containing adenine, participate in a large variety of cellular biochemical processes (Maurel and Décout, 1999).

Their best-known function is that of monomeric precursors of RNAs and DNAs. Nevertheless, derivatives of adenine are universal cofactors. They serve in biological systems as a source of energy (ATP), allosteric regulators of enzymatic activity and regulation signals (cyclic AMP). They are also found as acceptors during oxidative phosphorylation (ADP), as components of coenzymes (such as in FAD, NAD, NADP, coenzyme A), as transfer agents of methyl groups and of S-adenosylmethionine, as possible precursors of polyprenoids in C5 (adenosylhopane) (Neunlist et al., 1987), and – last but not least – adenine 2451 conserved within the large rRNA in the three kingdoms, would be involved in acid-base catalysis during the formation of the peptide bond (Muth et al., 2000). However, this role of adenine has been refuted based on mutagenesis studies and phylogenetic comparisons (Muth et al. 2001; Green and Lorsch, 2002).

On the other hand, biosynthesis of an amino acid, histidine, that would have appeared late in evolution, begins with 5-phosphoribosyl-1-phosphate that forms N'-(5-phosphoribosyl)-ATP by condensation with ATP. This reaction is akin to the initial reaction of purine biosynthesis. Finally, the ease with which purine bases are formed in prebiotic conditions[13] (Orò, 1960) suggests that these bases were probably essential components of an early genetic system. The first genetic system was probably capable of forming base pairs of the Watson–Crick type, Hoogsteen and other atypical associations, by hydrogen bonds as they still appear today in RNA. It probably contained a different skeleton from that of RNA, and no doubt also modified bases, thereby adding chemical functions, but also hydrophobic groups, and functions such as amine, thiol, imidazole, etc. Wächterhäuser (1988) also suggested novel pairings of the purine–purine type.

Originally, the principle probably rested on forced cooperation of genetic and functional components, rather than on selection by individual competition. It may have first entailed testing and improvements (learning by trial and error) of the informational content of the genes, i.e. linking the genotype (sequence) to the phenotype (shape). One can consider that in such a system the unforeseen was faced, so that the living organism would need to adapt favourably and rapidly.

6.4.3 Mimicking Darwinian Evolution

Most of the "rational" biochemical approaches consist of deducing the active sequence of a nucleic acid or protein from a primary sequence, or in synthesizing a defined compound by modelling and structural analysis. However, "real life", that of our ancestors as also that of our cells, does not proceed in this manner. The hunter-gatherers of prehistory survived only thanks to their extraordinary capacity to recognize objects. In addition, survival of a population in a new environment is often linked to the appearance of a few variants to which random mutations conferred the power to adapt and exploit the new situation to their advantage. Combinatorial methods, by modelling these observations[14], have now

---

[13] Purines have also been found in the Murchison meteorite

[14] And by giving access to many related molecules that can be sorted

become the alternative to the rational concept. Selection in vitro requires no information concerning the sequence of the molecules, and replaces the pre-established adjustments between the molecule and its target. What is needed, is to mimic the processes of evolution at the molecular level.

Indeed, it has been known since the experiments of Spiegelman (1971) and his colleagues (Kramer et al., 1974) that populations of different molecules capable of reproducing themselves in a hereditary manner, can evolve and adapt to an appropriate environment. Spiegelman, the inventor of non-natural selection indeed demonstrated in the 1960s, that RNA populations can evolve when they replicate with the help of an enzyme, the replicase of the bacteriophage $Q\beta$. A population of macromolecules can thus comply with the prerequisites of Darwinian theory, and must find a form adapted to recognition of the target in a sufficiently rich population. Coexistence in the same entity of shape and sequence, can favour the emergence of favourable candidates by means of a selection step (linked to the shape) and an amplification step (linked to the sequence) at the end of this molecular evolutionary process. A selection of this type could have occurred during early molecular evolution, some 3.5 billion years ago.

The original polymers more or less related to RNA and formed in the primitive world must have randomly contained the A, U, G and C bases. There are over one million possible sequences for a decanucleotide composed of 10 monomers A, U, G, C, and over 10^{12} sequences for a polynucleotide of 20 monomers[15]. Nature does not appear to have exploited all the possible combinations before having reached the remarkable functional unity of the living world, and given the immense number of possibilities it is also useless to try to explore experimentally, one by one, all the potentially functional sequences.

The SELEX method (systematic evolution of ligands by exponential enrichment) (Tuerk and Gold, 1990) is an efficient, quasi automatic method based on repeated cycles of reproductive selection of those individuals that are best adapted to a given function. Established in the 1990s, this method makes it possible to obtain new structures, aptamers, selected through their aptitude to recognize other molecules (Ellington and Szostak, 1990). Aptamers are capable to recognize targets as small as metal ions, or as large as cells. They can interact with a great variety of molecules that are important for primitive metabolism, like amino acids, porphyrines, nucleotide factors, coenzymes, small peptides and short oligonucleotides (Illangasekare and Yarus, 1997; Jadhav and Yarus, 2002; Joyce, 2002; McGinness et al., 2002; Reader and Joyce, 2002).

At the molecular level, the Darwinian behaviour requires that a method of selection (RNA-aptamers), of amplification of selected species, and of mutations (introduction of variants in the population by means of mutations) be established. Through several cycles of selection, amplification and mutations,

---

[15] For a nucleic acid of 200 nucleotides, 10^{120} different sequences are theoretically possible, and for a small protein containing 200 amino acids, 10^{280} arrangements are possible! Which also applies to the protein world (phage display and combinatorial synthesis of peptides)

populations of molecules are "pushed" to evolve towards novel properties. The molecules presenting the best "aptitudes" are selected and a new generation will thereby come out. Evolutionary processes performed experimentally thus make it possible for molecules to emerge that have not yet been produced by Nature, or allow the re-emergence of precursor molecules that have strongly diverged or naturally disappeared.

In practice, how does one proceed? A "library" of oligonucleotides is a conformations population containing at least one particular conformation able to regonize the molecular target we are interested in (Fig. 6.11). The protocol is composed of five steps: the creation of double-stranded DNA carrying the random "box"[16] flanked by regions required for amplification; transcription of this DNA into single-stranded RNA; selection; production of a DNA population by reverse transcription and PCR of the sequences retained during the selection step, then cloning and sequencing of the strands obtained after a certain number of selection and amplification cycles.

From a vast combination of nucleic acids, one can isolate aptamers that possess catalytic properties (RNA ligation, cleavage or synthesis of a peptide bond, transfer of an aminoacyl group, etc.). The first nucleic acids could possess independent domains, separated by flexible segments, creating reversible conformational motifs, dependent on ions and bound ligands. Thus, a peptide that is 10 amino-acids long can recognize fine structural differences within a micro-RNA helix (discrimination can be made between two closely placed microhelices). Just as protein and antibodies, RNA molecules can present hollows, cavities, or slits that make these specific molecular recognitions possible. RNAs must "behave as proteins". Whatever the chronology and the order of appearance of the various

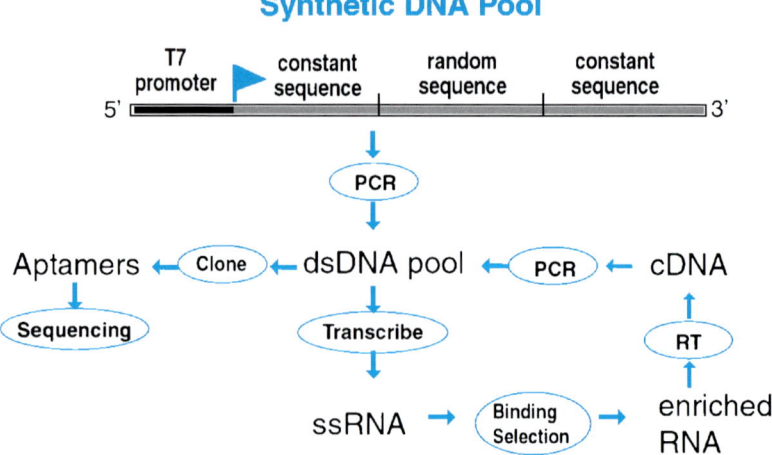

Fig. 6.11. The SELEX method (adapted from Wilson and Szostak, 1999)

---

[16] Region of defined length, for instance, of some 50 randomly aligned nucleotides

classes of molecules, the importance lies in the shape, the scaffolding and the architecture that have allowed functional associations.

Starting from a heterogeneous population of RNAs with 10^{15} variants (a population of 10^{15} different molecules) we have selected 5 populations of RNAs capable of specifically recognizing adenine after about ten generations (Meli et al., 2002). When cloned, sequenced and modelled, the best one among the individuals of these populations, has a shape reminiscent of a claw capable of grasping adenine. Is it the exact copy of a primitive ribo-organism that feeds on prebiotic adenine in prebiotic conditions? Functional and structural studies presently under way will highlight other activities, other conformations, etc.

Following this line of investigation we have selected two adenine-dependent ribozymes capable of triggering reversible cleavage reactions (Fig. 6.12). One of them is also active with imidazole alone. This result leads to very important perspectives (Meli et al., 2003).

A considerable amount of research has been focused on the selection of ribozymes in vitro. Recently, it was demonstrated that a ribozyme is capable of continuous evolution, adding successively up to 3 nucleotides to the initial molecule (McGuinnes, 2002). It is also possible to construct a ribozyme with only two different nucleotides, 2,6-diaminopurine and uracil (Reader and Joyce, 2002). Finally, Bartel and coworkers have selected a ribozyme-polymerase, capable of self-amplification (Johnston et al., 2001).

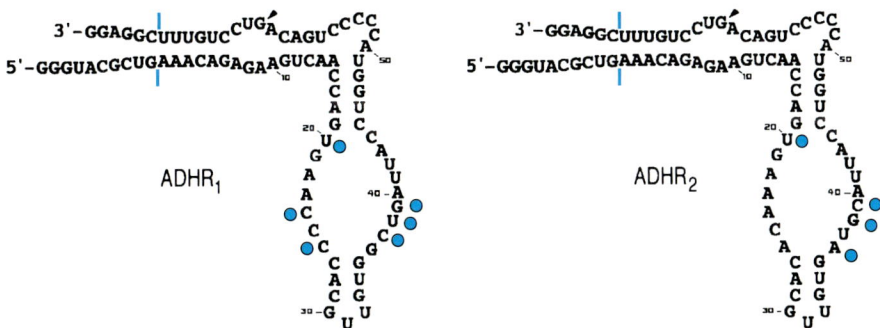

Fig. 6.12. Adenine-dependent hairpin ribozymes (ADHR). *Arrowheads*: cleavage sites; *Grey dots*: degenerated (mutated) sites; *Vertical bars*: separation between the primer binding region and the random sequence

6.4.4 Other Perspectives

Very little is known to date about the behaviour of macromolecules in "extreme" environments. How do structures behave? What are the major modifications observed? What are the conditions of structural and functional stability? How are the dynamics of the macromolecules and their interactions affected? What are the possibilities of conserving biological macromolecules in very ancient soils or in meteorites? Can we find traces of these macromolecules as molecular biosignatures, and if so in what form (Maurel and Zaccaï, 2001; Tehei et al., 2002)?

The selection of thermohalophilic aptamers, RNAs resistant to high temperatures (80°C) in the presence of salt (halites that are 30 million years old), undertaken in our laboratory, will perhaps allow us to answer some of these questions, that are fundamental for the search of past traces of life, and of life on other planets.

6.5 Conclusion

The RNA world thus contains innumerable perspectives. The combination of methods available today are the best adapted to explore the vast combinations of nucleic acids but also of peptides. Will they make it possible to reconstitute the first steps of the living world? Attractive simulations may emerge, opening new evolutionary paths that have not been envisaged or that Nature has not yet explored.

The RNA world, at whatever step we place it in the history of the living world, must be considered as a step in the history of life, an important step in the evolution of the contemporary cellular world. Because of its strong explanatory power, it also constitutes an important opening in the scientific study of the origin of life. Even if this concept does not explain how life appeared, it nevertheless promises a great number of experimental breakthroughs.

Acknowledgement

Figure 6.4 is reprinted from *Cell*, 2003,112, Barends S., Bink H.H.J., van den Worm S.H.E., Pleij C.W.A., Kraal B. Entrapping ribosomes for viral translation: tRNA mimicry as a molecular trojan horse. Copyright 2003, with permission from Elsevier. We also thank Dr. G.F. Joyce for his constructive comments on the manuscript.

References

Bachellerie J.P., Cavaillé J., Hüttenhofer A. (2002). The expanding snoRNA world. *Biochimie*, **84**, 775–790.
Ban N., Nissen P., Hansen J., Moore P.B., Steitz, T.A. (2000). The complete atomic structure of the large ribosomal subunit at 2.4 Å resolution. *Science*, **289**, 905–920.
Barends S., Bink H.H.J., van den Worm S.H.E., Pleij C.W.A., Kraal B. (2003). Entrapping ribosomes for viral translation: tRNA mimicry as a molecular trojan horse. *Cell*, **112**, 123–129.
Bartel D.P., Unrau P.J. (1999). Constructing an RNA world. *Trends Biochem. Sci.*, **24**, 9–13.

Benner S.A., Ellington A.D., Tauer A. (1989). Modern metabolism as a palimpsest of the RNA world. *Proc. Natl. Acad. Sci. USA*, **86**, 7054–7058.

Benner S.A., Cohen M.A., Gonnet G.H., Berkowitz D.B., Johnson K.P. (1993). Reading the palimpsest: contemporary biochemical data and the RNA World, in *The RNA World*, eds. Gesteland R.F., Atkins J.F., p. 27–70, Cold Spring Harbor Laboratory Press: Cold Spring Harbor NY.

Cairns-Smith A.G. (1966). The origin of life and the nature of the primitive gene. *J. Theor. Biol.*, **10**, 53–88.

Cairns-Smith A.G. (1982). *Genetic Takeover and the Mineral Origins of Life*, Cambridge University Press, Cambridge.

Cermakian N., Cedergren, R. (1998). Modified nucleosides always were: an evolutionary model, in *Modification and Editing of RNA*, ed. Grosjean H., Benne R., p. 535–541, ASM Press, Washington, D.C.

Chaput J.C., Szostak J.W. (2003). TNA synthesis by DNA polymerase. *J. Am. Chem. Soc.*, **125**, 9274–9275.

Cooper G., Kimmich N., Belisle W., Sarinana J., Brabham K., Garrel L. (2001). Sugar-related organic compounds in carbonaceous meteorites. *Nature*, **414**, 879–883.

Crick F.H. (1968). The origin of the genetic code. *J. Mol. Biol.*, **38**, 367–379.

Décout J-L., Vergne J., Maurel M-C. (1995). Synthesis and catalytic activity of adenine containing polyamines. *Macromol. Chem. Phys.*, **196**, 2615–2624.

Diener TO. (2001). The viroid: biological oddity or evolutionary fossil? *Adv. Virus. Res.* **57**, 137–184.

Eddy S.R. (2001). Non-coding RNA genes and the modern RNA world. *Nature Reviews Genetic*, **2**, 919–929.

Egholm M., Buchardt O., Christensen L., Behrens C., Freier S.M., Driver D.A., Berg R.H., Kim S.K., Norden B., Nielsen P.E. (1993). PNA hybridizes to complementary oligonucleotides obeying the Watson–Crick hydrogen-bonding rules. *Nature*, **365**, 566–568.

Ellington A.D., Szostak J.W. (1990). In vitro selection of RNA molecules that bind specific ligands. *Nature*, **346**, 818–822.

Eschenmoser A. (1994). Chemistry of potentially prebiological natural products. *Origins Life Evol. Biosphere*, **24**, 389–423.

Eschenmoser A. (1999). Chemical etiology of nucleic acid structure. *Science*, **284**, 2118–2124.

Fechter P., Rudonger-Thirion J., Florentz C., Giegé R. (2001). Novel features in the tRNA-like world of plant viral RNAs. *Cell. Mol. Life. Sci.* **58**, 1547–1561.

Ferris J.P. (1987). Prebiotic synthesis: problems and challenges *Cold Spring Harbor Symp. Quant. Biol.*, **LII**, 29–39.

Ferris J.P., Ertem G. (1992). Oligomerization of ribonucleotides on montmorillonite: reaction of the 5′ phosphorimidazolide of adenosine. *Science*, **257**, 1387–1389.

Ferris J.P., Hill A.R., Liu R., Orgel L.E. (1996). Synthesis of long prebiotic oligomers on mineral surfaces. *Nature*, **381**, 59–61.

Fuller W.D., Sanchez R.A., Orgel L.E. (1972). Studies in prebiotic synthesis. *J. Mol. Biol.*, **67**, 25–33.

Gesteland R.F., Cech T.R., Atkins J.F. (eds.) (1999). *The RNA World*, second edition, Cold Spring Harbor Laboratory Press, Cold Spring Harbor, NY.

Gilbert W. (1986). The RNA world. *Nature*, **319**, 618.

Green R., Lorsch J.R. (2002). The path to perdition is paved with protons. *Cell*, **110**, 665–668.
Grosjean H., Benne R. (eds.) (1998). *Modification and Editing of RNA*, ASM Press, Washington, D.C.
Grosshans H., Slack F.J. (2002). Micro-RNAs: small is plentiful. *J. Cell Biol.*, **156**, 17–21.
Guerrier-Takada C., Gardiner K., Marsh T., Pace N., Altman S. (1983). The RNA moiety of ribonuclease P is the catalytic subunit of the enzyme. *Cell*, **35**, 849–857.
Hill A.R., Orgel L.E., Wu T. (1993). The limits of template-directed synthesis with nucleoside-5'-phosphoro (2-methyl) imidazolides. *Orig. Life Evol. Biosphere*, **23**, 285–290.
Illangasekare M., Yarus M. (1997). Small-molecule substrate interactions with a self aminoacylating ribozyme. *J. Mol. Evol.*, **54**, 298–311.
Inoue T., Orgel L.E. (1983). A non-enzymatic RNA polymerase model. *Science*, **219**, 859–862.
Jacob F. (1970). *La Logique du Vivant*, Gallimard, Paris.
Jadahv V.R., Yarus M. (2002). Coenzymes as coribozymes. *Biochimie*, **84**, 877–888.
Johnston W.K., Unrau P.J., Lawrence M.S., Glasner M.E., Bartel D.P. (2001). RNA-catalyzed RNA polymerization: Accurate and general RNA-templated primer extension. *Science*, **292**, 1319–1325.
Joyce G.F., Orgel L.E. (1986). Non-enzymic template-directed synthesis on RNA random copolymers: poly (C,G) templates. *J. Mol. Biol.*, **188**, 433–441.
Joyce G.F., Schwartz A.W., Miller S.L., Orgel L.E. (1987). The case for an ancestral genetic system involving simple analogues of the nucleotides. *Proc. Natl. Acad. Sci. USA*, **84**, 4398–4402.
Joyce G.F. (1989). RNA evolution and the origins of life. *Nature*, **338**, 217–224.
Joyce G.F., Orgel L.E. (1999). Prospects for understanding the origin of the RNA world, in *The RNA World*, eds. Gesteland R.F., Cech T.R., Atkins J.F. p. 49–77, Cold Spring Harbor Laboratory Press: Cold Spring Harbor, NY.
Joyce G.F. (2002). The antiquity of RNA-based evolution. *Nature*, **418**, 214–221.
Kable M.L., Heidmann S., Stuart K.D. (1997). RNA editing: getting U into RNA. *Trends Biochem. Sci.*, **22**, 162–166.
Kong L.B., Siva A.C., Kickhoefer V.A., Rome L.H., Stewart P.L. (2000). RNA location and modeling of a WD40 repeat domain within the vault. *RNA*, **6**, 890–900.
Koppitz M., Nielsen P.E., Orgel L.E. (1998). Formation of oligonucleotide-PNA-chimeras by template-directed ligation. *J. Am. Chem. Soc.*, **120**, 4563–4569.
Kozlov I., Politis P.K., Pitsch S., Herdewijn P., Orgel L.E. (1999a). A highly enantioselective hexitol nucleic acid template for nonenzymatic oligoguanylate synthesis. *J. Am. Chem. Soc.*, **121**, 1108–1109.
Kozlov I., De Bouvere B., Van Aerschot A. Herdewijn P., Orgel L.E. (1999b). Efficient transfer of information from hexitol nucleic acids to RNA during nonenzymatic oligomerization. *J. Am. Chem. Soc.*, **121**, 5856–5859.
Kozlov I., Politis P.K., Van Aerschot A. Busson R., Herdewijn P., Orgel L.E. (1999c). Nonenzymatic synthesis of RNA and DNA oligomers on hexitol nucleic acid templates: the importance of A structure. *J. Am. Chem. Soc.*, **121**, 2653–2656.
Kozlov I., Zielinski M., Allart B., Kerremans L., Van Aerschot A, Busson R., Herdewijn P., Orgel L.E. (2000). Nonenzymatic template-directed reactions on altritol oligomers, preorganized analogues of oligonucleotides. *Chem. Eur. J.*, **6**, 151–155.

Kramer F.R., Mills D.R., Cole P.E., Nishihara T., Spiegelman S. (1974). *J. Mol. Biol.*, **89**, 719–736.

Lafontaine D.L., Tollervey D. (1998). Birth of the snoRNPs: the evolution of the modification-guide snoRNAs. *Trends Biochem. Sci.*, **83**, 383–388.

Lamond A.I. (1988). RNA editing and the mysterious undecovered genes of trypanosomatid mitochondria. *Trends Biochem. Sci.*, **13**, 283–284.

Maizels N., Weiner A.M., Yue D., Shi P.Y. (1999). New evidence for the genomic tag hypothesis: archaeal CCA-adding enzymes and DNA substrates. *Biol. Bull.*, **196**, 331–333.

Mattick J.S. (2003). Challenging the dogma: the hidden layer of non-protein-coding RNAs in complex organisms. *BioEssays*, **25**, 930–939.

Maurel M.-C., Ninio J. (1987). Catalysis by a prebiotic nucleotide analog of histidine. *Biochimie*, **69**, 551–553.

Maurel M.-C. (1992). RNA in evolution. *J. Evol. Biol.*, **2**, 173–188.

Maurel M.-C., Décout J.-L. (1999). Origins of life: molecular foundations and new approaches. *Tetrahedron*, **55**, 3141–3182.

Maurel M.-C., Zaccaï G. (2001). Why Biologists should support the exploration of Mars. *BioEssays*, **23**, 977–978.

McGinness K.E., Wright M.C., Joyce G.F. (2002). Continuous in vitro evolution of a ribozyme that catalyzes three successive nucleotidyl addition reactions. *Chem. Biol.*, **9**, 585–596.

Meli M., Albert-Fournier B., Maurel M.-C. (2001). Recent findings in the modern RNA world. *Int. Microbiol.* **4**, 5–11.

Meli M., Vergne J., Décout J.-L., Maurel M.-C. (2002). Adenine-aptamer complexes. A bipartite RNA site which binds the adenine nucleic base. *J. Biol. Chem.*, **277**, 2104–2111.

Meli M., Vergne J., Maurel M.-C. (2003). *In vitro* selection of adenine-dependent hairpin ribozymes. *J. Biol. Chem.*, **278**, 9835–9842.

Muth G.W., Ortoleva-Donnelly L., Strobel S.A. (2000). A single adenosine with a neutral pKa in the ribosomal peptidyl transferase center. *Science*, **289**, 947–950.

Muth G.W., Chen L., Kosek A.B., Strobel S.A. (2001). PH-dependent conformational flexibility within the ribosomal peptidyl transferase center. *RNA*, **7**, 1403–1415.

Neunlist S., Bisseret P., Rohmer M. (1987). The hopanoids of the purple non-sulfur bacteria *Rhodopseudomonas palustris* and *Rhodopseudomonas acidophila* and the absolute configuration of bacteriohopanetetrol. *Eur. J. Biochem.*, **87**, 245–252.

Nissen P., Hansen J., Ban N., Moore P.B., Steitz T.A. (2000). The structural basis of ribosome activity in peptide bond synthesis. *Science*, **289**, 920–930.

Nykanen A., Haley B., Zamore P.D. (2001). ATP requirements and small interfering RNA structure in the RNA interference pathway. *Cell*, **107**, 309–321.

Orgel L.E. (1968). Evolution of the genetic apparatus. *J. Mol. Biol.*, **38**, 381–393.

Orgel L.E. (1989). Was RNA the first genetic polymer? in *Evolutionary Tinkering in Gene Expression* eds. M. Grunberg-Manago et al., p. 215–224, Plenum Press, London.

Orgel L.E. (1992). Molecular replication. *Nature*, **358**, 203–209.

Orò J. (1960). *Biochem. Biophys. Res. Comm.*, **2**, 407–412.

Paecht-Horowitz M., Berger J., Katchalsky A. (1970). Prebiotic synthesis of polypeptides by heterogeneous polycondensation of amino acid adenylates. *Nature*, **7**, 847–850.

Reader J.S., Joyce G.F. (2002). A ribozyme composed of only two different nucleotides. *Nature*, **420**, 841–844.

Ricard J., Vergne J., Décout J.-L., Maurel M.-C. (1996). The origin of kinetic cooperativity in prebiotic catalysts. *J. Mol. Evol.*, **43**, 315–325.

Schmidt J.G., Nielsen P.E., Orgel L.E. (1997). Information transfer from peptide nucleic acid RNA by template-directed syntheses. *Nucl. Acids Res.*, **25**, 4797–4802.

Schöning K.-U., Scholz P., Guntha S., Wu X., Krishnamurthy R., Eschenmoser A. (2000). Chemical etiology of nucleic acid structure: the α-threo-furanosyl-(3'→2') oligonucleotide system. *Science*, **290**, 1347–1351.

Spiegelman S. (1971). An in vitro analysis of a replicating molecule. *Quaterly Review Biophys.*, **4**, 213–253.

Stuart K., Panigrahi A.K. (2002). RNA editing: complexity and complications. *Mol. Microbiol.*, **45**, 591–596.

Sutherland J.D., Whitfield J.N. (1997). Prebiotic chemistry: a bioorganic perspective. *Tetrahedron*, **53**, 11493–11527.

Tarn W.Y., Steitz J.A. (1997). Pre-mRNA splicing: the discovery of a new spliceosome doubles the challenge. *Trends Biochem. Sci.*, **22**, 132–137.

Tehei M., Franzetti B., Maurel M.-C., Vergne J., Hountondji C., Zaccaï G. (2002). Salt and the search for traces of life. *Extremophiles*, **6**, 427–430.

Terns M.P., Terns R.M. (2002). Small nucleolar RNAs: versatile transacting molecules of ancient evolutionary origin. *Gene Expr.*, **10**, 17–39.

Tuerk, C., Gold, L. (1990). Systematic evolution of ligands by exponential enrichment: RNA ligands to bacteriophage T4 DNA polymerase. *Science*, **249**, 505–510.

Valle M., Gillet R., Kaur S., Henne A., Ramakrishnan V., Frank J. (2003). Visualizing tmRNA entry into a stalled ribosome. *Science*, **300**, 127–130.

Wächtershäuser G. (1988). An all-purine precursor of nucleic acids. *Proc. Natl. Acad. Sci. USA*, **85**, 1134–1135.

Westhof E. (2002). Foreword. *Biochimie*, **84**, 687–689.

White H.B. (1976). Coenzymes as fossils of an earlier metabolic state. *J. Mol. Evol.*, **7**, 101–104.

Wild K., Weichenrieder O., Strub K., Sinning I., Cusack S. (2002). Towards the structure of the mammalian signal recognition particle. *Curr. Opinion Struct. Biol.*, **12**, 72–81.

Wilson D.S., Szostak J.W. (1999). In vitro selection of functional nucleic acids. *Annu. Rev. Biochem.*, **68**, 611–647.

Withey J.H., Friedman D.I. (2002). The biological roles of trans-translation. *Curr. Opinion Microbiol.*, **5**, 154–159.

Woese C.R. (1965). On the evolution of the genetic code. *Proc. Natl. Acad. Sci. USA*, **54**, 1546–1552.

Yusupov M.M., Yusupova G.Z., Baucom A., Lieberman K., Earnest T.N., Cate J.H., Noller H.F. (2001). Crystal Structure of the Ribosome at 5.5 Å resolution. *Science*, **292**, 883–896.

Zamore P.D. (2002). Ancient Pathways Programmed By Small RNAs. *Science*, **296**, 1265–1269.

7 Looking for the Most 'Primitive' Life Forms: Pitfalls and Progresses

Simonetta Gribaldo, Patrick Forterre

Two complementary approaches can be applied to try to understand how life appeared on our planet. The first, illustrated by Miller's seminal experiment, consists in trying to reproduce – either theoretically or/and practically – the conditions of primitive Earth, and to figure out which possible scenarios might have led from such conditions to the apparition of life as we know it today. This is the type of approach frequently adopted by chemists and physicists. The second approach, favoured by biologists, consists in looking at current organisms for hints of their most remote history. In this chapter we will mostly discuss this second approach.

Clearly, no barrier exists between researchers who look at the past to understand present time, and those who do the opposite. In fact, in order to reach a coherent view of the transition from unanimated matter to life, the scenarios suggested by both approaches must eventually converge. It is sensible to keep in mind this dichotomy of approaches, in order to be able to reach, in the end, their complementation. In fact, such a dichotomy raises a fundamental question that is rarely taken into consideration, i.e. at what time, over the history of life evolution on our planet, should the Darwinian notion of natural selection be called into play? In other words, when did chemistry give way to biology? The answer to this question is crucial to estimate the probability of apparition of life on other planets, in a starting environment compatible with life development.

In this chapter we will mostly focus on the quest, in the present-day biosphere, of "relic" cells that would have conserved the features harboured by the last common ancestor of all extant life forms, i.e. the LUCA (Last Universal Common Ancestor; Forterre, 1996a). Over the last several years, a few authors have suggested that life on our planet might have arisen at high temperatures and that LUCA was itself a hyperthermophile (i.e., thriving at temperatures close to the boiling point of water). These two hypotheses (a hot origin of life and a hot-loving LUCA) have become very popular amongst exobiologists. We will thus dedicate a large part of this chapter to discuss this issue. Moreover, given the current revolution in biology due to the sequencing of complete genomes from a growing number of diverse organisms, we will also focus on the most recent advances of comparative genomics, and whether they can disclose any detail on the nature of LUCA.

7.1 Simpler Doesn't Necessarily Mean Older!

One of the most traditional approaches of biologists interested in life origin issues is to search for the simplest extant cellular life forms, with the assumption that their features may date back to the first cells that appeared on our planet. This would make it easier to trace back the steps that drove to these first cells. In such a quest, simplicity and small size are often considered as ancestral characters (see, for example, the interest given to the supposedly primitive martian "nanobacteria"). However, this type of reasoning is risky and has already resulted in erroneous judgements; in fact, what is simple is not necessarily primitive, but can be the product of secondary simplification. For example, mycoplasma have been historically considered as candidates for primitive cells, since they are the smallest extant Bacteria. However, we now know that they are indeed ancient – and "complex" – Bacteria that have lost their cell wall and most of their biosynthesis abilities, due to adaptation to a parasitic lifestyle (Fig. 7.1). Nonetheless, mycoplasma still represent precious model organisms to both experimental and theoretical research on minimal genomes. Indeed, the smallest mycoplasma genome harbours less than 500 genes for a total of 500 000 pairs of nucleotides (0.5 mega bases, or MB), while the size of classical bacterial

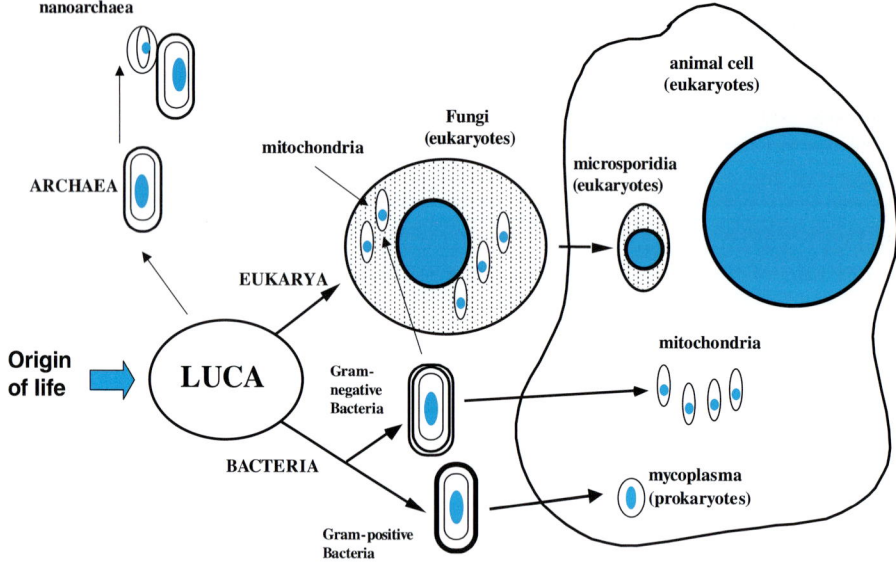

Fig. 7.1. Evolutionary mechanisms based on simplification and size reduction have played a major role in life's history. This sketch shows the origin of mycoplasma from Gram-positive Bacteria, the origin of mitochondria from Gram-negative Bacteria (note the loss of cell wall in both cases), and the origin of microsporidia from Fungi following mitochondrial loss. In all these cases, a reduction of the size of their genomes (*blue*) has occurred

genomes ranges from 1.5 to 10 MB. Recently, the interest in minimal genomes was renewed by the description of a nanosized hyperthermophilic archaeon from a submarine hot vent (Huber, 2002). This is the smallest archaeal genome – 0.5 MB – identified to date. Here again we are most likely in front of the product of a secondary simplification since this species leads a parasitic life onto the surface of a specific "normal-sized" archaeal host (Fig. 7.1). The same conclusions are also valid for eukaryal minimal genomes. In fact, it was recently shown that microsporidia, unicellular amitochondriate eukaryal parasites of animal cells, are not primitive Eukarya, as early molecular phylogenetic analyses seemed to indicate, but secondarily simplified fungi (Fig. 7.1) (Vivares, 2000). The smallest known microsporidial genome, sequenced at Génoscope, is of 2.9 MB, similarly to a medium-sized bacterium.

In conclusion, although the first cells were certainly very simple, the simplest extant cellular forms are instead their extremely evolved and specialized descendants; this hampers any attempt to reconstruct the past by starting from the present, and all the more any extrapolation of our knowledge on the present-day terrestrial biosphere to possible extraterrestrial ones.

7.2 Hyperthermophiles are not Primitives, but are Remnants from Thermophilic Organisms

7.2.1 Hyperthermophiles and the Hypothesis of a Hot Origin of Life

The hypothesis that life appeared in extremely hot environments (submarine hydrothermal vents) or directly within the Earth's mantle, either within or nearby oceanic ridges (interfaces between cold-fluid infiltrations and gas produced at their contact with the magma) is appealing to most exobiologists. Indeed, under such a hypothesis, the numbers of either solar or extra-solar planets that may have hosted life would rise enormously, given that initial atmosphere conditions and distance from a star no longer come into play. Moreover, if life arose in such conditions (i.e., far from Earth's surface) its first evolution steps would have been protected from the cataclysmic meteoritic bombardment that took place at the same time as planet formation. Consequently, hyperthermophiles (hot-loving micro-organisms thriving at temperatures from 80° to 113°C Forterre, 1999b) now replace mycoplasma as candidates for "primitive cells" in the majority of generic publications on life origins and textbooks (Madigan, 2000). A number of authors, such as Carl Woese and Norman Pace in the USA, and Carl Stetter and Gunther Wachtershäuser in Europe, have indeed supported the hypothesis of a hot origin of life up to LUCA (Pace 1991). According to these authors, LUCA would have appeared and lived on a still-heated Earth, and extant hyperthermophiles would be its direct descendants (see Ciba Foundation publications, 1996; and Wiegel, 1998). Indeed, terrestrial and submarine volcanic biotopes where microbiologists search for hyperthermophiles are environments traditionally associated to primitive Earth in the collective imaginary.

7.2.2 Hyperthermophiles are Complex Prokaryotes

The evidence that all extant hyperthermophiles are prokaryotes (cells without a nucleus) is a priori in agreement with a hyperthermophilic LUCA, since prokaryotes are generally considered as more representative of primitive cells with respect to Eukarya (we will get back on this important issue later). Indeed, most hyperthermophiles are Archaea (Fig. 7.2) – a definition evoking presumed ancient origins and an archaic nature.

However, molecular analyses have failed to highlight any primitive character in hyperthermophiles, but rather the opposite (Forterre, 1996b). For example, life at high temperatures needs specific lipids able to keep the cytoplasmic membrane impermeable to protons; these are giant lipids made up of two "classic" phospholipids covalently linked (tail-to-tail) by a sophisticated enzymatic system (see Sect. 7.2.3) (Alberts, 2000). Moreover, ribosomal and transfer RNAs from hyperthermophiles harbour modifications that increase their stability at high temperatures (Noon, 1998). Such modifications (frequently the addition of methyl groups) allow a larger range of hydrophobic interactions involved in RNA structuring that may protect RNA molecules from thermal degradation (see Sect. 7.2.4).

Giant lipids and RNA modifications do exist also in mesophilic (thriving at temperatures between 20–40°C and psycrophilic (cold-loving) organisms (Shouten, 2000). Instead, an exclusivity of hyperthermophiles is represented by reverse gyrase, an enzyme that introduces positive supercoiling in DNA molecules (Fig. 7.3) (Forterre, 1995a). In supercoiled DNA helices, the axis of the Watson and Crick double helix folds on itself in the same sense as the double helix (which is conventionally assumed to be positive). This positive supercoiling increases the number of topological links between the two strands of the DNA molecule, making more difficult the local opening-up of the helix, and consequently preventing the risk of denaturation that might occur at temperatures of 80°C and higher. Although the essential role of reverse gyrase at high temperature remains to be rigorously tested, this is strongly suggested by the systematic presence of this enzyme in hyperthermophiles. Indeed, comparative genomic analyses have shown that reverse gyrase is the only protein

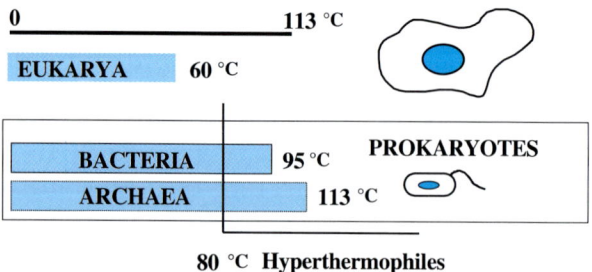

Fig. 7.2. Upper limits of life on Earth for the three major groups of cellular organisms

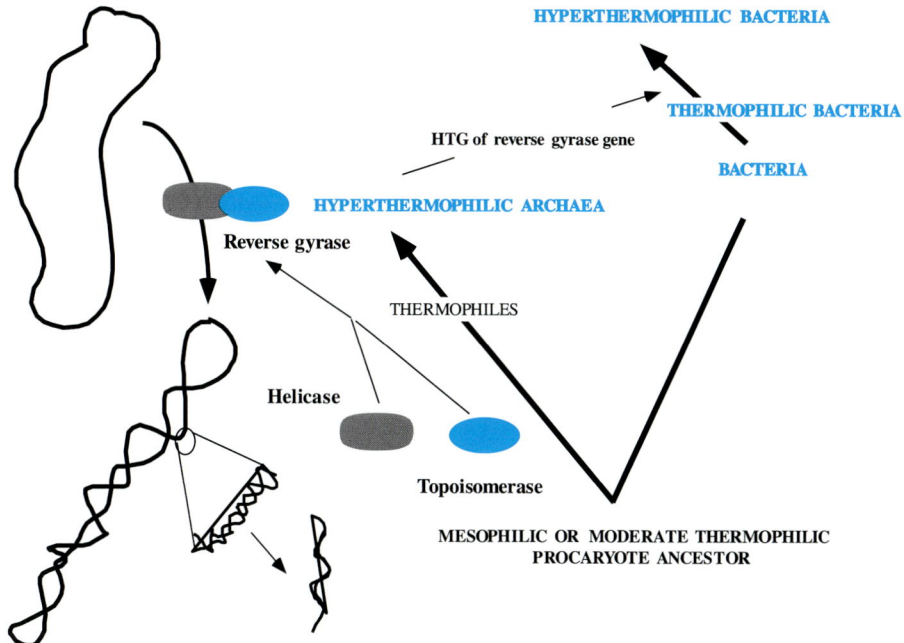

Fig. 7.3. A character specific to hyperthermophiles: the reverse gyrase. This enzyme, made up by the fusion of two 'classic' proteins, a helicase and a topoisomerase, transforms relaxed DNA into positive supercoiled DNA by the use of ATP energy. This enzyme might help to prevent denaturation of the double-helix structure at high temperature

displaying such a distribution (Forterre, 2002a), with one exception (see below). Molecular analyses have shown that reverse gyrase is a relatively modern enzyme, i.e., it appeared quite recently in evolution, by the fusion of the genes coding for two different proteins (a DNA topoisomerase and a DNA helicase) (Fig. 7.3), each belonging to a protein family already specialised at the time of the fusion (Confalonieri, 1993). Thus, if reverse gyrase is crucial to the functioning of DNA at temperatures higher than 80°C, hyperthermophiles would have had a late emergence, subsequent to the "invention" of this enzyme (Forterre, 1995a). Reverse gyrase likely appeared in a thermophilic organism thriving between 70 and 80°C, and whose proteins were thus already adapted to high temperatures. This is consistent with the recent evidence of a gene coding for reverse gyrase in the genome of *Thermoanaerobacter tencongensis*, a thermophilic bacterium. A late emergence of hyperthermophiles would exclude a direct link between extant hot-loving organisms and a hot origin of life. However, the possibility that life arose in warm environments (50–80°C) remains a possible scenario.

7.2.3 Origin of Hyperthermophily

As far as we know, the apparition of reverse gyrase in all bacterial and archaeal hyperthermophiles seems to have been at the origin of adaptation to hot environments. Molecular phylogenies have allowed retracing the evolution of this protein (Forterre, 2000b). Reverse gyrase most likely appeared in Archaea and was later transferred to some Bacteria by independent lateral gene transfers from different archaeal sources (Fig. 7.3). Indeed, comparative genomics has shown that such transfers are very frequent both between and within the two prokaryotic groups, and that they boost adaptation to new biotopes. In hyperthermophilic Bacteria such as *Thermotoga maritima* (upper growth temperature 96°C) and *Aquifex aeolicus* (upper growth temperature 95°C) a large fraction of genes are indeed much more similar to their archaeal counterparts rather than to their bacterial ones (Aravind, 1998; Nelson, 1999), indicating that they were likely acquired from Archaea by horizontal transfer. This evidence, together with the fact that reverse gyrase in hyperthermophilic Bacteria is likely of archaeal origin, suggests that Archaea might have been the first organisms to overcome the 80°C barrier and that they transferred this ability to Bacteria. Nonetheless, Bacteria did not succeed in adapting to temperatures as high as those of Archaea (maximal growing temperature 113°C) (Fig. 7.2).

Indeed, all extant Archaea might have a hyperthermophilic origin. This hypothesis is suggested by a few lines of evidence. For instance, the unique structure of archaeal glycerolipids indicates that they possibly arose as an adaptation for survival at high temperatures. Archaeal glygerolipids are glycerol and phytanol di- or tetraethers (highly hydrophobic long isoprenoid chains). In hyperthermophilic Archaea such tetraethers form a monolayer at the level of the cytoplasm membrane (Fig. 7.4). Tetraether lipids are particularly adapted to extremely hot environments, since ether links are very stable and the formation of monolayers prevents high-temperature-driven proton leakage through the membrane, a key feature to maintain a ATP-producing proton gradients. Tetraether lipids are also found in some mesophilic and psycrophilic Archaea, suggesting that they can maintain membrane fluidity over a large temperature spectrum. This type of lipids might thus have appeared in mesophiles and permitted adaptation to hyperthermophily. Alternatively, they might be an ancestral character dating back to a hyperthermophilic archaeal ancestor. This latter hypothesis, i.e. a hyperthermophilic origin of Archaea, is supported by rRNA-based phylogenies, where hyperthermophilic lineages emerge as the most basal offshoots (reviewed in Forterre, 2002b). A recent archaeal phylogeny based on ribosomal proteins also showed a similar basal placement of hyperthermophilic species (Matte-Taillez, 2002). By contrast, hyperthermophily in Bacteria appears to be a secondary adaptation. Bacterial glycerolipids, such as eukaryal ones, are generally glycerol and fatty acid esters. However, those of hyperthermophilic Bacteria tend to mimic the archaeal ones. Some of them are based on ether links that can form monolayer structures

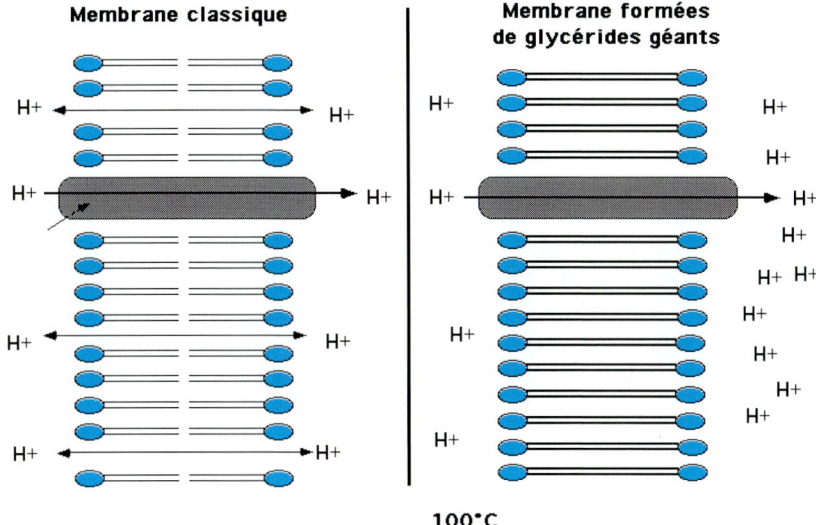

Fig. 7.4. Proton permeability of cell membranes at high temperature (100 °C). Lipid polar heads are drawn in *blue*. The protein pump creates a proton gradient that allows the production of cellular energy. At 100 °C this gradient cannot be created since bilayer classic membranes become permeable to protons at this temperature (**A**). In hyperthermophiles, the nonpermeability of the membrane to protons at high temperature is granted by a monolayer glycerid structure and/or by higher hydrophobicity of lipid tails (**B**)

(Schouten, 2000), while some others are based on tetraesters, as suggested by the identification in *Thermotoga maritima* of long fatty acid chains harbouring an acid function at each end ("diabolic acids"). Despite such archaeal-like features, hyperthermophilic bacterial glycerolipids remain strongly bacterial in essence. In fact, they do not harbour any isoprenic chain, and the absolute configuration of the asymetric carbon atom of the glycerolipids are different in mesophilic Bacteria and Eukarya, on the one hand, in Archea, one the other hand. Significantly, lipids from hyperthermophilic Bacteria are different between lineages. This strongly suggests an independent adaptation to hyperthermophily in different bacterial lineages, consistently with the fact that they likely recruited their reverse gyrases from different archaeal sources (Forterre, 2000a).

7.2.4 LUCA was Probably not a Hyperthermophile

Adaptation to temperatures higher than 80 °C is thus a character that likely appeared in Archaea and was punctually transferred to some Bacteria. If the root of the life tree lies in the bacterial branch, this implies a nonhyperthermophilic LUCA. Indeed, this is consistent with a simulation analysis that reconstructed the hypothetical ancestral ribosomal RNA of LUCA and obtained a molecule whose content of guanine and cytosine (G-C content) is characteristic of extant organisms thriving at temperatures lower than 70°C (Galtier, 1999). Ribosomal RNAs of hyperthermophiles are in fact very rich in G-C, a feature that stabilises their numerous minihelixes since three hydrogen bonds exist between G-C base pairs and only two between A-U base pairs (Fig. 7.5). Since the reconstructed ancestral ribosomal RNA is not very G-C rich, LUCA may have rather been a mesophilic or moderately thermophilic organism. As opposed to their RNA, it should be noted that the genomic DNA of hyperthermophiles is not especially G-C rich. This is not surprising, given that cellular DNA is "topologically closed", i.e. its two strands cannot rotate freely one over the other; such a DNA is stable up to a temperature of at least 107 °C (Marguet, 1994).

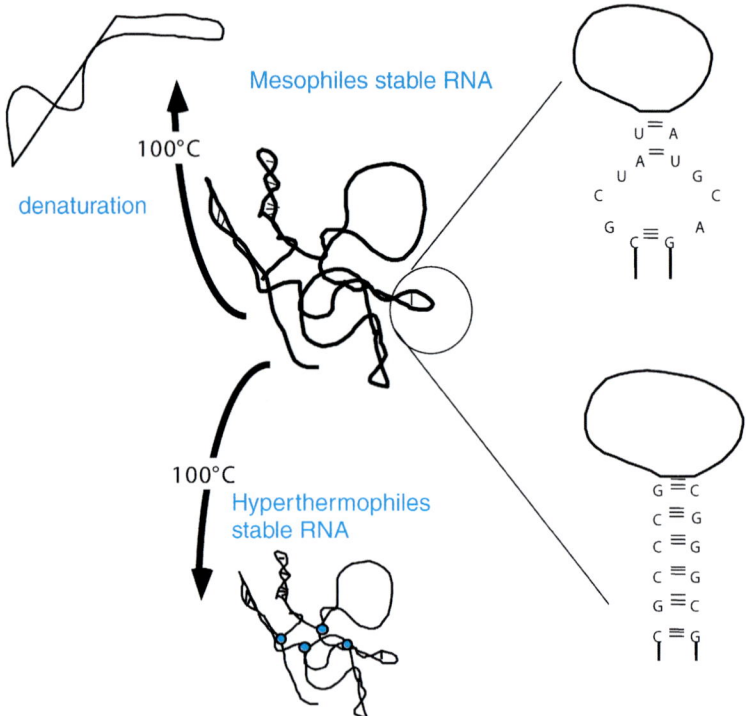

Fig. 7.5. Stable RNA protection at high temperature. Stable RNA molecules (ribosomal RNA, transfer RNA, etc.) are composed of a single polynucleotide chain folded on itself to form a number of 'hairpins' corresponding to stem-loop structures. In hyperthermophiles, these hairpins are stabilized by longer stems and a higher content of guanine (G) and cytosine (C). The overall globular structure can be stabilised by the addition of proteins and/or chemical groups at specific positions (*blue circles*)

7.2.5 Temperature and the RNA World

The fact that LUCA was not a hyperthermophile does not necessarily exclude the possibility of a hot origin of life, since a very long period of evolution occurred between the apparition of life and that of LUCA (Forterre, 1995a; Lazcano, 1996). Life might thus have appeared at very high temperatures, gone through a cold evolution period, and diversified at the whole range of temperatures from 0 to 110°C! It is likely that the step corresponding to a "RNA world" occurred at relatively low temperatures (Joyce, 1988; Forterre, 1995b). Indeed, RNA is a very fragile molecule due to the reactive oxygen at position 2′ of ribose (absent in DNA, deoxyribonucleic acid) that can cause the breaking of nucleotide bonds (Fig. 7.6). This reaction, faster at high temperature, as are all chemical reactions, leads to rapid degradation of RNA by hydrolysis at temperatures close to 100°C. Moreover, this thermodegradation is highly stimulated *in vitro* by magnesium chloride, a cofactor necessary to the activity of ribozymes (RNA enzymes), which themselves were essential components of metabolism in a RNA world.

However, a hot RNA world cannot yet be totally excluded. Actually, it has been shown that RNA thermodegradation can be inhibited by strong concentrations of monovalent ions (> 100mM potassium chloride) (Hetcke, 1999; Tehei, 2002). Thus, a hot RNA world might have existed in combination with monovalent salts and/or unknown RNA-stabilizing molecules. Given its thermosensi-

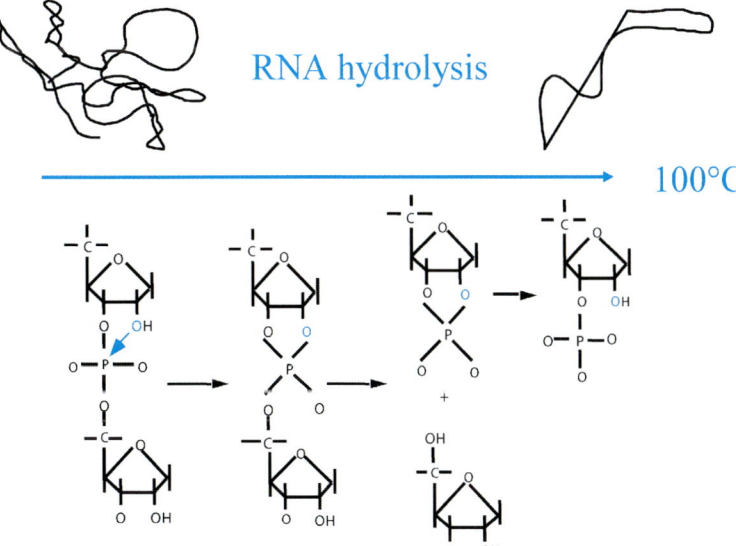

Fig. 7.6. RNA hydrolysis at high temperature. The link between two ribose molecules can be cleaved at the level of the phosphate by the ribose oxygen at position 2′ (*blue*)

bility, this scenario implies also the absence of magnesium chloride, and consequently the presence of ribozymes depending only on monovalent ions. It would be interesting to test such dramatic constraints by in vitro RNA selection tests.

In extant hyperthermophiles, stable RNAs (transfer RNA, ribosomal RNA) are protected against thermodegradation by methylation of a number of ribose 2' oxygens and by association with proteins (Fig. 7.5). Messenger RNA is already *per se* highly unstable in prokaryotes (half-life of about one minute). mRNA is, in fact, used by ribosomes for protein synthesis (translation). RNA instability at high temperatures might perhaps explain why most thermophilic Eukarya (such as some fungi) are unable to grow beyond 60°C (Forterre, 1995b) (Fig. 7.1). In fact, in Eukarya, messenger RNA has to be stable enough to survive over its transfer from the nucleus (transcription site) to the cytoplasm (translation site). It was even proposed that selection pressure for adaptation to thermophily might have been the driving force for the appearance of the prokaryotic phenotype (e.g., rapid macromolecular turnover; elimination of the nuclear membrane to allow coupling between transcription and translation) – the "thermoreduction hypothesis" (Forterre, 1995b).

7.3 Comparative Genomics: a Novel Approach to Retrace Our Most Distant Past

7.3.1 Simple or Complex LUCA? A RNA or a DNA Genome?

While the study of mycoplasma or hyperthermophiles does not provide many insights into the nature of LUCA, comparative genomics seems to be a much more promising approach (Koonin, 2000). At the beginning of 2004, 169 genomes (131 bacterial, 17 archaeal, and 21 eukaryal) have been completely sequenced and are publicly available in genomic data banks (http://wit.integratedgenomics.com/GOLD/). Their comparison has permitted the identification of approximately a hundred genes that are universally distributed in all these genomes and thus were likely present in LUCA. The majority of the proteins encoded by these genes are involved in either protein or RNA synthesis (indeed mostly ribosomal proteins) (they can be easily viewed on the COG website at http://www.ncbi.nlm.nih.gov/COG/).

In general, these proteins are much closer between Eukarya and Archaea than between Bacteria and either of the other two domains (Olsen, 1997). Moreover, a certain number of translation and transcription proteins are shared only between Eukarya and Archaea. Finally – and remarkably – all bacterial proteins involved in DNA replication are very different from their analogues in the other two domains; conversely, they are very similar between Archaea and Eukarya (Mushegian, 1996; Olsen, 1997; Edgell, 1997; Myllykallio, 2000).

These observations have been interpreted by invoking two opposite scenarios. In the first, LUCA would have been much simpler than any extant organism, i.e. it would have been a progenote, an organism still belonging to an RNA world (or to the very dawn of a DNA world, prior to the apparition of DNA replication) (Fig. 7.7A) (Mushegian, 1996; Leipe, 1999). This scenario is consistent with the rooting of the universal tree of life between Bacteria on one side and Archaea/Eukarya on the other (Woese, 2000). This rooting was proposed

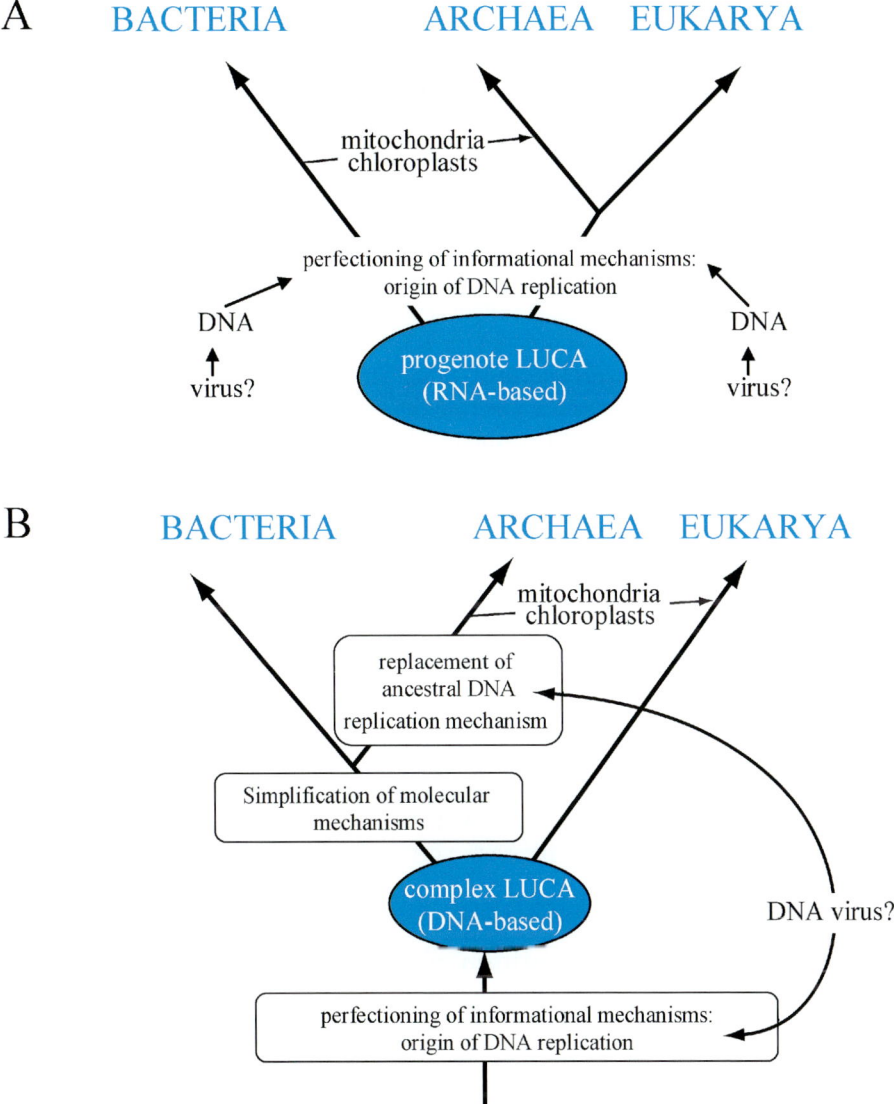

Fig. 7.7. Two alternative hypotheses to explain evidence from comparative genomics: a simple (**A**) or complex (**B**) Last Universal Common Ancestor (LUCA)

about 15 years ago, on the basis of phylogenetic analyses of paralogue protein couples (proteins coded by duplicated genes) present in all three domains of life. Since these couples are the products of a gene duplication that occurred prior to the separation of the three lines of descent, i.e. in LUCA, the tree built on the first member of the duplicated couple provides the outgroup to the tree built on the second one, and vice versa, thus producing two rooted universal phylogenies (Iwabe, 1989; Gogarten, 1989). However, such analyses should be treated with caution, due to the methodological problems affecting the reconstruction of such ancient relationships (Gribaldo, 2002).

In opposition to the idea of a primitive LUCA, a few authors have put forward the hypothesis of a complex ancestor, whose molecular biology was closer to that of extant Eukarya rather than that of extant prokaryotes (Fig. 7.7B) (Forterre, 1999c; Penny, 1999). This hypothesis rests on the presence in Eukarya of a number of RNA-based mechanisms that may possibly be relics of an RNA world. This is particularly the case of a eukaryal mechanism producing functional messenger RNAs by the cutting off of noncoding regions (exons) and the joining of coding ones (introns). This mechanism, called «splicing», is performed by a molecular complex (the spliceosome) composed of RNA and proteins. No counterpart of the spliceosome exists in prokaryotes. Since such a structure is similar to that of the spliceosome, the ribosome itself might be an ancient ribozyme. In this alternative scenario, the root of the universal tree of life would rather lie between Eukarya and Archaea/Bacteria, rendering prokaryotes monophyletic and indicating all characters common to Eukarya and Archaea as ancestral. In this case, the functional differences between bacterial informational proteins and their archaeal/eukaryal counterparts would be explained by a dramatic acceleration of evolution of these proteins in Bacteria, or even by their replacement by proteins of distinct origins (particularly with regards to DNA replication enzymes), incluting viral ones (Forterre, 1999a). Indeed, the genomes of many DNA viruses code for replication proteins that are very different from their cellular analogues as well as from other viruses, suggesting that viral and cellular DNA replication proteins diverged a long time ago.

For the time being, no evidence allows favouring either a complex or a simple LUCA. The answer might be provided by a more accurate analysis of the ribosomal machinery, since it presents very well preserved features in all three domains of life.

7.3.2 A Key Step: the Apparition of DNA

Another important issue is represented by the period comprised between the apparition of proteins and that of DNA. This period might be called the "RNA world's second age", to distinguish it from the first age, dominated by ribozymes. This age witnessed the progressive assemblage of the complex systems of protein synthesis and of the genetic code in LUCA, together with the apparition of RNA viruses, composed of proteins and RNA. Large metabolic pathways of synthesis and degradation might also date back to this period, together with the apparition of bioenergetic mechanisms and cellular membranes, as we know them today.

DNA might have appeared in two steps towards the end of this period. The first step would have been the apparition of ribonucleotide reductase (RNR), an enzyme that produces deoxyribonucleotides (substrates for DNA polymerases) starting from ribonucleotides (substrates for RNA polymerases) (Fig. 7.8). Ancestral RNRs were already most likely proteins, since the mechanism of ribose reduction involves highly active radical groups that would have not been stable in an RNA molecule (Freeland, 1999). The second step would have been the appari-

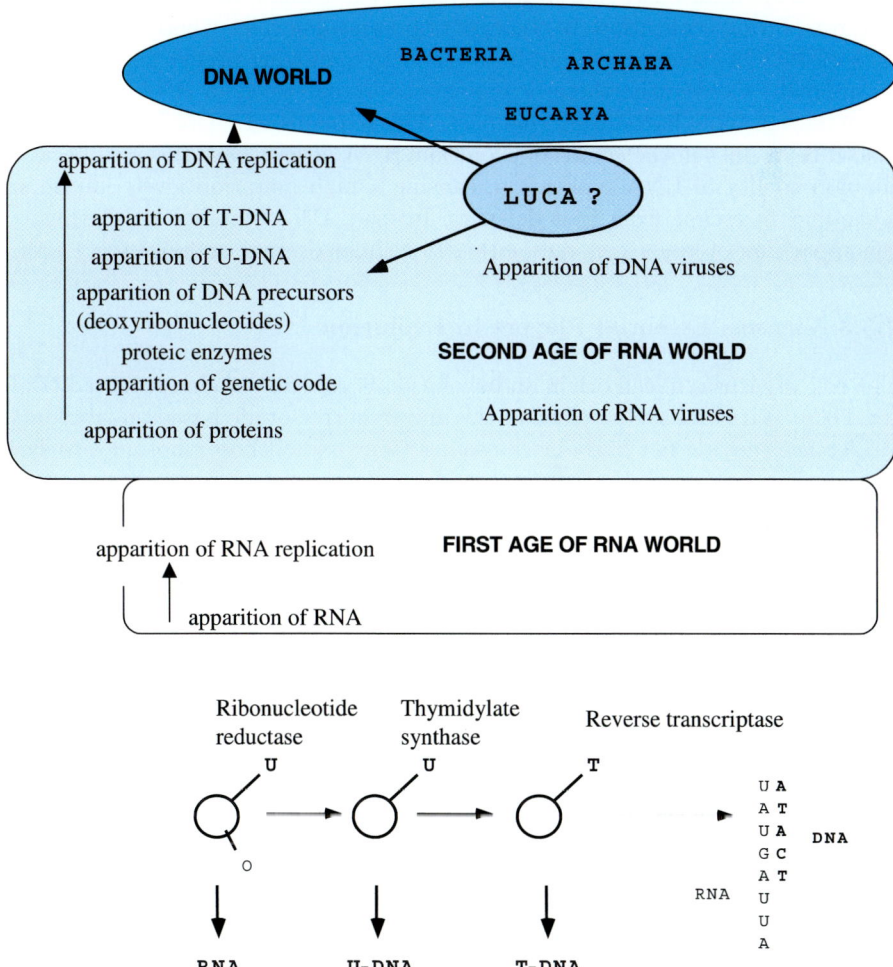

Fig. 7.8. The first steps of life evolution on Earth based on molecular biology data: the transformation of a ribonucleotide into a deoxyribonucleotide and the appearance of DNA from RNA

tion of enzymatic proteins necessary to deoxyribonucleotide polymerisation (reverse transcriptase and DNA polymerase), and the formation of deoxythymidine monophosphate (dTMP) starting from uracyl monophosphate (dUMP), leading to the replacement of uracyl by thymine in DNA. The fact that this transition occurred at the level of deoxynucleotides (base+desoxyribose+phosphate) shows that the apparition of U-DNA (uracyl-containing DNA) must have predated that of present-time T-DNA (thymine-containing DNA) (Fig. 7.8).

DNA is more stable than RNA due to the absence of oxygen at position 2′ of ribose. Moreover, it can be more reliably replicated than RNA due to the replacement of uracyl by thymidine. Indeed, DNA and RNA cytosines tend to transform spontaneously into uracyl in vivo by a deamination reaction, leading to errors during replication. In the case of present-day cells, whose genome consists of T-DNA, such errors can be avoided by removing uracyl, recognized as an anomaly. Conversely, these errors cannot be corrected in RNA or U-DNA genomes, normally harbouring this base. This explains why DNA cells have replaced RNA ones in the course of evolution. RNA viruses have survived because the non-fidelity of RNA replication, causing a high mutation level, can be an advantage to escape from host defences. Instead, DNA stability has permitted the apparition of larger genomes, either in cellular or viral organisms.

7.3.3 Viruses: Essential Players in Evolution

The role of viruses in cell origin and evolution is generally underestimated (Balter, 2000). Viruses have no place in the universal tree of life based on ribosomal RNA since they do not harbour ribosomes (they exploit host ribosomes to synthesise their proteins). Nonetheless, it is well known that viruses have played, and still play, a crucial role in gene transfer, so frequent in prokaryotes. Viruses have also played an important role in the apparition of new proteins, thanks to their well-developed ability to recombine and to their particularly high mutation rate. It might also be argued whether viruses have indeed "invented" DNA (Forterre, 2002c). In fact, some viruses do modify their DNA to make it resistant to nucleases produced by the cells they infect, and DNA can be indeed considered as a modified RNA. As mentioned above, the advantages of DNA over RNA (higher stability and higher replication fidelity) perfectly explain why DNA cell populations outcompeted RNA ones, although they cannot account for the initial selective advantage of the first DNA organism. The apparition of DNA in a virus would have provided an immediate selective advantage, giving a satisfactory answer, in terms of selective pressure, to the apparition of this molecule. If viruses invented DNA and DNA-replicating proteins, these proteins might have had the possibility to diverge very early over the course of evolution. This would explain why viral-replication proteins are often very divergent from virus to virus, and very different from their cellular homologues. In such a scenario, different cellular lineages may have acquired very different DNA replication proteins from different viral sources. This would indeed explain an important evolutionary puzzle, i.e., the fact that, despite being very similar in terms of overall mechanisms and enzymatic steps, several essential proteins involved in identical steps of cellular DNA replication are evolutionary unrelated in Archaea and Bacteria. Moreover, the rising number of complete genomes in public databases is revealing the existence of a number of other evolutionary unrelated (or very distantly related) protein families performing essential and common steps in DNA replication and metabolism in the three domains of life, Archaea, Bacteria, and Eukarya. This is, for instance, the case of topoisomerases, essential enzymes involved in the handling of DNA topological constraints that arise in a number of basic cellular processes (i.e., replication, transcription, recombination, chromatin remodelling). In fact, two types of DNA topoisomerases, TopoIV in Bacteria and TopoVI in Archaea/Eukarya, perform essentially the same function, despite being evolutionary unrelated. This peculiar distribution is consistent with a possible viral origin of these two enzymes (Gadelle, 2003). Another stunning example is that of tymidylate synthase (ThyA), a key enzyme that catalyzes the transformation of deoxyuracyl monophosphate (dUMP) into deoxythymidine monophosphate (dTMP), one of the four fundamental components of DNA. Recent investigations have evidenced that a number of bacterial species harbour an evolutionary unrelated analogue (ThyX), which catalyzes the same reaction (Myllykallio, 2002). Here again, the presence of numerous ThyX and ThyA viral homologues leaves open the possibility of a viral origin for these enzymes. The possibility that an enzyme involved into a fundamental step of DNA metabolism may be of viral origin is consistent with the hypothesis that T-DNA might have first appeared in viruses (Forterre, 2002c).

Returning to the two scenarios for the origin of LUCA discussed above, the introduction of viral DNA into cells might have occurred by two different ways (Fig. 7.8). The first would have been a double independent transfer of DNA and its replication machinery from a viral source to an RNA-based LUCA, once in the line leading to Bacteria, and another in that leading to Archaea and Eukarya. This scenario, although implying a bacterial rooting for the tree of life, is nonetheless consistent with an RNA-based LUCA, either a simple prokaryotic-like organism, or a fairly complex protoeukaryal one, containing perhaps RNA genes with introns and exons. In the latter case, present-day Bacteria would be the product of a radical secondary simplification (Forterre, 1995b). The second scenario would posit that a transfer of DNA and its machinery occurred from a viral source prior to LUCA, leading to a DNA-based LUCA. Another transfer would have followed either in the bacterial branch, or in the branch common to Archaea and Eukarya, leading to a radical replacement of the ancestral replication machinery by an analogue and nonhomologue system (Forterre, 1999a).

7.3.4 The Origin of the Nucleus: a Further Puzzle

Evidently, a number of unknown variables are involved in retracing the history of life on our planet. One of them is the origin of the eukaryal cell nucleus. We now know that mitochondria and chloroplasts are remnants from Bacteria that engaged endosymbioses with ancestral eukaryal cells (Fig. 7.1). By analogy, a few authors have suggested that the nucleus also originated from the endosymbiosis of an archaeon within a bacterial host, in order to explain the archaeal nature of eukaryal proteins involved in the expression and replication of the genomic material (reviewed in Lopez-Garcia, 1999). In this model, a total disappearance of the ancestral archaeal membrane has to be invoked. Indeed, the actual nuclear membrane corresponds to an expansion of the endoplasmic reticulum (the net of vesicles present in all eukaryal cells) and does not contain any archaeal lipid. The big concern with such a model is the origin of nuclear pores, whose structure is utterly complex and has no counterpart in prokaryotes. Moreover, it is difficult to understand why the evolution of the nucleus would have occurred by a complexification of the archaeal endosymbiont, when endosymbionts normally go through a reductive evolution of most of their genome (simplification). Finally, this model does not explain the disappearance of the whole genome of the bacterial host, together with all its informational proteins. The recent completion of the first eukaryotic genomes has triggered comparative genomics approaches aimed at testing alternative hypotheses on the origin of the eukaryotic cell. The fact that eukaryotic proteins functionally related to nuclear functions share higher similarity with their archaeal homologues, while those related to metabolisms and cytoplasmic activities are more similar to their bacterial counterparts has been used as a strong support for an archaeal endosymbiotic origin of the nucleus in a bacterial host (Horiike, 2001). However, the reliability of such an approach has been seriously questioned (Poole, 2001; Rotte, 2001). More recently, searches for sequence homologues across complete genomes from members of the three domains allowed the identification of 347 proteins that are unique to Eukarya (Hartman, 2002). These proteins were thus assumed to represent genuine 'eukaryal signatures' that would have evolved in a protoeukaryotic anucleate ancestor that was named "chronocyte" (Hartman, 2002). The cronocyte would have thus been a complex cell already provided with many important novelties, including a cytoskeleton apparatus. The nucleus would have originated by the engulfment of an archaeal symbiont. The authors invoke an RNA-based host in order to explain the origin of the nuclear envelope due to the need to physically isolate the RNA genome from the archaeal symbiont DNA genome. The complement of eukaryal proteins with bacterial origin would thus have been introduced into the eukaryal genome subsequently, by LTG from mitochondria to the nucleus.

However, these theories on the origin of the nucleus from an archaeal symbiosis (either in a bacterial or a protoeukaryal host) are at odds with the evidence that a substantial number of eukaryal nuclear proteins are not of archaeal origin (e.g., eukaryotic type II DNA topoisomerases or some replicative DNA polymerases). In some cases, these proteins have homologues in viral genomes instead (Gadelle, 2003; Villarreal, 2000). This has led some authors to propose that viruses may have played an important role in the assembling of eukaryotic-specific molecular processes, up to the very origin of the nucleus ("viral eukaryogenesis"; Bell, 2001; Takemura, 2001). Indeed, some stunning affinities exist between the process of nuclear envelope dissolution during eukaryal cellular division, and the cycle of some DNA viruses such as Pox viruses. Moreover, nuclear pores are similar to those that some RNA viruses utilise to excrete their viral messenger RNA into the cellular cytoplasm of the host. It will be interesting to test this possibility by including viral sequences into genomic approaches and attempting to dissect the impact (and timing) of a viral contribution to the eukaryotic genome.

7.4 Conclusions and Perspectives

7.4.1 More Data are Needed

In conclusion, the quest for primitive cells in the present-day biosphere has led, up to date, only to wrong conclusions. It seems that no cells preceding LUCA have survived, and that all extant cells are equally evolved. In particular, the study of hyperthermophiles has not unveiled any obvious primitive life forms. However, it has largely contributed to the debate on a hot origin of life and to the clarification of the steps leading to hyperthermophily. Conversely, the identification of three domains in the cellular world and the systematic comparison of these three domains at the genomic and molecular level is extremely instructive. Molecular mechanisms common to all three domains allow to retrace a part of life history prior to the apparition of LUCA, and to answer a few precise questions on both the nature of LUCA and the mechanisms that have led to present-day cellular types.

The sequencing of an ever-growing number of genomes should provide an answer to all these questions. It would be particularly interesting to analyse the genomes of a number of protists (unicellular eukaryotes), and of novel archaeal and bacterial lineages. In most cases, the existence of these lineages is known only from ribosomal RNA sequences amplified randomly by PCR (polymerase chain reaction) from either natural or artificial biotope samples. One of the major future challenges of biology will be indeed to cultivate these organisms in order to be able to perform physiological and genomic analyses. It would be equally crucial to sequence a number of viral genomes in order to test the hypotheses discussed in this chapter on the role of viruses during the first steps of the DNA world. For example, some viruses harbour genomes larger than those of some Bacteria (La Scola, 2003). The sequencing of such genomes might reveal many surprises. Viruses represent the majority of the biosphere (each cellular species being the target of many different viruses) and we know only a tiny fraction of this diversity.

7.4.2 To not Forget Darwin!

In this chapter, we implicitly assumed that all evolutionary steps prior to LUCA and involving RNA, proteins (via RNA), and DNA (via proteins) occurred in a cellular world. We reckoned it essential since, to us, natural selection acts on individuals, either cellular or viral, competing one with each other (the core of the hypothesis on a viral origin of DNA). Nonetheless, this notion is not evident to all researchers involved in the study of life origins and the first steps of evolution. To some of them, evolution could have occurred in a noncellular medium up to the emergence of a RNA world, viruses, or even LUCA! Is it really possible to imagine evolution by natural selection in similar conditions? Is this not a return to pre-Darwininan concepts? Similarly, it is very fashionable these days to discuss of a primitive and nondifferentiated cellular world where continuous gene transfers prevented any speciation (Woese, 2000). Again, is this scenario compatible with a natural selection process?

It would thus be sensible to bring Darwin back in our discussions on life origin, particularly for researchers dealing with the domain of evolution but not having a specific preparation on the subject (the great majority indeed). This remark is valid not only for non-biologists, but also for molecular biologists and geneticists (Kupic, 2000).

References

General

Ciba Foundation, *Evolution of Hydrothermal Ecosystems on Earth (and Mars?)* (1996). J. Wiley and Son, Chichester

Madigan M.T., Martinko J.M, Parker J., *Brock Biology of Microorganisms* (2000) 9th edition, Prentice Hall, USA

Wiegel J., Adams W.W., *Thermophiles, the Keys to Molecular Evolution and the Origin of Life?* (1998) Taylor and Francis, London

Specialised

Alberts, S.V., Van de Vossenberg, J.L., Driessen, A.J., Konings, W.N. (2000) Adaptation of archaeal cell membrane to heat stress. *Front. Biosci.* **5**, D813–820

Aravind, L., Tatusov, R.L, Wolf, Y.L., Walker, D.R., Koonin, E.V. (1998) Evidence for massive gene exchange between archaeal and Bacterial hyperthermophiles. *Trends Genet.* **15**, 298–299

Bell, P.J.L. (2001) Viral eukaryogenesis: was the ancestor of the nucleus a complex DNA virus? *J. Mol. Evol.* **53**, 251–256

Ciba Fondation, *Evolution of Hydrothermal Ecosystems on Earth (and Mars?)* 1996, J. Wiley and Son, Chichester

Confalonieri, F., Elie, C., Nadal, M., Bouthier de la Tour, C., Fortere, P., Duguet, M. (1993) Reverse gyrase: a helicase-like domain and a type I topoisomerase in the same polypeptide. *Proc. Natl. Acad. Sci. USA* **90**, 4753–4757

Edgell, D.R., Doolittle, W.F. (1997) Archaea and the origin(s) of DNA replication proteins. *Cell* **89**, 995–998

Forterre, P., Confalonieri, F., Charbonnier, F.,Duguet, M. (1995a) Speculations on the origin of life and thermophily *Origin of Life and Evolution of the Biosphere* **25**, 235–249

Forterre, P. (1995b) Thermoreduction, a hypothesis for the origin of prokaryotes, *C. R. Acad. Sci. Paris* **318**, 415–422

Forterre, P. (1996a) A la recherche de LUCA. Compte rendu du colloque de la fondation des Treilles, http://www-archbac.u-psud.fr/Meetings/LesTreilles/Treilles_frm.html

Forterre, P. (1996b) A hot topic: the origin of hyperthermophiles. *Cell* **85**, 789–792

Forterre, P. (1999a) Displacement of cellular proteins by functional analogues from plasmids or viruses could explain puzzling phylogenies of many DNA informational proteins. *Mol. Microbiol.* **33**, 457–465

Forterre, P. (1999b) Les hyperthermophiles sont-ils nos ancêtres? *La Recherche* **317**, 36–43

Forterre, P., Philippe, H. (1999c) Where is the root of the universal tree of life? *Bioessays.* **10**, 871–879

Forterre, P., Bouthier de la Tour, C., Philippe, H., Duguet, M. (2000b) Reverse gyrase from hyperthermophiles, probable transfer of a thermoadaptation trait from Archaea to Bacteria. *Trends Genet.* **16**, 152–154

Forterre, P. (2002a) A hot story from complarative genomics: reverse gyrase is the only hyperthermophile-specific protein. *Trends Genet.* **18**, 236–238

Forterre, P., Brochier, C., Philippe, H. (2002b) Evolution of the Archaea. *Theor. Popul. Biol.* **61**, 409–422

Forterre, P. (2002c) The origin of DNA genomes and DNA replication proteins. *Curr. Opin. Microbiol.* **5**, 525–32

Freeland, S.J., Knight, R.D., Landweber, L.F. (1999) Do proteins predate DNA? *Science* **286**, 690–691

Gadelle, D., Filée, J., Buhler, C., Forterre P. (2003) Phylogenomics of type II DNA topoisomerases. *Bioessays* **25**, 232–242

Galtier, N., Tourasse, N., Gouy, M. (1999) A non hyperthermophilic common ancestor to extant life forms. *Science* **283**, 220–221

Gribaldo, S., Philippe, H. (2002) Ancient phylogenetic relationships. *Theor. Popul. Biol.* **61**, 391–408

Gogarten, J.P., Kibak, H., Dittrich, P., Taiz, L., Bowman, E.J., Bowman, B.J., Manolson, M.F., Poole, R.J., Date, T., Oshima, T. (1989) Evolution of the vacuolar H^+-ATPase: implications for the origin of Eukarya. *Proc. Natl. Acad. Sci. USA* **86**, 6661–6665

Hartman, H. and Fedorov, A. (2002) The origin of the eukaryotic cell: a genomic investigation. *Proc. Natl. Acad. Sci. USA* **99**, 1420–1425

Hethke, C., Bergerat, A., Hausner, W., Forterre, P., Thomm, M. (1999) Cell-free transcription at 95 degrees: thermostability of transcriptional components and DNA topology requirements of *Pyrococcus* transcription *Genetics* **152**, 1325–1333

Horiike, T., Hamada, K., Kanaya, S., Shinozawa, T. (2001) Origin of eukaryotic cell nuclei by symbiosis of Archaea in Bacteria is revealed by homology-hit analysis. *Nat. Cell. Biol.* **3(2)**, 210–214

Huber, H., Hohn, M.J., Rachel, R., Fuchs, T., Wimmer, V.C., Stetter, K.O. (2002) A new phylum of Archaea represented by a nanosized hyperthermophilic symbiont. *Nature* **417**, 63–67

Iwabe, N., Kuma, K.-I., Hasegawa, M., Osawa, S., Miyata, T. (1989) Evolutionary relationship of Archaea, euBacteria and Eukarya inferred from phylogenetic trees of duplicated genes. *Proc. Natl. Acad. Sci.* **86**, 9355–9359

Joyce, G. (1988) Hydrothermal vents too hot? *Nature* **334**, 564–565

Koonin, E.V., Aravind, L., Kondrashov, A.S. (2000) The impact of comparative genomics on our understanding of evolution, *Cell* **101**, 573–6

Kupic, J.P., Sonigo, P. (2000), *Ni Dieu ni Maître*. Editions Odile Jacob, Paris

La Scola, B., Audic, S., Robert, C., Jungang, L., de Lamballerie, X., Drancourt, M., Birtles, R., Claverie, J.M., Raoult, D. (2003) A giant virus in amoebae. *Science* **299**, 2033

Lazcano, A., Miller, S.L. (1996) The origin and early evolution of life: prebiotic chemistry, the pre-RNA world, and time. *Cell* **14**, 793–798

Leipe, D.D., Aravind, L., Koonin, E.V. (1999) Did DNA replication evolve twice independently? *Nucleic Acids Res.* **27**, 3389–3401

Lopez-Garcia, P., Moreira, D. (1999) Metabolic symbiosis at the origin of Eukarya, *Trends Biochem. Sci.* **24**, 88–93

Matte-Tailliez, O., Brochier, C., Forterre, P., Philippe, H. (2002) Archaeal phylogeny based on ribosomal proteins. *Mol. Biol. Evol.* **19**, 631–639

Marguet, E., Forterre, P. (1994) DNA stability at temperatures typical for hyperthermophiles. *Nucl. Acid Res.* **22**, 1681–1686

Myllykallio, H., Lipowski, G., Leduc, D., Filée, J., Forterre, P., Liebl, U. (2002) An alternative flavin-dependent mechanism for thymidylate synthesis. *Science* **297**, 105–107

Nelson, K.E. et al. (1999) Evidence for lateral gene transfer between Archaea and bacteria from genome sequence of Thermotoga maritima. *Nature* **399**, 323–329.

Noon, K.R., Bruenger, E., McCloskey, J.A. (1998) Post-transcriptional modifications in 16S and 23S rRNAs of the archaeal hyperthermophile *Sulfolobus solfataricus*. *J. Bacteriol.* **80**, 2883–2888

Madigan, M.T., Martinko, J.M., Parker, J. (2000*) Brock Biology of Micro-organisms*, 9th edition, Prentice Hall, USA

Mushegian, A., Koonin, E.V. (1996) A minimal gene set for cellular life derived by comparison of complete Bacterial genomes. *Proc. Natl. Acad. Sci.* **93**, 10 268–10 273

Myllykallio, H., Lopez, P., Lopez-Garcia, P., Hailig, R., Saurin, W., Zivanovic, Y., Philippe, H., Forterre, P. (2000) *Science* **288**, 2212–2215

Nelson, K.E., et al. (1999) Evidence for lateral gene transfer between archaea and Bacteria from genome sequence of *Thermotoga maritima*. *Nature* **399**, 323–329.

Olsen, G., Woese, C.R. (1997) Archaeal genomics: an overview. *Cell* **89**, 991–994

Pace, N. (1991) Origin of life-facing up the physical setting. *Cell* **65**, 531–533

Penny, D., Poole, A.M. (1999) the nature of the universal ancestor. *Curr. Opin. Genet. Devel.* **9**, 672–677

Poole, A., Penny, D. (2001) Does endo-symbiosis explain the origin of the nucleus? *Nat. Cell. Biol.* **3(8)**, E173–174

Rotte, C., Martin, W. (2001) Does endo-symbiosis explain the origin of the nucleus? *Nat. Cell. Biol.* **3(8)**, E173–174

Schouten, S., Hopmans, E.C., Pancost, R.D., Damsté, J.S.S. (2000) Widespread occurrence of structurally diverse tetraether membrane lipids: evidence for the ubiquitous presence of low-temperature relatives of hyperthermophiles. *Proc. Natl. Acad. Sci. USA* **97**, 14 421–14 426

Takemura, M. (2001) Poxviruses and the origin of the eukaryotic nucleus. *J. Mol. Evol.* **52**, 419–425

Tehei, M., Franzetti, B., Maurel, M.C., Vergne, J., Hountondji, C., Zaccai, G. (2002) The search for traces of life: the protective effect of salt on biological macromolecules. *Extremophiles* **5**, 427–430

Villarreal, L.P., DeFilippis, V.R. (2000) A hypothesis for DNA viruses as the origin of eukaryotic replication proteins. *J. Virol.* **74(15)**, 7079–7084

Vivares, C.P., Méténier, G. (2000) Towards the minimal eukaryal parasitic genome. *Curr. Opin. Microbiol.* **5**, 463–467

Wiegel, J., Adams, W.W. (1998) *Thermophiles, the keys to molecular evolution and the origin of life?* Taylor and Francis eds. 339 p London

Woese, C.R. (2000) Interpreting the universal phylogenetic tree. *Proc. Natl. Acad. Sci. USA* **97**, 8392–8396

8 The Universal Tree of Life: From Simple to Complex or From Complex to Simple

Henner Brinkmann, Hervé Philippe

The idea of Zuckerkandl and Pauling put forward in 1965 (Zuckerkandl and Pauling 1965), that the macromolecules, which are the genetic inheritance of cells, could conserve traces of the evolutionary history of the extant organisms, profoundly changed our view of the universal tree of life. In the late 1970s Carl Woese promoted the use of the ribosomal RNA as a molecular marker, which allowed the definite confirmation of the endosymbiotic origin of mitochondria and plastids, and to define the major lines of microbial evolution. The prime discovery that Prokaryotes are composed of two very distinct groups, Eubacteria and Archaebacteria, was completely unexpected and triggered a great number of new research fields.

In this chapter we will first explain the principles of molecular phylogenetic reconstruction methods. Then, we will present the predominant view of the universal tree of life (often called the tree of Woese) and the main conclusions about the ancient evolution, which were deduced from it in the early 1990s. The end of the 1990s was dominated by something that is best described as the crisis of molecular phylogenetic reconstructions, because two major problems appeared almost simultaneously: (1) the proof of serious tree reconstruction artefacts and (2) a much higher than expected frequency of lateral gene transfers/duplications, which implied that the phylogeny of a given gene does not necessarily reflect the evolutionary history of the organism. However, a clear progress in the methods of molecular tree reconstruction together with a high number of sequences mostly produced by genome-sequencing projects did permit several important errors present in the original tree to be corrected. These corrections have naturally consequences for the hypotheses about the ancestral/early evolution (e.g., the origin of Eukaryotes) and also to a certain degree for the origin of life.

8.1 Principles of Tree-Reconstruction Methods

Phylogenetic inference consists of using what one can observe from organisms (e.g., either DNA sequences of the extant organisms or morphology for both extant and fossil) in order to infer their relationships. The first step is to find comparable characters, a task that had taken morphologists several centuries to obtain a rigorous approach. In 1848 Richard Owen defined the concept of homology, by applying the principle of connectivity of Geoffroy Saint-Hilaire:

two structures (or organs using the terminology of that epoch) in two different species are homologous (and therefore comparable) irrespective of their forms, if they are connected in the same way to identical structures. The same principle is used for DNA sequences. In fact, during the alignment step, one tries to maximise the number of identical nucleotides between two (or n) sequences (equivalent to the organs of Owen) for finding the homologous nucleotides (differing between the two or n species compared, but linked in the same way to identical nucleotide positions). The alignment step represents a problem that can be easily formalized and it is the object of an intense algorithmic research (Hickson et al. 2000). Theoretically, this objectivity may look like progress compared to the phylogeny based on morphology (the famous art of the taxonomists), but in reality the molecular phylogenies remain quite subjective. The current approach consists, in fact, in using an automatic alignment program, manually refining the result and then choosing the nonambiguously aligned regions (also manually). Even if certain programs allow formalizing the last step, e.g., (Castresana 2000), there is still a large theoretical and algorithmic progress to be realized in order to render the sequence alignment step efficient and objective.

We will assume in the following that we possess several perfectly aligned sequences and that we want to reconstruct their phylogenetic relationships. Figure 8.1 shows a tree, based on the evolution of eleven amino-acid positions from four species. Starting from the aligned sequences of the four extant species (Fig. 8.1b), how can we find the correct phylogeny? A first simplification is introduced by the fact that we only search for an unrooted tree (see Fig. 8.1c–e), which means a tree where we do not know the location of the last common ancestor of the four species (the specific point of the tree that is called the root, see Fig. 8.1a). In order to know the position of the root, we need to know in which direction the characters have evolved (this is called character polarisation). If, in certain cases, morphologists are able to polarize characters thanks to palaeontological data, the same approach is impossible with molecular data (or only possible under very strong assumptions on the evolutionary process (Gu and Li 1998)). The position of the root in a molecular phylogeny is thus normally determined by the addition of the sequence of an outgroup, about which we know that it is not part of the group under study. This allows us to polarize the evolution of characters, and thus to reconstruct the order of emergence of the ingroup species. For example, we can use birds to root the mammalian tree, or, in the case of Fig. 8.1, species number four can be used as an outgroup if we have additional external information supporting this fact.

By only considering unrooted trees, the number of possible trees is lowered: with four (n) species there are only three unrooted trees ($[2n-5] \times [2n-7] \times \ldots \times 3 \times 1$) but 15 rooted trees ($[2n-3] \times [2n-5] \times \ldots \times 5 \times 3$). In the case of Fig. 8.1, we have to find among the three possible unrooted trees the right one: tree X, which unites species 1 and 2, tree Y, which unites species 1 and 3, or tree Z, which unites species 2 and 3 (Fig. 8.1c). If we look at the first amino acid position, all species have a lysine (K), except species 4 that has a histidine

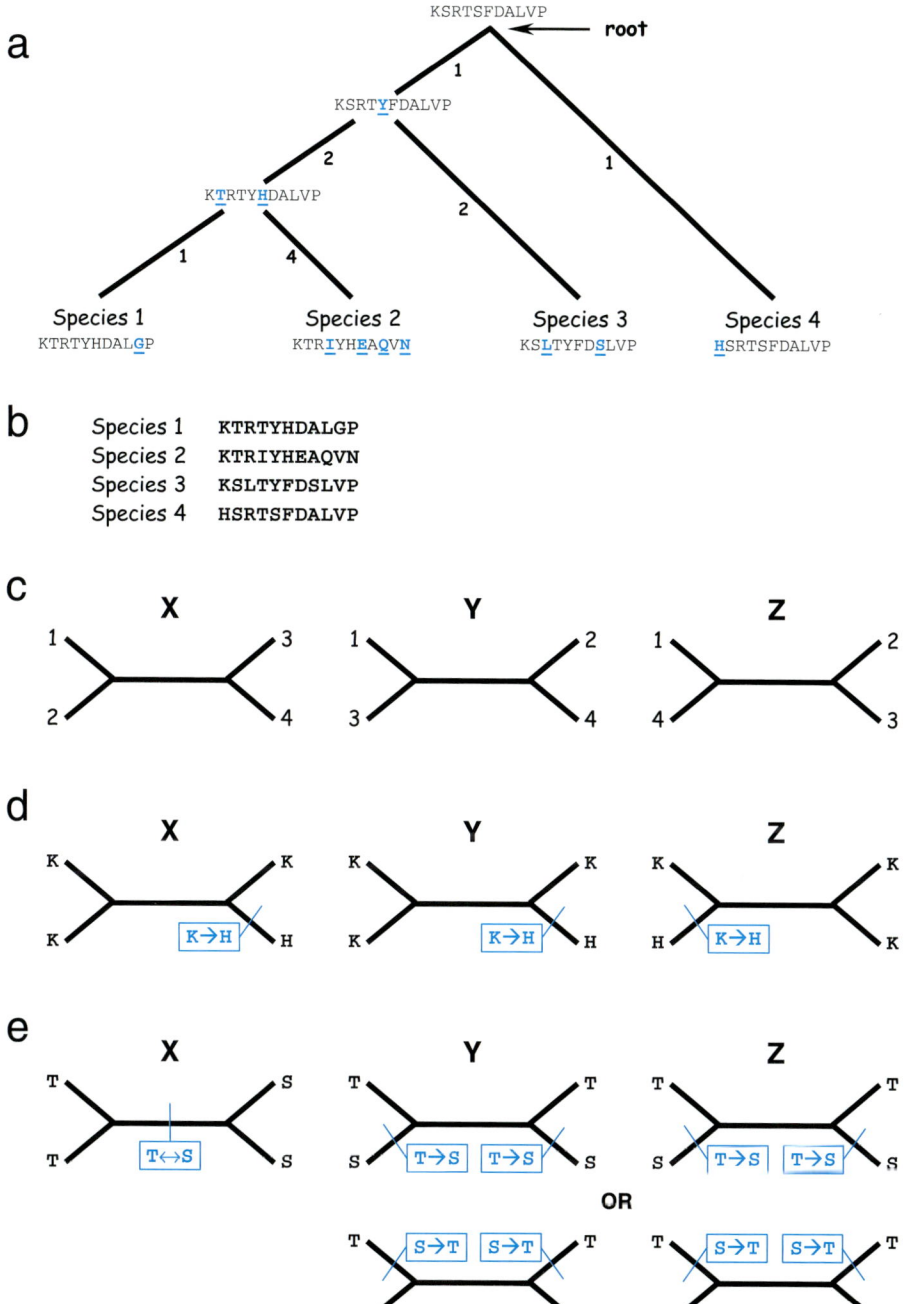

Fig. 8.1. Scheme illustrating the principles of phylogenetic reconstruction methods. An ideal case is presented, because each position changed once and only once

(H). To explain the actual distribution, it is sufficient to postulate that there was a nucleotide substitution leading to the change of amino acid K into H on the branch that leads to species 4; this change is independent of the topology (Fig. 8.1d). This position is therefore not informative, because it does not give any clue that allows a choice to be made between the three alternative tree topologies, but it has an impact on the branch lengths of the tree by adding an additional step to the branch of this species. It should be noted that other explanations are possible (e.g. a change from K to H in the internal branch and another one from H to K in the branch leading to species 2), but, according to the rule known as Occam's razor, only the most parsimonious interpretation is retained.

We will shortly discuss the additional difficulties introduced by the genetic code. The 20 different amino acids used in normal proteins are encoded by a triplet code (three nucleotides), with a total of 64 different codons (4^3, there are four different bases). The genetic code is redundant, since the number of codons is over three times higher than that of amino acids. Most mutations of nucleotides do not lead to a change of the encoded amino acid (synonymous mutations). For the replacement of an amino acid, there may be anything from a single to up to three nucleotide changes necessary. There are changes corresponding to a single nucleotide change (e.g., glycine, GGN versus alanine, GCN), or two changes (e.g., threonine, ACN versus valine, GTN), or even three changes (e.g., tryptophan, TGG versus histidine, CAC/T). However, empirical experience indicates that for the amino-acid-based phylogenetic reconstructions, the way in which substitutions are realized at the nucleotide level is not of great consequence. Furthermore, for the inference of distant evolutionary relationships based on proteins, amino acids are more appropriate than codons (nucleotides).

The second position of the sequences is informative (Fig. 8.1e), because a T (threonine) is present in the species 1 and 2 and an S (serine) is found in the species 3 and 4. Taking the tree X, it would be sufficient to postulate a single change between T and S in the internal branch to explain the distribution of characters for this position in the given species. We actually do not know whether the change is from T to S or from S to T, because it is not possible to polarize the evolution of the sequences. If we try to explain the current character distribution with the trees Y and Z, we would need two independent substitution events. For example for tree Y, we need to assume one T to S substitution in the branch leading to species 3 and one T to S substitution in the branch leading to species 4 (or two S to T substitutions in the branches leading to species 1 and 2). In conclusion, position 2 needs one substitution to fit the tree X and two substitutions in the trees Y and Z. According to the principle of economy of hypotheses (also known under the name of Occam's razor or parsimony), one has to choose the tree X. A phylogenetic method, named maximum parsimony, chooses the tree that needs the lowest number of changes (or steps) over all positions. In the case of the sequences of Fig. 8.1b, the tree X, which has a total length of eleven steps, is the most parsimonious and is therefore favoured over the trees Y and Z, which need 13 steps to explain the data. This tree corresponds to the true tree of Fig. 8.1a. In fact, there are only two informative characters (positions number two and six); they correspond to changes in the central branch of the unrooted tree. Another important point is that the correct tree is found despite quite different evolutionary speeds among the species (e.g., species 2 evolves four times faster than species 1).

8.2 The Universal Tree of Life According to Woese

Woese and Fox published an analysis about the oligonucleotide catalogues of the ribosomal RNA (rRNA) of the so-called 16S rRNA, which is part of the small subunit (30S) of the prokaryotic ribosome (S=Svedberg coefficient, refers to the sedimentation coefficient, is dependent of shape and weight) from a large ensemble of organisms (Woese and Fox 1977). The most important result of their work was the discovery that a group of poorly characterised prokaryotes (thermoacidophiles, methanogenes and extreme halophiles) were so distantly related to both Eukaryotes and Eubacteria, that they had to be placed in a third major group, called the Archaebacteria. This separation (Fig. 8.2) was later confirmed by the analyses of the rRNA sequences of the small subunit (SSU) of the ribosome: the Archaebacteria did represent an independent branch on the universal rRNA tree of Life (Woese 1987). The root was placed on the bacterial branch in the universal tree of life (Fig. 8.2) based on the analyses of anciently duplicated genes (Gogarten et al. 1989; Iwabe et al. 1989). This led to a strong rejection of the traditional dichotomy between Eukaryotes and prokaryotes and subsequently a new tripartite distribution of the biodiversity was established, assigning the extant organisms to one of three domains: Eukaryotes (Eucarya), Archaebacteria (Archaea), and Eubacteria (Bacteria) (Woese et al. 1990).

As discussed above, establishing the order of descent in a group of organisms starting from a common ancestor necessitates the use of an external group, which allows rooting the phylogeny. The fact that there is obviously no external group to all extant life forms has hindered for a long time the rooting of the tree of life. However, this problem was going to be solved by the use of anciently duplicated (paralogous) genes (Schwartz and Dayhoff 1978). If a pair of paralogous genes is universally distributed (which means that the two copies are found in all representatives of the three domains), we can deduce that the duplication, from which they originated, must have taken place in the last universal common ancestor (LUCA) or more likely before. In this context, one paralogous gene furnishes the external group for the phylogeny established by the second copy and vice versa. The first two identified pairs of universal paralogous sequences (paralogs) were the translation elongation factors of EF-Tu/1α and EF-G/2 (Iwabe et al. 1989) and the V- and F-type ATPases (Gogarten et al. 1989).

The four resulting phylogenies support the rooting along the bacterial branch of the tree of life, thereby proposing a sister-group relationship between Archaea

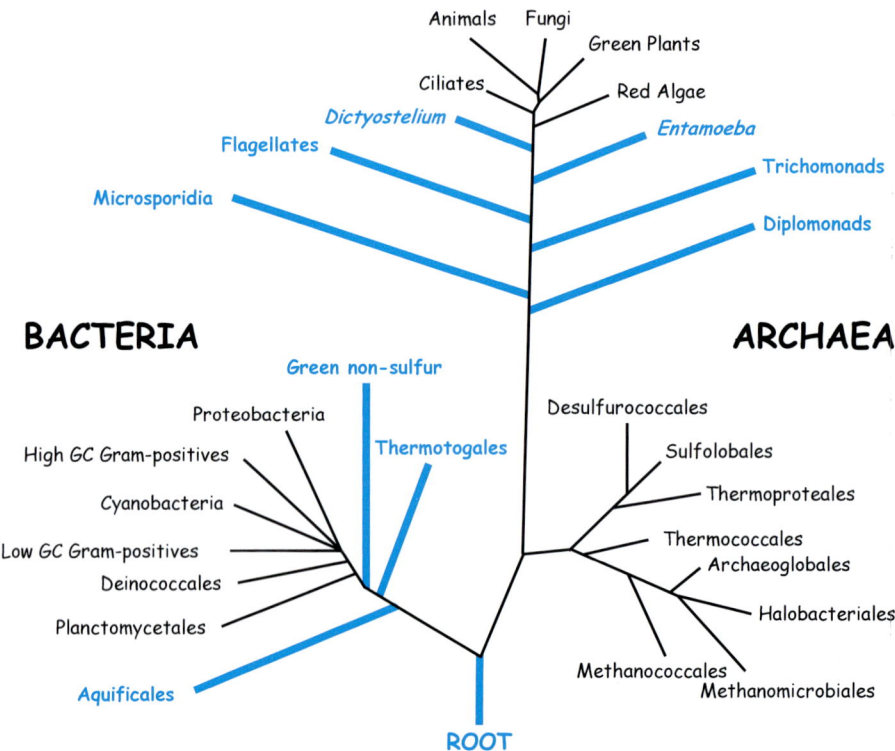

Fig. 8.2. The classical view of the universal tree of life. The topology was inspired by (Stetter 1996), and is principally based on the comparison of the rRNA sequences. Branches that could be affected by the long-branch attraction artefact (e.g., the position of the root on the bacterial branch or the early emergence of the hyperthermophilic taxa among the Bacteria) are indicated by *thick blue lines*

and Eukaryotes (Fig. 8.2). Unfortunately, the pairs of universal paralogs remain very rare. Until now only five additional anciently duplicated gene pairs have been found: the tRNA synthetases (tRS) for the amino acids valine and isoleucine (Brown and Doolittle 1995) and for tyrosine and tryptophane (Brown et al. 1997), an internal duplication in the subunit B of the carbamoyl-phosphate synthetase (Lawson et al. 1996), two components of the pathway directing nascent translation products into the endoplasmic reticulum (ER), SRP54/SRα (Gribaldo and Cammarano 1998), and the histidine biosynthetic genes *hisA/hisF* (Charlebois et al. 1997). All of the reciprocally rooted phylogenies based on these markers confirmed the bacterial rooting, with the only exception being the *hisA/hisF* pair. The rRNA phylogeny rooted between the Bacteria and the group Archaea/Eukaryotes became in consequence the universal tree of life shown in all the textbooks even for pupils (Fig. 8.2). Currently the classification of the life in three domains and the deduction that LUCA had a prokaryotic-type organisation is representing a paradigm that is largely accepted by the scientific community with only a few exceptions (Gupta 1998; Margulis 1996; Mayr 1998). Two other major evolutionary inferences were deduced from the tree of Fig. 8.2.

First, the phylogenies based on rRNA data suggest that the most deeply emerging branches in both prokaryotic domains correspond to hyperthermophilic organisms (Fig. 8.2). The members of the two known hyperthermophilic bacterial phyla – Aquificales and Thermotogales – are representing the most basal branches, followed by the moderate thermophilic bacteria, and finally the remaining species emerge (mostly mesophilic Bacteria). This observation linked to the traditional bacterial rooting of the universal tree of life and an early emergence of the hyperthermophilic Archaea was leading to the hypothesis that LUCA was a hyperthermophilic organism (Stetter 1996). Furthermore, the fact that the branches of the hyperthermophilic Bacteria are shorter than those of the mesophilic taxa was taken as an indication that a reconstruction artefact could not be responsible for their early branching (see below) and that the hyperthermophilic Bacteria are "living fossils", meaning that they had retained many ancestral features since LUCA. The idea of a hyperthermophilic LUCA was tightly linked to the hypothesis of a "hyperthermophilic Eden", which is suggesting that life originated and evolved in an extremely hot environment (more precisely a hydrothermal vent), and that the ancestral organisms remained hyperthermophilic until the appearance of the last common ancestor of the three domains (Gogarten-Boekels et al. 1995; Imai et al. 1999; Nisbet and Sleep 2001; Olsen et al. 1994; Pace 1991; Stetter 1996; Woese 1987).

Secondly, the phylogeny of Eukaryotes based on rRNA data (Fig. 8.2) can be divided into two parts, a basal part with several successively emerging amitochondriate lineages (such as, e.g. microsporidia, diplomonads and trichomonads), followed by mitochondriate groups (e.g. Euglenozoa or amoebae), and finally the remaining groups containing all late emerging lineages, that form a poorly resolved huge radiation, called the crown (Knoll 1992; Sogin 1991). The early appearance of amitochondriate protists in the rRNA tree had been interpreted as proof of the Archezoa hypothesis (Cavalier-Smith 1987), which suggested that several eukaryotic lineages originated before the mitochondrial endosymbiosis (microsporidia, diplomonads, trichomonads and Archaemoeba). Archezoa are often regarded as "living fossils", which furnish us information about the ancestral Eukaryotes (Knoll 1992; Sogin 1991).

The general idea of the evolutionary scenarios that may be deduced from this tree is that simple organisms preceded complex organisms (even if the definition of simplicity is not clear). The hypothesis that the "simple" Prokaryotes would be the ancestors of the "complex" Eukaryotes is a typical illustration. Furthermore, the simple Eukaryotes without mitochondria preceded the complex Eukaryotes with mitochondria. In fact, the idea that evolution is going from simple to complex is almost the direct inheritance of the scale of beings (scala naturae) from Aristotle. The latter was classifying the objects, living or inanimate, only in function of their complexity (from mineral to man), of course without assuming that the organisms have the capacity to change. The evolutionary theory was adapted to this philosophic frame simply by supposing those simple organisms are "living fossils", which ultimately gave rise to complex organisms. This apparent congruence between the modern results (ribosomal tree) and solidly rooted philosophical concepts (Aristotle) probably largely explains the fast and considerable acceptance of this new paradigm.

Nevertheless, this view of the universal tree of life is based on a very limited sampling of the genome (ca. 1000 nucleotides for the 16S rRNA, and ca. 1000 amino acids for the anciently duplicated paralogs). With the establishment of several whole genome sequences from the three domains, it rapidly became obvious that the trees based on alternative markers were largely contradicting the rRNA-based phylogeny and also each other, thus undermining the general consensus (Pennisi 1998). The reasons for these contradictions reside often in biological phenomena, like, e.g., lateral/horizontal gene transfers (LGT) or in the comparison of paralogous instead of orthologous genes (hidden paralogy) (Doolittle 1999c). However, mathematical features (tree-reconstruction artefacts) are more and more identified as an additional and important source of incorrect results in molecular phylogenetic reconstruction (Philippe and Laurent 1998).

8.3 Reconstruction Artefacts

8.3.1 Multiple Substitutions Generate Reconstruction Problems

After explaining the principles of phylogenetic reconstruction, we will now inspect the limitations of these methods. In Fig. 8.3a, the same phylogenetic tree as in Fig. 8.1a is shown, displaying the same number of substitutions; the only difference is that it is not exactly the same positions that have changed, as is indicated by the arrows. In the tree of Fig. 8.1a each position only changed once, whereas in Fig. 8.3 three positions (the second, the sixth and the eleventh) encountered two changes/substitutions. Let us see the influence of these multiple substitutions on the phylogenetic reconstruction methods. The second position was subject to a reversion (return to the previous state) that changed the T to S in the branch leading to the species 2. Therefore a position that was originally supporting the tree X (see Fig. 8.1e) is now noninformative. In the sixth position, which was also supporting tree X, there is a second substitution from F to L located on the branch leading to species 3 and it is also becoming noninformative, because we need two changes to explain the observed distribution for all three possible tree topologies (Fig. 8.3c). Even if the species 1 and 2 share the same amino acid (H), this does not necessarily mean that they are specifically related, a fact that was well explained by Willy Hennig (Hennig 1966). One has therefore to be critical towards the intuitive assumption that the more similar two organisms the more closely related they are.

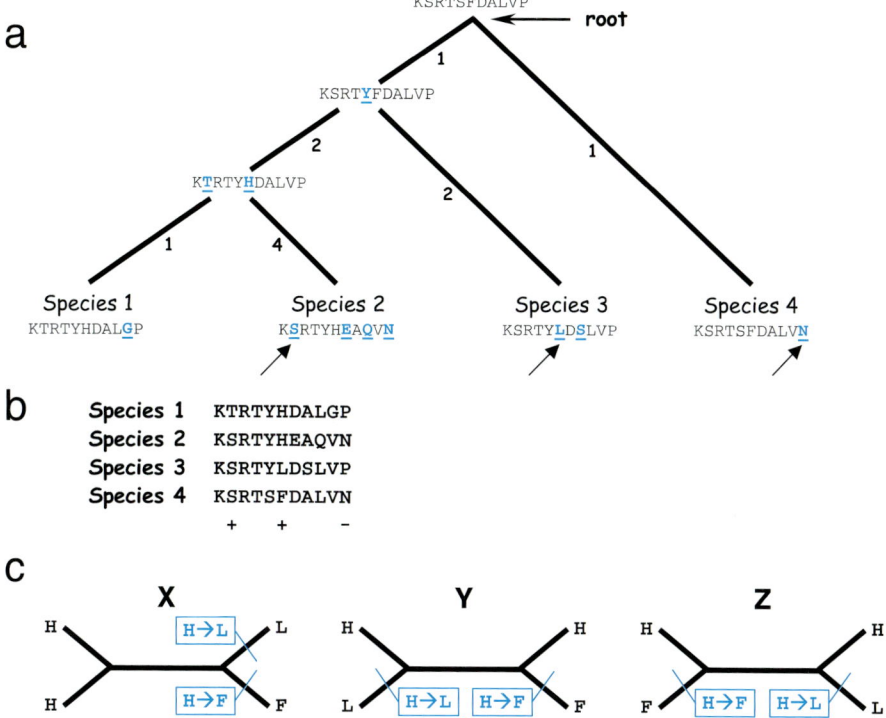

Fig. 8.3. Schema showing the limits of the molecular phylogenies. It is almost exactly the same tree as in Fig. 8.1, the only difference is that certain positions changed several times. These multiple substitutions generate noise that is excluding the reconstruction of the right tree

The eleventh position did experience a convergence on the branch leading to the species 2 and 4; this means the same amino acid, N, originated independently at the same position twice. Unfortunately, this position is now becoming informative, but in favour of the wrong tree (tree Y). In conclusion, the multiple substitutions destroyed the phylogenetic signal of two positions that support the right tree (indicated by a+ in Fig. 8.3b) and did create an erroneous informative position (indicated by a− in Fig. 8.3b). In this case, the most parsimonious tree based on the alignment is tree Y, thus leading to the inference of the wrong tree. Therefore the substitutions, which are the basis of phylogenetic information, are also responsible for its disappearance if they affect the same position several times.

8.3.2 Mutational Saturation Versus Resolving Power

As we saw before, positions that changed only once furnish reliable information for inferring bifurcations in the phylogenetic trees (e.g. the change from K to Y in site 1 of Fig. 8.4). If we have a sufficient number of these positions, we are able to reconstruct any given evolutionary history (Swofford et al. 1996). In practice, in order to resolve a phylogeny of about 50 species we will need more than 150 reliable positions (at least three changes for each node). Unfortunately, substitutions involve most of the time only a subsample of the total number of positions during history (Fitch and Markowitz 1970), generating more noise than signal for molecular phylogenetics (e.g. site 2, with 25 changes in Fig. 8.4). The positions that contain essential information (changes on the internal branches) often reach a state of mutational saturation, because they have lost all valuable information (site 2 in Fig. 8.4), e.g. the signal they contain is becoming random. Although the resolving power of molecular phylogenies is directly proportional to the length of the data set and to its evolutionary speed, it is diminishing if the

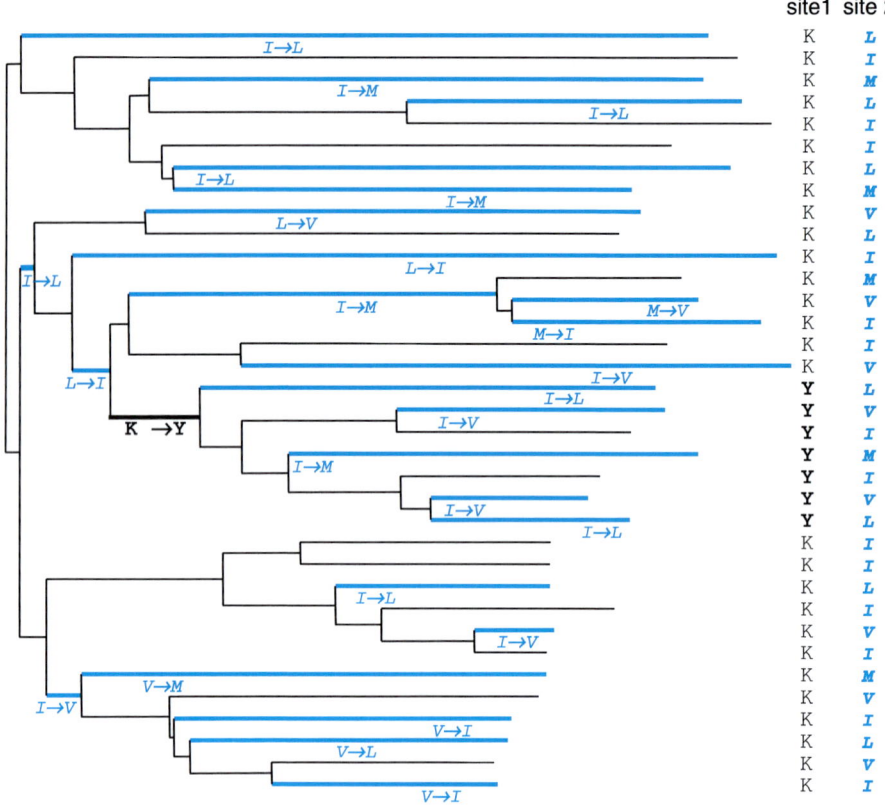

Fig. 8.4. Archetype of "good" and "bad" positions. Position 1 changed once (lysine versus tyrosine) and it is conserving a weak signal. In contrast, position 2 underwent multiple changes between several hydrophobic residues, which have completely intermingled the phylogenetic signal; more exactly it reached the point of mutational saturation, and in consequence it will essentially add more noise rather than phylogenetic information

speed of a given position is too high. In conclusion, there is the following major paradox: an informational overload is finally leading to the loss of the original phylogenetic signal.

Although an important number of the positions is highly conserved (and allows the alignment to be created), most of the other positions normally had changed much more than once. In order to demonstrate this fact, we analysed an alignment of 97 valyl/isoleucyl-tRNA synthetases (tRS) using the maximum parsimony (MP) method (Fig. 8.5). Five or less changes were inferred for only about 20% of the positions (68 of 344, with 15 constant sites). The information given by these positions consists of 144 changes (generally called steps). In contrast, for the 20% of the most variable positions (67 positions with more than 35 substitutions) a total of 2929 changes are inferred. Because the noisy positions are contributing tremendously more than the slowly evolving reliable ones to the choice of the best tree (2929 vs. 144 steps), it is of the utmost importance to develop methods that allow the efficient discrimination between signal and noise (see below).

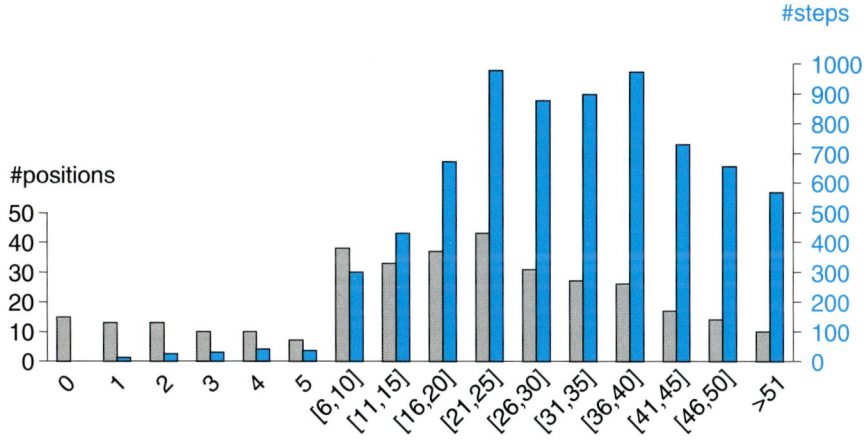

Fig. 8.5. Mutational saturation of the Ile/Val-tRNA synthetase sequences. On the X axis, the positions are regrouped according to their number of changes, calculated by the maximum parsimony method. On the Y axis, the number of positions that belong to the different classes is indicated on the *left* columns (in *light grey*) and the total number of steps that they are adding to the most parsimonious tree on the *right* columns (*blue*). The information added by the positions that only accepted a few changes (e.g., the positions with up to five changes furnished 144 steps), is largely outnumbered by those introduced by the positions that frequently changed (for example, the positions with more than 35 inferred changes furnished 2929 steps). The more noisy a position is, the more important is its contribution to the choice of the tree, thus leading us to the question of the validity of the bacterial rooting of the universal tree, observed for this gene (Brown and Doolittle 1995)

Saturation can also be identified by the presence of a plateau in a diagram, where the number of multiple substitutions calculated based on the tree is compared to the observed differences, for pairwise comparisons of sequences (Philippe et al. 1994). Mutational saturation represents a central issue of the reconstruction of ancient phylogenies, because all markers used for the inference of these trees are highly saturated (Philippe and Adoutte 1998; Philippe and Forterre 1999; Roger et al. 1999). Completely saturated data should not contain any signal and, in principle, should produce a star-like topology (without any resolution). Nevertheless, biases of multiple natures can lead in an erroneous way to resolved trees, because character states can be shared by chance between evolutionary distant species. This is linked to the fact that these biases are not correctly treated by the current tree-reconstruction methods. Three principal biases had been identified: (1) compositional bias, (2) variable rates of evolution and (3) heterotachy (also entitled covarion structure).

8.3.3 Compositional Bias

It was earlier shown that the G + C content of the rRNA was fundamentally heterogeneous among species (spreading between 30 and 70%) and could lead to an erroneous grouping of distantly related species that share a similar content (Loomis and Smith 1990; Weisburg et al. 1989). For example, thermophilic organisms have an elevated G + C content (Galtier and Lobry 1997); this produces a systematic bias. Modifications of the three principal tree-reconstruction methods were developed in order to take this problem into account (e.g. transversion-based parsimony (Woese et al. 1991), log-det distance (Lockhart et al. 1994) and nonstationary maximum likelihood (Galtier and Gouy 1995)). Despite some observable improvements (Lockhart et al. 1994; Woese et al. 1991), these methods did not considerably modify the inferred phylogenies. The corrections may even lead to less correct results, as exemplified by the enhanced statistical support for the early emergence of microsporidia, if the G + C content is taken into account (Galtier and Gouy 1995).

Furthermore, contrary to the first affirmations (Hasegawa and Hashimoto 1993), protein sequences can also be sensitive to the influence of compositional bias (Foster and Hickey 1999). For example, in a recent phylogeny based on a concatenation of ribosomal proteins, taxa with G + C rich genomes (e.g. *Mycobacterium* and *Deinococcus*) were artefactually put together (Brochier et al. 2002). Nevertheless, despite these few counterexamples, the impact of compositional biases is reasonably well treated by the current methods and cannot therefore be considered as a major problem, especially because it is quite easy to identify.

8.3.4 Long-branch Attraction

Another well-known bias, which can lead to errors in phylogenetic reconstructions, if the true phylogenetic signal is attenuated, is the one introduced by the variation of the evolutionary speed among species. For example, the evolutionary rates of the rRNA differ by a factor of 100 among planktonic foraminifers (Pawlowski et al. 1997). The artefact known as long-branch attraction (LBA) leads to the association of the longest branches, that correspond to the fastest evolving species, irrespective of their real phylogenetic affiliations (Felsenstein 1978). In the case of phylogenies rooted by a distantly related external group (outgroup), independently fast-evolving sequences of the studied taxa (ingroup) will emerge separately as the deepest branches, because they are attracted by the long branch of the distant outgroup (Philippe and Laurent 1998). Microsporidia that were proposed to be an early branching eukaryotic lineage (Vossbrinck et al. 1987), represent a spectacular example of LBA, because they are in reality highly evolved fungi (Keeling 2001b). The LBA artefact actually represents a major problem for molecular phylogenies, because it may affect the position of practically every deeply diverging lineage. In consequence, the position of many taxa, shown in bold in the universal tree of Fig. 8.2, should be considered as highly suspect, especially in the case of Eukaryotes, and at a lesser degree also for Bacteria.

Three main approaches were used to detect and/or reduce the impact of the LBA artefact: (1) enhanced or modified taxon sampling, (2) focusing on slowly evolving positions, and (3) better models of sequence evolution.

It was suggested very early during the development of molecular systematics that the addition of species reduces the impact of the LBA artefact, via breaking long branches (Hendy and Penny 1989). The eukaryotic phylogeny based on the elongation factor EF-1α furnishes a good example of this approach, even if the resolution of the tree is becoming limited concomitantly with the rising number of species (Moreira et al. 1999). Even if the theoretical basis is still under study (Poe and Swofford 1999), the practical approach is proving that this is an efficient method. Alternatively, if several representatives are available for a given group, it is important to do a more specific choice of taxa conserving only those that are more slowly evolving, rather than increasing naively the number of species. For example, the monophyly of the moulting animals (Ecdysozoa) was established by using a slowly evolving nematode species (Aguinaldo et al. 1997). The branch length of a given species taken from the root to the tip of a phylogenetic tree is roughly reflecting its evolutionary rate (e.g., short branches correspond to slowly evolving sequences).

The phylogenetic inference can also be improved by using the more slowly evolving positions (Olsen 1987), because they are less likely to contain noise (which means multiple substitutions, see Figs. 8.3 and 8.4) and will therefore more likely maintain the ancient phylogenetic signal (Felsenstein 2001). Our group recently developed a method, named SF (slow/fast) method, in order to detect these positions (Brinkmann and Philippe 1999). In short, the number of changes for each position is calculated by a maximum parsimony analysis independently for several predefined monophyletic groups. The evolutionary rate of a position is estimated as the sum of the number of substitutions observed in all groups; this estimate is therefore independent of the phylogenetic relationships among the groups. The number of substitutions at which the noise is exceeding the signal is not *a priori* known. Different data sets, containing all positions with less than n changes, are successively generated, covering the complete range of the relationships between signal and noise, and the evolution of the phylogenetic inference is also followed at the same time. If a rapidly evolving group is badly placed by the standard methods of phylogenetic reconstruction because of an LBA artefact, its true position should be found by using the most slowly evolving positions, and it will emerge progressively earlier when the more rapidly evolving positions are taken increasingly into account. Even if the reliable signal is rare and the resolution is often weak (see below), the SF method is efficient in detecting groups that are most likely incorrectly located because of LBA and may help to find the true affinity (Brinkmann and Philippe 1999; Philippe et al. 2000).

The LBA artefact is due to the fact that the observed number of substitutions is an underestimate of the real number, because of multiple substitutions that had taken place in the same position (Fig. 8.4); this is even more pronounced when a taxon is rapidly evolving (Olsen 1987). This underestimation happens if the model of sequence evolution used in the reconstruction process is not sufficiently reflecting the reality. The first model proposed for the detection of multiple substitutions assumes that all types of substitution are equivalent and that all sites have the same probability of change (Jukes and Cantor 1969). A progress was made by differentiating the types of substitutions (e.g., transitions vs. transversions) (Swofford et al. 1996). However, the most significant improvement was obtained when the unrealistic assumption of an equal substitution rate for all sites had been rejected (Yang 1996). For instance, the presence of invariant positions, although noninformative, violates the assumption of an equal substitution rate for all sites, and their identification and subsequent elimination improves the phylogenetic reconstruction (Lockhart et al. 1996). The support for the early emergence of the microsporidia in the phylogenies based on the elongation factors considerably decreased if the constant positions were removed from the analysis (Hirt et al. 1999). Actually, the most widely used method is based on the implementation of a model with variable rates between the positions (rate-across sites or RAS), modelling the heterogeneity of the substitution rates between the sites in the form of a continuous distribution of a Γ law type. This kind of approach, for example, helped to demonstrate that the microsporidia are highly derived fungi (Van de Peer et al. 2000).

8.3.5 Heterotachy

Models of the RAS type treat the variation of the substitution rate along the entire molecule, but they assume that the rate of a given position remains the same during its entire history (i.e. on all the branches of a phylogenetic tree). We named this hypothesis homotachy, for the same speed in Greek (Lopez et al.

2002). In fact homotachy has been shown to be unrealistic since the 1970s, with the pioneering work of Fitch (Fitch 1971), who demonstrated, for the cytochrome c, that the evolutionary rate of positions across the molecule are distributed differently between fungi and animals. A simple model of sequence evolution, the covarion model, was suggested to explain these observations. This model assumes that, at a given moment, only a fraction of the positions (the concomitantly variable codons, or covarions) is free to accept substitutions, with a constant probability for each of them (Fitch and Markowitz 1970). The rejection of the homotachy is therefore often described under the name of "covarion-like" phenomenon (Galtier 2001; Lockhart et al. 1996; Penny et al. 2001). Because large taxonomic samples are necessary for testing the hypothesis of homotachy (Lopez et al. 1999), the current massive increase of the data bases for molecular sequences permits a more thorough study of this question. We recently analysed about 2000 vertebrate cytochrome b sequences and demonstrated that 95% of the variable positions are significantly heterogeneous with respect to their substitution rates with time (Lopez et al. 2002). We used the term heterotachy instead of "covarion-like" to describe this major phenomenon, because the covarion model is only one among many other possible heterotachous models, and in addition it does not describe satisfactorily the evolution of the cytochrome b sequences (Lopez et al. 2002).

Heterotachy can generate reliable phylogenetic signals in sequence datasets (Lopez et al. 1999), in fossilizing ancient substitutions (e.g. a position that changed once several billions years ago became invariable rapidly after that change, thus excluding the possibility that multiple subsequent substitutions can erase this signal). Sometimes, heterotachy can represent an important source of trouble. It was shown that the presence of invariable positions in alignments can lead to an erroneous tree reconstruction (including the maximum likelihood methods, which are usually the most robust) (Lockhart et al. 1996). The analysis of the 16S rRNA and the elongation factor tufA of a number of non-photosynthetic and of oxygen-liberating photosynthetic prokaryotes (cyanobacteria and their endosymbiotic descendents chloroplasts) for example, has proved that the sharing of a similar ensemble of covarions (sites free to vary) can bias the estimate of the origin of chloroplasts (Lockhart et al. 1998). Because programs that take into account the phenomenon of heterotachy were not available, an easy method for improving the fit of the RAS model to the data is simply to eliminate the heterotachous positions from the alignment. This approach had shown, based on a concatenation of the 18S and 28S rRNA sequences, that a profoundly revised phylogeny of the Eukaryotes (compare Figs. 8.2 and 8.7) is not significantly rejected by these markers (Philippe and Germot 2000). The reconstruction of ancient phylogenies could especially benefit from the implementation of heterotachous models. The evolutionary processes of the heterotachous sequences could be responsible for the conservation of an ancient phylogenetic signal even if the mutational saturation would have almost totally eradicated all evolutionary messages (e.g., if a substitution did happen in an early emerging lineage and remained unchanged ever since) (Lopez et al. 1999; Penny et al. 2001). To model the heterotachy is very complex, but there is an increasing effort to achieve its implementation in the phylogenetic reconstruction methods (Galtier 2001; Huelsenbeck 2002; Penny et al. 2001).

8.3.6 Rare Genomic Events as an Alternative Approach?

In order to avoid the pitfalls described above, the approach called "hennigian" (the use of characters less sensitive towards saturation) could be powerful (Philippe and Laurent 1998). The comparison of rare genomic changes (e.g. insertions/deletions, or indels) could represent valuable markers supporting the monophyly of a group (Rokas and Holland 2000). Several promising results were obtained by using, for example, the integration of retroposons in the genome for mammals (Lum et al. 2000), the positions of introns for vertebrates (Venkatesh et al. 1999), and the order of mitochondrial genes for animals (Boore and Brown 1998). For the ancient phylogenies, this approach is less convincing. However, four indels had been used to support the opisthokont monophyly (animals and fungi) (Baldauf and Palmer 1993). One of these indels was also present in microsporidia, in agreement with the proximity of this group to the fungi (Hashimoto and Hasegawa 1996).

Small-sized indels could be particularly misleading as phylogenetic markers, because they are easily subject to convergence. In fact, they are usually found in regions of the protein (e.g. in loops exposed to the surface) that can easily accept the insertion or the deletion of segments representing several hydrophilic amino acids. For example, the reliability of an indel of two amino acids in the enolase, suggested as a support in favour of the early emergence of the trichomonads in the eukaryotic tree (Keeling and Palmer 2000), was seriously questioned (Bapteste and Philippe 2002; Hannaert et al. 2000). Assuming indels to have arisen several times independently by convergent evolution (and therefore being misleading) seems reasonable in the case of a few positions, but this problem should be less relevant in the case of large indels. But can they in consequence be considered to be perfect characters?

The eukaryotic gene coding for the valyl tRNA synthetase (ValRS), whose product is present in the cytoplasm and in the mitochondria of fungi, is considered to be probably of mitochondrial origin. In the phylogenies based on the ValRS, the eukaryotic sequences form a monophyletic group with the Proteobacteria (Brown and Doolittle 1995). This affiliation was further supported by the presence of a highly conserved, large insertion of 37 amino acids, shared by the eukaryotic and the γ-proteobacterial sequences, which was completely in agreement with the phylogeny based on this gene (Hashimoto et al. 1998). Nevertheless, because α-Proteobacteria are most closely related to the mitochondria, the mitochondrial origin of the ValRS was not completely convincing. A lateral gene transfer (LGT) of the ValRS gene from an archaebacterium to

Rickettsia, the first available sequence from an α-Proteobacteria, unfortunately did not allow this test to be performed (Woese et al. 2000). We decided to create an updated alignment of the ValRS consisting of 185 sequences, containing numerous α-Proteobacteria generated in the meantime. A phylogenetic analysis of the ValRS (Fig. 8.6) indicates that the α and the β/γ-Proteobacteria form two monophyletic groups, which are sister-groups, as expected. However, Eukaryotes did not show any specific affinity to the α-Proteobacteria, but rather emerge earlier, most likely because of a LBA artefact. Furthermore, the large insertion although present in all β/γ-Proteobacteria and in all Eukaryotes, is surprisingly absent in the α-Proteobacteria (see Fig. 8.6). At first sight, this observation seems to exclude the mitochondrial origin of the eukaryotic ValRS gene and would rather suggest a lateral gene transfer (LGT) from the β/γ-Proteobacteria to Eukaryotes. But, another well-conserved insertion of six amino acids is shared by a monophyletic group of four γ-Proteobacteria and by two representatives of the α subdivision (*Sinorhizobium* and *Agrobacterium*). These organisms are nevertheless solidly grouped with the other α-Proteobacteria, which do not have this insertion (Fig. 8.6). The same insertion, although longer and less conserved, is also present in one of the members of the high GC Gram-positives (*Mycobacterium*). In consequence, because the two large indels in the sequence of the ValRS are in conflict with each other and also in conflict with the phylogeny, they must be considered as ambiguous characters, and did not give any answer to the question of the origin of the eukaryotic ValRS.

Furthermore, rare genomic events are equally sensitive to LGT, as standard phylogenetic inference. For example, a large indel in the amino-terminal part of the HSP70 proteins (heat shock protein, 70 kDa) was for a long time considered as evidence to unite the Archaea and the Gram-positive bacteria, reinforced by the claim that some HSP70-based phylogenies seem to support this point of view (Gupta and Singh 1994). This indel was subsequently used for supporting a new hypothesis about a chimerical origin of the Eukaryotes (Gupta and Singh 1994). However, by using a larger taxonomic sampling, it was shown that (1) several Gram-negative bacteria did have a HSP70 gene with the same deletion that was supposed to be specific for the Gram-positive bacteria (Philippe et al. 1999) and (2) the majority of the archaeal species do not have a HSP70 gene, suggesting that the Archaea originally did not possess any HSP70 gene at all (Gribaldo et al. 1999). In fact, the taxonomic distribution of HSP70 genes in the complete genomes actually accessible (its absence in all Crenarchaeota, in the Pyrococci and in *Methanococcus*), strongly suggests a single LGT event with a Gram-positive bacterium, after the emergence of *Methanococcus* (see Fig. 2 in (Forterre et al. 2002)). In conclusion, rare genomic changes could be only useful markers in the reconstruction of phylogenies if they are used in addition to sequence data (Rokas and Holland 2000), but their direct use as an indicator of relatedness between species (Gupta and Johari 1998) should always be regarded very carefully.

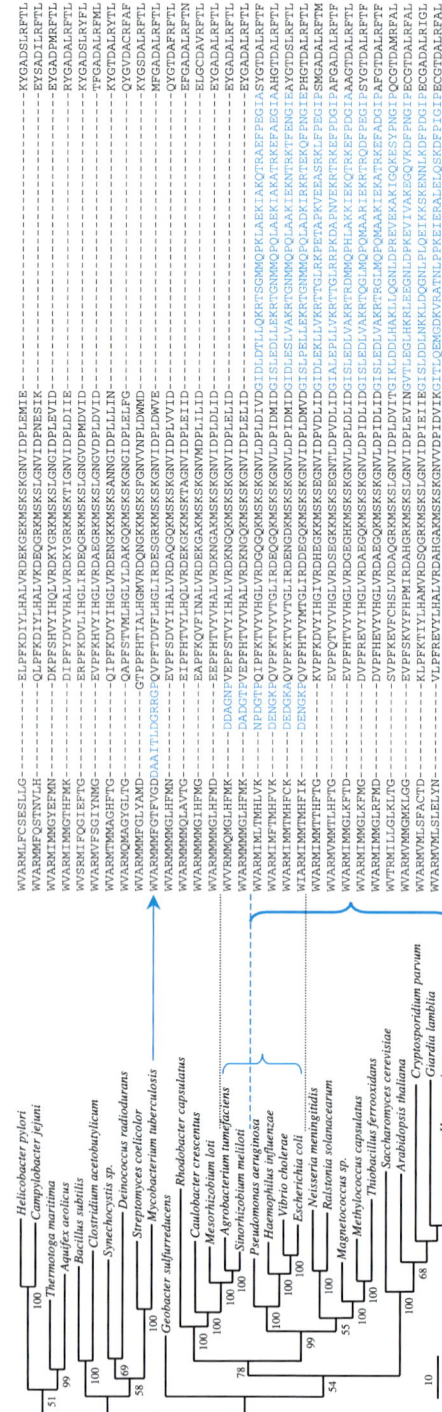

Fig. 8.6. Rare genomic events and the evolution of the Ile/Val-tRNA synthetase. (**A**) a representative sample of 30 sequences has been selected starting from a dataset containing 185 sequences of ValRS covering a huge taxonomic diversity. The tree has been inferred with a maximum likelihood method (Adachi and Hasegawa 1996). The proteobacteria of the α and β/γ subdivisions form two monophyletic groups, which are sistergroups, as expected. Surprisingly, Eukaryotes did not show any affinity for the α-Proteobacteria. (**B**) an extract of the ValRS alignment, which is focused on the region surrounding two highly conserved insertions. The large insertion of 37 residues has been proposed as a strong support in favour of the mitochondrial origin of the eukaryotic ValRS (Hashimoto et al. 1998), although this insertion is present in all β/γ-Proteobacteria and Eukaryotes, it is surprisingly absent from the α-Proteobacteria. Furthermore, the phylogenetic distribution of another well-conserved insertion of six amino acids is clearly in conflict with the phylogeny presented in (**A**). It is in fact shared by four γ-Proteobacteria and two β-Proteobacteria, which do not form a common group at all. In addition, the taxonomic distribution of this indel is also in conflict with the larger indel. Although, longer and less conserved, an insertion at the same position than the six-amino-acid insertion is also present in several high G + C Gram-positives (more precisely, present in *Mycobacterium*, but absent in *Streptomyces*)

8.4 Lateral Gene Transfer and the Quest for a Phylogeny of the Organisms

The sequencing of complete genomes has confirmed the high frequency of LGTs in prokaryotes (Koonin et al. 2001; Ochman et al. 2000) and of genes duplications in Eukaryotes (Lynch and Conery 2000). Therefore it is potentially dangerous to deduce the phylogeny of the organisms starting from trees that are only based on a single gene (Page and Charleston 1997). The presence of multiple copies of a homologous gene in a single genome opens the way to recombination. But, if a gene becomes involved in recombination events, the most appropriate mathematical representation of its history would be a network, rather than a tree (i.e. a connected graph without cycles). Difficult to identify, the intragene recombination was recently demonstrated through the incongruent taxonomic distribution of indels in the enolase and in the inositol monophosphate dehydrogenase (IMPDH), rendering the phylogenetic reconstruction of Eukaryotes based on these genes elusive (Bapteste and Philippe 2002; Keeling and Palmer 2001). In fact, recombination between paralogous or xenologous genes is probably more frequent than was previously thought (Archibald and Roger 2002), rendering the phylogenetic inference even more problematic (Posada and Crandall 2002).

The methods used to identify the LGTs (so-called indirect methods) are principally based on the non-normal nucleotide composition and the fact that the most similar sequences in the data banks did not belong to closely related species (similarity is normally estimated by a BLAST search). But these approaches are only poor indicators (Guindon and Perriere 2001; Koski and Golding 2001; Koski et al. 2001; Wang 2001). This is illustrated by the fact that four alternative methods, detected different groups of genes (Ragan 2001). The most likely explanation is that these indirect methods could potentially detect LGT of different ages (Lawrence and Ochman 2002; Ragan 2001). In consequence, albeit slow, the thorough phylogenetic approach represents the only direct detection method that should be used systematically (see below).

The underlying assumption of this method is that a phylogeny of organisms exists and can be represented in the form of a bifurcating tree. Because LGTs seem to be common and affect all genes (Koonin et al. 2001), it was suggested that "the history of life could not be correctly represented in the form of a tree" (Doolittle 1999c). Other authors have less radical opinions on the damaging impact of LGTs on the reconstruction of the evolutionary history. For example, it was suggested that LGTs could have played an important role only during a precellular stage of the evolution because it was hypothesised that "primitive entities were basically modular (loosely coupled) in construction", thus facilitating LGT events (Woese 2000). Alternatively, only a subdivision of the genetic materials could be affected by this phenomenon, excluding a veritable core of nontransferable genes (Jain et al. 1999). In any case, it seems that the LGTs likely represent an important evolutionary force in the diversification of the prokaryotes (Doolittle 1999b; Martin 1999; Ochman et al. 2000).

In order to examine the sheer existence of the phylogeny of organisms, several groups applied genomic approaches (Fitz-Gibbon and House 1999; Korbel et al. 2002; Lin and Gerstein 2000; Tekaia et al. 1999; Wolf et al. 2001). These methods are normally based on gene content (presence or absence) or the order of genes. The resulting phylogenies are more or less similar to the rRNA-based tree. For example, the monophyly of the three domains and of several other major groups (like animals, spirochaetes, Proteobacteria) is generally recovered. However, unexpected relationships, like that between *Thermoplasma* (Euryarchaeota) and the Crenarchaeotes (Korbel et al. 2002), demonstrate biases introduced in these approaches by frequent LGTs between phylogenetically distant taxa, but that are coexisting in the same environment (Ruepp et al. 2000). Another striking example is given by the hyperthermophilic bacteria, which probably acquired up to 24% of their genes from the Archaea (Aravind et al. 1998; Nelson et al. 1999). Furthermore, the convergent loss/gain of a comparable ensemble of genes due to similar physiological conditions, e.g. the adaptation to intracellular parasitism, can equally introduce biases. In conclusion, the methodology based on complete genomes should rather be considered as phenetic rather than as phylogenetic approaches (Doolittle 1999a; Wolf et al. 2001). Large-scale studies based on the concatenation of a great number of genes (Brown et al. 2001), despite their great potential for the inference of phylogenetic trees, can also be influenced by LBA artefacts, especially if they are applied to prokaryotes. In fact, if too many of the laterally transferred genes are included in the data sets, the results could correspond to a simple mean of the LGTs frequencies between the organisms, rather than being a correct indicator of the phylogeny of the organisms. Circular and therefore specious methodology could be applied in order to recover the tree in which one believes. For example, Brown and collaborators (2001) observed incongruities between their tree based on several fused genes and the one based on the rRNA. They obtained an increase of the congruence (e.g. the basal emergence of the hyperthermophilic bacteria) by eliminating the nine genes that did not recover the monophyly of the bacterial clade out of a total of 23. Such an approach based on only 14 genes does not represent convincing evidence in favour of the existence of a phylogeny of the organisms.

In order to avoid these limitations, our group recently developed a method that evaluates congruence among genes without a reference phylogeny (Brochier et al. 2002). Starting with 45 complete bacterial genomes, 59 markers implicated in translation (57 proteins and the 16S and 23S) were chosen, which were supposed to be excluded from LGTs (Jain et al. 1999), as well as 39 markers well known for being transferred (essentially tRNA synthetases (Woese et al. 2000), but also some ribosomal proteins (Brochier et al. 2000)). To estimate the congruence, each marker was described by a vector containing its likelihood values for an ensemble of representative topologies. These vectors were subsequently examined by a principal-component analysis (PCA). If two genes share the same evolutionary history, they furnish comparable support for the same topologies and will therefore group in the same area of the PCA. After eliminating stochastic effects, 46 of the 59 markers a priori supposed to be nontransferable, and six of the 39 a priori supposed transferable formed a compact cloud of points. In consequence, the identification of a genuine core of 52 genes, that belong a posteriori to the nontransferred fraction of the genome among the 45 analysed species, argues strongly in favour of a phylogeny of organisms for the Bacteria. Furthermore, the phylogeny based on these proteins is in good agreement with that based on the two concatenated rRNA genes (Brochier et al. 2002). The PCA method also allowed detection of the very likely cases of LGTs that were not reported earlier (e.g. for the translation elongation factor EF-G).

We applied the same approach to 14 archaeal species and obtained very similar results: a core of nontransferred genes could be detected and the phylogeny based on their concatenation gives a good approximation of the phylogeny of the species (e.g. obtained by the concatenated rRNAs) (Matte-Tailliez et al. 2002). Without minimising the importance of the LGTs in Prokaryotes, these results nevertheless demonstrate that a universal tree of organisms really exists and that its inference represents the primary challenge of the molecular phylogenies. We will now focus on the recent progress made to resolve this challenge.

8.5 A New Evaluation of the Universal Tree of Life

In light of the tree-reconstruction problems described above, especially the LBA artefact, it is legitimate to question whether the standard view of the ancient phylogenies is really sound. Especially, many of the basal branches (shown in bold on Fig. 8.2) could potentially be incorrect.

8.5.1 The Root of the Universal Tree of Life

The inference of the root of the universal tree is fundamentally the most difficult duty for a molecular phylogeneticist, because it corresponds to the most ancient evolutionary event that one can study. It is somewhat surprising that the vast majority of the researchers take the bacterial rooting for granted (Fig. 8.2), because this had been deduced from a few genes (less than ten), each of them having only a few alignable positions (ca. 100), and a very limited taxon sampling. This is even more striking because the methods used to reconstruct the trees were rather rudimentary, in general distance-based methods without correction for the variation of the evolutionary rates across sites. This approach is known to be very sensitive to LBA artefacts (Swofford et al. 1996), particularly with a very limited species sampling (Hillis 1996; Lecointre et al. 1993; Philippe and Douzery 1994). In a reanalysis with an improved taxon sampling, the majority of the phylogenies used for the inference of the root of the universal tree of life (Brown and Doolittle 1995; Brown et al. 1997; Gogarten et al. 1989; Gribaldo and Cammarano 1998; Iwabe et al. 1989; Lawson et al. 1996) indicated a much more complex evolutionary pattern (Philippe and Forterre

1999). For example, in the phylogeny based on the isoleucyl tRNA synthetase the two prokaryotic domains are so extremely intermingled that it is very difficult, even impossible, to know which sequences are the true representatives of the Archaea and of the Bacteria, respectively. More fundamentally, all genes revealed very high levels of mutational saturation (Philippe and Forterre 1999). In fact, these highly conserved genes contain by essence a certain number of invariant positions. But saturation is free to accumulate at variable positions, which furnish noise rather than phylogenetic information. In consequence, instead of informative positions (like the position 2 in Fig. 8.4), noisy positions (like the position 2 in Fig. 8.4) furnish the principal information to determine the location of the root, as illustrated by the case of the duplicated genes of the isoleucyl/valyl tRNA synthetases (Fig. 8.5). In that instance, the most slowly evolving positions (less than ten changes) furnish a negligible number of steps (442) in comparison to those (2032) generated by the rapidly evolving positions (more than 40 changes). Encountering such a high level of saturation, the classical methods of reconstruction have only a moderate performance and are susceptible to artefacts.

Furthermore, among the three domains of life, the bacterial branch is always the longest one for all of the genes used to root the universal trees of life (Philippe and Forterre 1999). Because this root is inferred by using an anciently duplicated paralogous gene, which *per se* represents a very long branch, a LBA artefact between these two long branches is predictable because of the mutational saturation. Therefore, the bacterial rooting could be simply the result of an artefact. Since branch length equals λt, with λ being the substitution rate and t the absolute time, the long bacterial branch can be schematically explained in two opposite ways: the origin of the bacterial domain is really more ancient (and t has a very elevated value) or the bacterial genes evolve more rapidly than their archaeal and eukaryotic homologs (and λ has a very high value). In the first case, the bias created by the LBA artefact is acting in favour of the true phylogeny, whereas in the second case it will hinder its discovery. A better way to distinguish between these two hypotheses (high value of t or λ) is to apply methods that focus on the slowly evolving positions, because they are less sensitive to the LBA artefact (Olsen 1987).

Such a method was developed by our group and applied to the paralogous pair EF1α/EF2 (Lopez et al. 1999). By using the more slowly evolving positions, the two opposite topologies (Bacteria, (Archaea, Eukaryotes)) – the current paradigm – and ((Bacteria, Archaea), Eukaryotes) – prokaryotic monophyly – could not be distinguished. This result is in agreement with an elevated λ (substitution rate) and not an elevated time t. The most concrete point of this analysis was the very long branch that separates EF1α and EF2, that renders it very difficult to avoid LBA artefact. We subsequently focused our attention on the SRP54/SRα sequences, because these two genes are by far the most similar pair among the anciently duplicated paralogs (Brinkmann and Philippe 1999). When the positions, containing the most reliable phylogenetic signal were extracted (positions with less than six changes), the Bacteria displayed a very long branch, but were nevertheless a sister-group of the Archaea (indicating a eukaryotic rooting of the universal tree). This fact constitutes a strong argument in favour of the hypothesis assuming a large λ. The utilisation of the positions with seven substitutions is provoking a shift from the topology ((B,A),E) to (B,(A,E)). Quite interestingly, this shift is not due to any signal in favour of the topology (B,(A,E)), but rather to a rejection of the ((B,A),E) by noisy char-

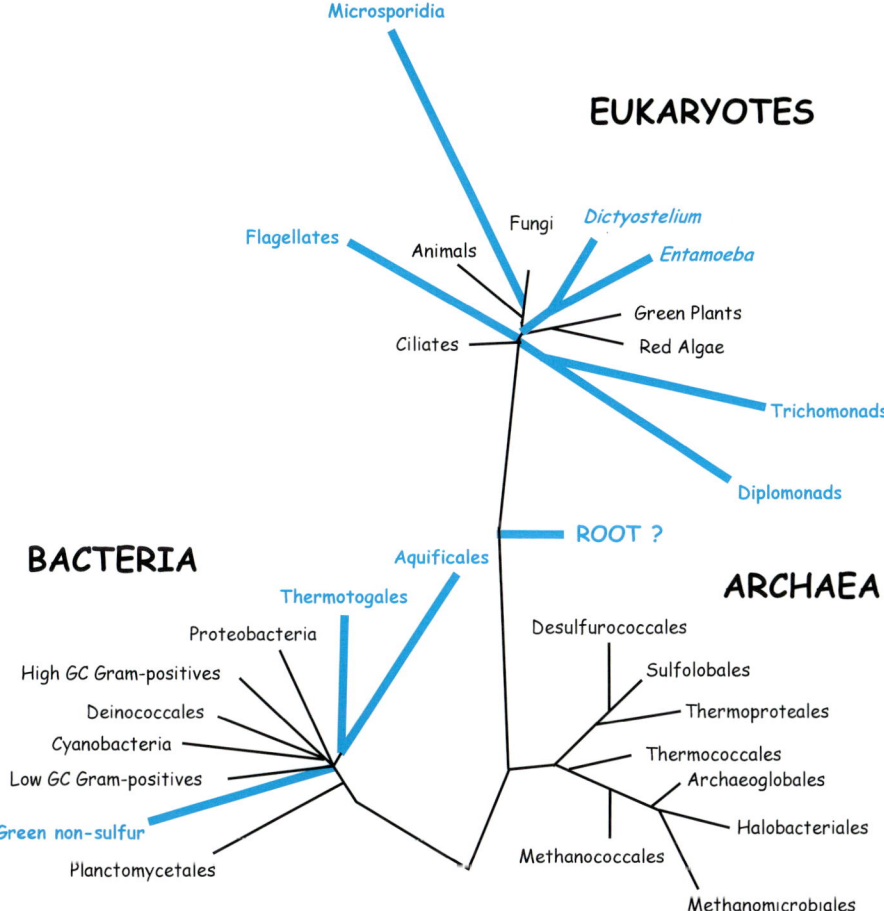

Fig. 8.7. A new vision on the universal tree of life. Taxa that are incorrectly placed in Fig. 8.2 because of the LBA artefact are replaced on this phylogeny, based either on the reanalysis of the rRNA (e.g., hyperthermophilic bacteria) and of the anciently duplicated genes (for example the root), or based on the analyses of new markers (for example, the grouping of *Dictyostelium* with *Entamoeba*). It should be noted that, while some of these revisions are robust (e.g. microdsporidia with fungi), several others still remain open (e.g. the root)

acters, because of a LBA artefact. In conclusion, the study of slowly evolving positions provides a serious proof that the rooting in the bacterial branch is due to an elevated λ(LBA) and not an elevated t (the true phylogeny). The statistical support for the eukaryotic rooting (the monophyly of Prokaryotes, Fig. 8.7) is weak, because the slowly evolving positions correspond to too small a number. The implications of the rooting in the eukaryotic branch have been discussed in detail elsewhere ((Forterre and Philippe 1999a; Forterre and Philippe 1999b); Penny and Poole 1999; Philippe and Forterre 1999), and will only be mentioned very briefly below. The most important consequence is that such a rooting prevents knowledge of whether LUCA was a Prokaryote or a Eukaryote. It seems therefore that the rooting of the universal tree of life should be considered as an open question.

8.5.2 Prokaryotic Phylogeny

The fact that the branches leading to the hyperthermophilic bacteria in the tree based on the rRNA are shorter than those leading to the mesophiles had been taken as an indication that a LBA artefact was not responsible for their early emergence, but rather that the hyperthermophilic bacteria were "living fossils". The observations about the prokaryotic phylogeny and the rooting in the eukaryotic branch also led to the hypothesis of thermoreduction (Forterre 1995), that propose that the prokaryotic organisation of the cells emerged originally as an adaptation of life to high temperatures.

The first branches of the bacterial phylogeny are potentially misplaced for the reason of the LBA, because the external group (Archaea and Eukaryotes) is very distant (Fig. 8.2). We recently applied the SF-method, which focuses on the most slowly evolving positions (Brinkmann and Philippe 1999), to a large dataset of the rRNA containing the representatives of the majority of the prokaryotic phyla and we demonstrated that the early emergence of hyperthermophilic bacteria (Aquificales and Thermotogales) is due to a LBA artefact (Brochier and Philippe 2002). This fact argues in favour of a secondary adaptation to high temperatures in these lineages. This deduction is in agreement with an estimate of the G + C content of the rRNA of LUCA (54–56%)(Galtier et al. 1999), a value that is incompatible with a growth temperature higher than 80°C. The complete genomes of *Aquifex* and *Thermotoga* showed that numerous genes found in these species (respectively 16% and 24%) have been acquired by LGTs from the Archaea (Aravind et al. 1998; Deckert et al. 1998; Nelson et al. 1999). Such a high flux of genes could have helped the adaptation to a life at high temperature in hyperthermophilic bacteria. An example of this type of genes is the reverse gyrase, which generates a positive superturn (supercoil) and very likely serves in the stabilisation of the DNA at high temperature. A detailed phylogenetic analysis of this gene supports the independent acquisitions of this gene by *Aquifex* and *Thermotoga* from Archaea (Forterre et al. 2000). These recent results therefore strongly argue against the hypothesis of a hyperthermophilic ancestor of Bacteria and advocate towards a nonhyperthermophile (in contrast, the ancestor of Archaea is most likely a hyperthermophile (Forterre et al. 2002)).

Surprisingly, the SF-method applied to the bacterial rRNA suggests, with a reasonable statistical support (bootstrap value of 75%), that the Planctomycetales may be the earliest emerging bacterial phylum (Brochier and Philippe 2002). This phylum is an important bacterial division where the members share several remarkable characteristics, like the absence of peptidoglycan in their cell wall and reproduction via budding (Fuerst 1995). Another particularly intriguing trait is the existence of an internal membrane, simple or double, around the bacterial chromosome in the genera *Gemmata* and *Pirellula*, respectively. Although this structure had been compared to the nuclear envelope of the Eukaryotes (Fuerst 1995), the evolutionary homology with the nuclear membrane has not been proven. Unfortunately, only a few studies were made on this group, despite their special characteristics; yet their implication in the anaerobic oxidation of ammonium has been recently demonstrated (Strous et al. 1999). If the early emergence of the Planctomycetales favoured in our reanalysis of the rRNA is confirmed with additional phylogenetic markers, the origin of Bacteria and the nature of LUCA should be seriously re-evaluated. With the exception of the basal branches, the bacterial phylogeny based on the rRNA has been globally confirmed by the analyses of complete genomes, such as the trees based on the gene order or on the gene content (Korbel et al. 2002; Wolf et al. 2001), or by the simultaneous analyses of numerous genes (Brochier and Philippe 2002; Brown et al. 2001; Daubin and Gouy 2001; Wolf et al. 2001). Surprisingly, the resolution between the major bacterial phyla has been slightly improved despite the use of very large datasets (Fig. 8.7). Only three new supergroups have been suggested by the phylogenetic analysis of concatenated genomic datasets; these are Chlamydiales/Spirochaetes, Thermotogales/Aquificales and high G + C Gram-positive bacteria/Deinococcales/Cyanobacteria (Brochier et al. 2002; Brown et al. 2001; Daubin and Gouy 2001; Wolf et al. 2001). However, it is not impossible that these affiliations could have been influenced by one of the biases discussed above; i.e. a very elevated evolutionary rate (due to parasitism), the amino-acid composition (linked to hyperthermophily) and the genomic G + C content, respectively.

In conclusion, since new data generated in numerous genome-sequencing projects are released rapidly, approaches based on large concatenations of genes represent the most promising perspective for resolving the prokaryotic phylogeny. It is, nevertheless, crucial to work on the methodological improvement of tree reconstruction, in order to extract the maximum information from this enormous quantity of sequences. One example of a methodological improvement will be given in the following section.

8.5.3 Eukaryotic Phylogeny

The Archaezoan hypothesis, proposing eukaryotic lineages that appeared before the mitochondrial endosymbiosis, has been criticised. In fact, it is possible to know whether a species does not have mitochondria because they have been lost, or because they were never present. Indeed, after the mitochondrial endosymbiosis, the essential part of the genetic inheritance of the α-Proteobacteria has been either lost or transferred to the nucleus (Lang et al. 1999). The proteins encoded by these genes are often imported into the mitochondria, but are also located in other compartments (cytoplasma, nucleus, etc.). Therefore, if a species is losing their mitochondria, or at least the oxidative respiration that constitutes its principal function, during the adaptation to an anaerobic environment, it will likely conserve some genes of mitochondrial origin. The discovery of genes of mitochondrial origin in all of the eukaryotic lineages without mitochondria (especially, the early emerging ones like the diplomonads and the microsporidia) strongly rejected the Archaezoan hypothesis (Embley and Hirt 1998). Furthermore, the early emergence of these lineages, which represented the second basis of this hypothesis, is highly suspicious, given the very long branch leading to the prokaryotic outgroup (Fig. 8.2). In fact, numerous independent analyses have shown that the basal branches in rRNA-based phylogenies are incorrectly placed because of the LBA artefact (for a review see (Philippe 2000)).

The challenge is now to find enough correct phylogenetic positions of all the lineages for which rRNA sequences are rapidly evolving. Analyses based on a single gene shed some light on these questions. For example, microsporidia have been grouped together with the fungi thanks to phylogenies based on tubulin (Edlind et al. 1996), the mitochondrial HSP70 (Germot et al. 1997), and the RNA polymerases II (Hirt et al. 1999). The trees based on actin (Keeling 2001a) and an indel found in polyubiquitin (Archibald et al. 2003) have suggested the grouping of foraminifers and Cercozoa, a group of highly heterogeneous protists. The principal drawback of phylogenies based on a single gene is a very limited resolution of the relationships between the principal eukaryotic lineages (Budin and Philippe 1998). In consequence, the necessity to increase the resolving power imposes the simultaneous use of several genes. The first attempts to reconstruct the eukaryotic phylogeny by using a large number of data were done starting in the mid-1990s (Kuma et al. 1995; Nikoh et al. 1994). Unfortunately, the taxon sampling was very limited (four or five species), rendering these phylogenetic inferences not very reliable. Several attempts have subsequently been made with a larger number of species, although this reduced the number of genes used, invariably leading to poorly resolved phylogenies (Budin and Philippe 1998; Germot and Philippe 1999).

One notable exception is a phylogeny based on concatenated mitochondrial genes (cob1, cox1, cox2, and cox3), which provided a strong statistical support for the monophyly of Opisthokonta (animal+fungi) and of Plantae (red algae and green plants) (Burger et al. 1999). The recent explosion of the sequence data for a much larger taxonomic variety, led to an increase of analyses based on multiple genes, in particular for angiosperms (for review see (Chase and Fay 2001)), the mammals (e.g. (Madsen et al. 2001; Murphy et al. 2001)), but also for the Eukaryotes ((Arisue et al. 2002a; Arisue et al. 2002b); Baldauf et al. 2000; Bapteste et al. 2002; Fast et al. 2002; Moreira et al. 2000).

Several recent results have clarified important questions of eukaryotic evolution (e.g. the acquisition and the loss of organelles). The analyses based on multi-gene fusions (Burger et al. 1999; Moreira et al. 2000) furnished significant support for a single primary endosymbiosis with a cyanobacterium at the origin of chloroplasts, because they indicated the monophyly of Plantae (green plants, red algae and glaucocystophytes). Furthermore, the grouping of the two principal phyla that harbour a chloroplast enveloped by four membranes, the alveolates and the stramenopiles (Arisue et al. 2002b; Baldauf et al. 2000; Bapteste et al. 2002; Fast et al. 2001) strongly suggested that secondary endosymbioses (the phagocytotic uptake of a photosynthetic Eukaryote by another Eukaryote) are much rarer than was previously thought. In fact, only two secondary endosymbiotic events could have taken place, in contrast to seven as was previously assumed (Cavalier-Smith 1999). Nevertheless, in order to definitely solve this question, additional data from other lineages of photosynthetic Eukaryotes (especially, haptophytes, cryptophytes, chlorarachniophytes and the euglenoids) are needed. The discovery of two new eukaryotic super-groups of organisms without mitochondria, *Entamoeba* and *Mastigamoeba* on the one hand (Arisue et al. 2002a; Bapteste et al. 2002), trichomonads and diplomonads on the other hand (Baldauf et al. 2000; Embley and Hirt 1998; Henze et al. 2001), also reduced the expected number of independent secondary losses of mitochondria. Anyhow, the independent losses of mitochondria and chloroplasts are likely much more frequent than their acquisition by endosymbiosis (see also below).

8.6 The Importance of an Evolution by Simplification and by Extinction

The idea that evolution is going from simple to complex, which is inherited from the philosophy of Aristotle (as already mentioned), deserves to be discussed here in more detail in light of the recent results from molecular phylogenies. First of all, it is certain that starting from the origin of life (and even before) up to the emergence of LUCA, a very important rise in complexity of living organisms happened. But, there is no proof that this rise in complexity arrived in a linear way and that there had not also been episodes of secondary simplification. There is often a tendency to extrapolate the phenomenon of a steady rise in complexity to the period after LUCA. But in particular, the often-committed error is to associate the simplicity of extant species with primitiveness, even often considering them as inferior (e.g., the lower animals to designate the invertebrates).

The concept of the "living fossil" is intimately mixed with this problem. Hence, the microsporidia, which are simple Eukaryotes (without flagella, Golgi apparatus and mitochondria) were often considered as primitive species, the "living fossils" from an era antedating the mitochondrial endosymbiosis. Unfortunately, the notion of a "living fossil" is highly arguable. In fact, all extant species are the descendants from a single common ancestor, LUCA, and did evolve exactly during the same period of time: they are therefore all approximately equally evolved (not taking into account different rates of evolution). To say that an extant species is a fossil implies the assumption that it is less evolved than the others. However, a major issue concerning this concept is that its application is highly subjective. For example, the coelacanth, probably the most famous example of a "living fossil", has a morphology that is globally similar to that of certain several hundred million years old fossils, but it is very likely that its physiology, and in consequence also its life style, is very different. Furthermore, it is obvious that the genome of the coelacanth is as evolved as the genome of all other vertebrates. One can propose an objective measure for the degree of evolution of an organism, and it would be the number of generations since the last common ancestor. This measure is perfectly reasonable because the influence of the natural selection rises with the number of generations (or of DNA replications). Based on these grounds, the less-evolved species would be those that have the longest generation times (e.g., the elephant, man or the turtle) and the most highly evolved would be those that have the smallest generation times (e.g., the Bacteria *Escherichia coli* or *Bacillus subtilis*). In fact, the sheer concept of evolved or primitive species is without any interest if one considers extant species and is only correct for palaeontologists. Let us go back to the problem of the evolution from simple to complex without further considering the useless complications introduced by the use of expressions like "primitive species" or "lower organisms".

The molecular phylogenies have indeed demonstrated that in numerous cases the species having numerous simple characteristics often emerge late in the trees. For example, the Bacteria (*Mycoplasma*) and the Archaea (*Thermoplasma*) without a cell wall arose among species with a cell wall, thus strongly suggesting secondary loss of cell walls. Eukaryotes without flagella, like red algae and the majority of fungi, emerge in between organisms that have well-developed flagella, suggesting that this complex organelle (composed of several hundred different proteins) was lost at least twice. More surprisingly, chloroplasts were lost in several independent lineages. In certain cases, the photosynthetic function was lost, but the structure was conserved, like for example in *Plasmodium falciparum*, the intracellular parasite that causes malaria (Waller et al. 1998). In other cases, like for example ciliates (Fast et al. 2001), diplomonads, trichomonads and the trypanosomes (Andersson and Roger 2002; Hannaert et al. 2003), it seems that the original chloroplast was completely lost and that only some genes of cyanobacterial origin were conserved in the nucleus. Finally, as already discussed before, mitochondria were lost several times. To be more precise, the respiratory function based on oxygen as well as the genome were lost, but a mitochondrial structure (limited by two membranes) is conserved. In particular, thanks to monoclonal antibodies directed against the chaperonin proteins of mitochondrial origin, it was possible to demonstrate the presence of highly reduced mitochondria in *Entamoeba histolytica* (Mai et al. 1999; Tovar et al. 1999), in the microsporidia

(Williams et al. 2002) and in diplomonads (Tovar et al. 2003). It is conceivable that the complete loss of mitochondrial structures is impossible, because the mitochondria are implicated in numerous, nonrespiratory functions central to eukaryotic cells, like the iron-sulphur protein maturation (Katinka et al. 2001; Tovar et al. 2003).

In contrast to what seems to indicate the first results (Fig. 8.2), molecular phylogenies have revealed that the evolution by secondary simplification represents a major process. They indeed simply allowed recovering a quite old result. In 1943, André Lwoff published a book exclusively dedicated to the importance of the evolution from the complex towards the simple, as its title suggests, "L'évolution physiologique. Etude des pertes de fonctions chez les microorganismes" (The physiological evolution: Studies of functional losses in microorganisms) (Lwoff 1943). The introduction of this book, which is still perfectly up to date and whose lecture we strongly recommend, explains very well the reasons of the psychological rejection that we have of regressive evolution. In particular, because of positivism, there exists a very strong association between evolution and progress on the one hand and between progress and complexity on the other hand. Therefore, an evolution by simplification is strongly opposite to the myth of progress. More recently, the palaeontologist Stephen Jay Gould renewed that discussion and added a very simple but also very convincing argument (Gould 1996). Let us assume that evolution corresponds to a random walk, going sometimes to the complex, sometimes to the simple. Since living organisms cannot exceed a minimal limit of simplicity (especially, if one does not consider parasitic species), the random walk will generate more and more complex organisms, even if the majority of species is relatively simple. Therefore, the existence of complex organism like animals and plants does not at all mean that there is a tendency of the evolution towards complexity, but can very well be explained by a stochastic process.

In conclusion, it is certain that in order to explain the evolution of life one has to take into account both the evolution from simple to complex and the evolution from complex to simple (Forterre and Philippe 1999a). In fact, it would be the best to completely re-evaluate the problem. The fundamental point is that the evolutionary steps have to be simple, in order that the innovations can be maintained by natural selection, either for simplifying or for making the organisms more complex. It is therefore useless to look for simple organisms, in order to inform us about the ancient evolution. In contrast, it is indispensable to study the evolutionary process and to develop evolutionary scenarios about the origin of the major groups of organisms, which suppose simple evolutionary stages. A provoking proposition inspired by this idea is that LUCA would have resembled more a Eukaryote than a prokaryote. The major argument employed is that Eukaryotes have much more RNA-based reactions than the prokaryotes (Forterre and Philippe 1999a; Poole et al. 1999). If one accepts the hypothesis about the RNA world for the origin of life (see also Maurel and Haenni Chap. 6, Part II), the replacement of RNA-based reactions by protein-based reactions is essentially an irreversible phenomenon. In this sense, the scarcity of catalytic RNA in prokaryotes supports their late emergence (or of their derived status). Even if this hypothesis remains controversial, one should remember that the picture that we have about the ancient evolution is under major reconstruction (compare Figs. 8.2 and 8.7).

8.7 Exobiology, a Procession of Extinctions?

We have not discussed a major problem that is limiting the efficiency of the methods based on the comparison of extant species. This is the question of extinctions. For example, if one takes a look at the mammals, the last common ancestor of the extant species has an age of about 130 million years, whereas the divergence between mammals and the other vertebrates happened about 310 million years ago. This indicates that, on a branch corresponding to 180 million years of evolution, only a single species has survived; these circumstances exclude the access of biologists to any living material that originated within this period. It is extremely difficult to understand the appearance of the mammalian characteristics only based on the extant species (like for example the bones that form the internal ear). In this special case, rich fossil data allowed us to elucidate numerous problems (like the progressive movement of the jaw bones towards the inner ear). Unfortunately, if one is interested in more ancient evolution, palaeontology is barely useful because of the generally bad fossilisation of micro-organisms. More precisely, the fossils usually provide only morphological information, which is of extremely limited taxonomic use, because of their mostly very simple rod-shaped or coccoid morphology. It is even difficult to recognize these objects as true fossils since minerals can mimic the simple morphologies of a large part of the micro-organisms, as exemplified by the debate about the archaean Cyanobacteria (Brasier et al. 2002; Garcia-Ruiz et al. 2003; Schopf et al. 2002). The existence of long internal branches, where not a single extant (e.g. surviving) species emerges (see the basis of Bacteria and Eukaryotes in Fig. 8.7), is therefore a major limitation on the inferences. It is possible that a better description of the terrestrial biodiversity, especially due to the molecular ecological methods, will allow several additional important lineages to be discovered. In fact, numerous species, even the majority, have escaped to the classical culture methods, in both prokaryotes (Pace 1997) and Eukaryotes (Lopez-Garcia et al. 2001). If one extrapolates the observations made based on the fossil record of multicellular organisms, it is expected that the vast majority of these species, especially those that could have witnessed the simple stages of ancient evolution have disappeared. These extinctions explain the difficulties that can arise if one studies the evolution of life since LUCA despite the knowledge of the complete sequence of more than one hundred genomes. The exobiology can therefore be of great help to us, by discovering other living systems. In light of the importance of the historical contingency (Gould 1996), these systems would very likely have evolved in different ways and would have suffered from completely independent extinction events. This would constitute, in some respect, a parade of extinctions that would have taken place on Earth.

Acknowledgement

We thank Bernard Barbier, Christophe Douady, Hervé Martin, and David Moreira for many useful comments and Muriel Gargaud for her kindness, her patience, and her never-ending interest in the subject. This work was supported by the Canada research chair program.

References

Adachi J, Hasegawa M (1996) MOLPHY version 2.3: programs for molecular phylogenetics based on maximum likelihood. *Comput. Sci. Monogr.* **28**:1–150

Aguinaldo AM, Turbeville JM, Linford LS, Rivera MC, Garey JR, Raff RA, Lake JA (1997) Evidence for a clade of nematodes, arthropods and other moulting animals. *Nature* **387**:489–493

Andersson JO, Roger AJ (2002) A cyanobacterial gene in nonphotosynthetic protists – an early chloroplast acquisition in Eukaryotes? *Curr. Biol.* **12**:115–119

Aravind L, Tatusov RL, Wolf YI, Walker DR, Koonin EV (1998) Evidence for massive gene exchange between archaeal and bacterial hyperthermophiles. *Trends Genet.* **14**:442–444

Archibald JM, Longet D, Pawlowski J, Keeling PJ (2003) A novel polyubiquitin structure in cercozoa and foraminifera: evidence for a new eukaryotic supergroup. *Mol. Biol. Evol.* **20**:62–66

Archibald JM, Roger AJ (2002) Gene duplication and gene conversion shape the evolution of archaeal chaperonins. *J. Mol. Biol.* **316**:1041–1050

Arisue N, Hashimoto T, Lee JA, Moore DV, Gordon P, Sensen CW, Gaasterland T, Hasegawa M, Müller M (2002a) The phylogenetic position of the pelobiont *Mastigamoeba balamuthi* based on sequences of rDNA and translation elongation factors EF-1a and EF-2. *J. Eukaryot. Microbiol.* **49**:1–10

Arisue N, Hasshimoto T, Yoshikawa H, Nakamura Y, Nakamura G, Nakamura F, Yano T-A, Hasegawa M (2002b) Phylogenetic position of *Blastocystis hominis* and of stramenopiles inferred from multiple molecular sequence data. *J. Eukaryot. Microbiol.* **49**:42–53

Baldauf SL, Palmer JD (1993) Animals and fungi are each other's closest relatives: congruent evidence from multiple proteins. *Proc. Natl. Acad. Sci. USA* **90**:11558–11562

Baldauf SL, Roger AJ, Wenk-Siefert I, Doolittle WF (2000) A kingdom-level phylogeny of Eukaryotes based on combined protein data. *Science* **290**:972–977

Bapteste E, Brinkmann H, Lee JA, Moore DV, Sensen CW, Gordon P, Durufle L, Gaasterland T, Lopez P, Muller M, Philippe H (2002) The analysis of 100 genes supports the grouping of three highly divergent amoebae: *Dictyostelium*, *Entamoeba*, and *Mastigamoeba*. *Proc. Natl. Acad. Sci. USA* **99**:1414–1419

Bapteste E, Philippe H (2002) The potential value of indels as phylogenetic markers: position of trichomonads as a case study. *Mol. Biol. Evol.* **19**:972–977

Boore JL, Brown WM (1998) Big trees from little genomes: mitochondrial gene order as a phylogenetic tool. *Curr. Opin. Genet. Dev.* **8**:668–674

Brasier MD, Green OR, Jephcoat AP, Kleppe AK, Van Kranendonk MJ, Lindsay JF, Steele A, Grassineau NV (2002) Questioning the evidence for Earth's oldest fossils. *Nature* **416**:76–81

Brinkmann H, Philippe H (1999) Archaea sister group of Bacteria? Indications from tree reconstruction artifacts in ancient phylogenies. *Mol. Biol. Evol.* **16**:817–825

Brochier C, Bapteste E, Moreira D, Philippe H (2002) Eubacterial phylogeny based on translational apparatus proteins. *Trends Genet.* **18**:1–5

Brochier C, Philippe H (2002) Phylogeny: a non-hyperthermophilic ancestor for bacteria. *Nature* **417**:244

Brochier C, Philippe H, Moreira D (2000) The evolutionary history of ribosomal protein RpS14: horizontal gene transfer at the heart of the ribosome. *Trends Genet.* **16**:529–533

Brown JR, Doolittle WF (1995) Root of the universal tree of life based on ancient aminoacyl-tRNA synthetase gene duplications. *Proc. Natl. Acad. Sci. USA* **92**:2441–2445

Brown JR, Douady CJ, Italia MJ, Marshall WE, Stanhope MJ (2001) Universal trees based on large combined protein sequence data sets. *Nat. Genet.* **28**:281–285

Brown JR, Robb FT, Weiss R, Doolittle WF (1997) Evidence for the early divergence of tryptophanyl- and tyrosyl-tRNA synthetases. *J. Mol. Evol.* **45**:9–16

Budin K, Philippe H (1998) New insights into the phylogeny of Eukaryotes based on ciliate Hsp70 sequences. *Mol. Biol. Evol.* **15**:943–956

Burger G, Saint-Louis D, Gray MW, Lang BF (1999) Complete sequence of the mitochondrial DNA of the red alga *Porphyra purpurea*. Cyanobacterial introns and shared ancestry of red and green algae. *Plant Cell* **11**:1675–1694

Castresana J (2000) Selection of conserved blocks from multiple alignments for their use in phylogenetic analysis. *Mol. Biol. Evol.* **17**:540–552

Cavalier-Smith T (1987) Eukaryotes with no mitochondria. *Nature* **326**:332–333

Cavalier-Smith T (1999) Principles of protein and lipid targeting in secondary symbiogenesis: euglenoid, dinoflagellate, and sporozoan plastid origins and the Eukaryote familly tree. *J. Eukaryot. Microbiol.* **46**:347–366

Charlebois RL, Sensen CW, Doolittle WF, Brown JR (1997) Evolutionary analysis of the hisCGABdFDEHI gene cluster from the archaeon *Sulfolobus solfataricus* P2. *J. Bacteriol.* **179**:4429–4432

Chase MW, Fay MF (2001) Ancient flowering plants: DNA sequences and angiosperm classification. *Genome Biol.* **2**:REVIEWS1012

Daubin V, Gouy M (2001) Bacterial molecular phylogeny using supertree approach. Genome Inform Ser Workshop *Genome Inform.* **12**:155–164

Deckert G, Warren PV, Gaasterland T, Young WG, Lenox AL, Graham DE, Overbeek R, Snead MA, Keller M, Aujay M, Huber R, Feldman RA, Short JM, Olsen GJ, Swanson RV (1998) The complete genome of the hyperthermophilic bacterium Aquifex aeolicus. *Nature* **392**:353–358

Doolittle WF (1999a) Lateral gene transfer, genome surveys, and the phylogeny of prokaryotes. Technical comment on M. Huynen et al. *Science* **286**:1443

Doolittle WF (1999b) Lateral genomics. *Trends Cell Biol.* **9**:M5–8

Doolittle WF (1999c) Phylogenetic classification and the universal tree. *Science* **284**: 2124–2129

Edlind TD, Li J, Visvesvara GS, Vodkin MH, McLaughlin GL, Katiyar SK (1996) Phylogenetic analysis of beta-tubulin sequences from amitochondrial protozoa. *Mol. Phylogenet. Evol.* **5**:359–367

Embley TM, Hirt RP (1998) Early branching Eukaryotes? *Curr. Opin. Genet. Dev.* **8**:624–629

Fast NM, Kissinger JC, Roos DS, Keeling PJ (2001) Nuclear-encoded, plastid-targeted genes suggest a single common origin for apicomplexan and dinoflagellate plastids. *Mol. Biol. Evol.* **18**:418–426

Fast NM, XUE L, Bingham S, Keeling PJ (2002) Re-examining alveolate evolution using multiple protein molecular phylogenies. *J. Eukaryot. Microbiol.* **49**:30–37

Felsenstein J (1978) Cases in which parsimony or compatibility methods will be positively misleading. *Syst. Zool.* **27**:401–410

Felsenstein J (2001) Taking Variation of Evolutionary Rates Between Sites into Account in Inferring Phylogenies. *J. Mol. Evol.* **53**:447–455

Fitch WM (1971) The nonidentity of invariable positions in the cytochromes c of different species. *Biochem. Genet.* **5**:231–241

Fitch WM, Markowitz E (1970) An improved method for determining codon variability in a gene and its application to the rate of fixation of mutations in evolution. *Biochem. Genet.* **4**:579–593

Fitz-Gibbon ST, House CH (1999) Whole genome-based phylogenetic analysis of free-living microorganisms. *Nucleic Acids Res.* **27**:4218–4222

Forterre P (1995) Thermoreduction, a hypothesis for the origin of prokaryotes. *C.R. Acad. Sci. III.* **318**:415–422

Forterre P, Bouthier De La Tour C, Philippe H, Duguet M (2000) Reverse gyrase from hyperthermophiles: probable transfer of a thermoadaptation trait from archaea to bacteria. *Trends Genet.* **16**:152–154

Forterre P, Brochier C, Philippe H (2002) Evolution of the Archaea. *Theor. Popul. Biol.* **61**:409–422

Forterre P, Philippe H (1999a) The last universal common ancestor (LUCA), simple or complex? *Biol. Bull.* **196**:373–375; discussion 375–377

Forterre P, Philippe H (1999b) Where is the root of the universal tree of life? *BioEssays* **21**:871–879

Foster PG, Hickey DA (1999) Compositional bias may affect both DNA-based and protein-based phylogenetic reconstructions. *J. Mol. Evol.* **48**:284–290

Fuerst JA (1995) The planctomycetes: emerging models for microbial ecology, evolution and cell biology. *Microbiology* **141**:1493–1506

Galtier N (2001) Maximum-likelihood phylogenetic analysis under a covarion-like model. *Mol. Biol. Evol.* **18**:866–873

Galtier N, Gouy M (1995) Inferring phylogenies from DNA sequences of unequal base compositions. *Proc. Natl. Acad. Sci. USA* **92**:11 317–11 321

Galtier N, Lobry JR (1997) Relationships between genomic G + C content, RNA secondary structures, and optimal growth temperature in prokaryotes. *J. Mol. Evol.* **44**:632–636

Galtier N, Tourasse N, Gouy M (1999) A nonhyperthermophilic common ancestor to extant life forms. *Science* **283**:220–221

Garcia-Ruiz JM, Hyde ST, Carnerup AM, Christy AG, Van Kranendonk MJ, Welham NJ (2003) Self-assembled silica-carbonate structures and detection of ancient microfossils. *Science* **302**:1194–1197

Germot A, Philippe H (1999) Critical analysis of eukaryotic phylogeny: a case study based on the HSP70 family. *J. Eukaryot. Microbiol.* **46**:116–124

Germot A, Philippe H, Le Guyader H (1997) Evidence for loss of mitochondria in Microsporidia from a mitochondrial- type HSP70 in *Nosema locustae*. *Mol. Biochem. Parasitol.* **87**:159–168

Gogarten JP, Kibak H, Dittrich P, Taiz L, Bowman EJ, Bowman BJ, Manolson MF, Poole RJ, Date T, Oshima T, Konishi J, Denda K, Yoshida M (1989) Evolution of the vacuolar H+-ATPase: implications for the origin of Eukaryotes. *Proc. Natl. Acad. Sci. USA* **86**:6661–6665

Gogarten-Boekels M, Hilario E, Gogarten JP (1995) The effects of heavy meteorite bombardment on the early evolution – the emergence of the three domains of life. *Orig. Life Evol. Biosph.* **25**:251–264

Gould SJ (1996) Full House: *The Spread of Excellence From Plato to Darwin*. Harmony Books

Gribaldo S, Cammarano P (1998) The root of the universal tree of life inferred from anciently duplicated genes encoding components of the protein-targeting machinery. *J. Mol. Evol.* **47**:508–516

Gribaldo S, Lumia V, Creti R, de Macario EC, Sanangelantoni A, Cammarano P (1999) Discontinuous occurrence of the hsp70 (dnaK) gene among Archaea and sequence features of HSP70 suggest a novel outlook on phylogenies inferred from this protein. *J. Bacteriol.* **181**:434–443

Gu X, Li WH (1998) Estimation of evolutionary distances under stationary and nonstationary models of nucleotide substitution. *Proc. Natl. Acad. Sci. USA* **95**:5899–5905

Guindon S, Perriere G (2001) Intragenomic base content variation is a potential source of biases when searching for horizontally transferred genes. *Mol. Biol. Evol.* **18**:1838–1840

Gupta RS (1998) What are Archaebacteria: life's third domain or monoderm prokaryotes related to Gram-positive bacteria? A new proposal for the classification of prokaryotic organisms. *Mol. Microbiol.* **229**:695–708

Gupta RS, Johari V (1998) Signature sequences in diverse proteins provide evidence of a close evolutionary relationship between the deinococcus-thermus group and cyanobacteria. *J. Mol. Evol.* **46**:716–720

Gupta RS, Singh B (1994) Phylogenetic analysis of 70 kD heat shock protein sequences suggests a chimeric origin for the eukaryotic cell nucleus. *Curr. Biol.* **4**:1104–1114

Hannaert V, Brinkmann H, Nowitzki U, LeeJ.A., Albert M-A, Sensen CW, Gaasterland T, Müller M, Michels P, Martin W (2000) Enolase from *Trypanosoma brucei*, from the amitochondriate protist *Mastigamoeba balamuthi*, and from the chloroplast and cytosol of *Euglena gracilis*: pieces in the evolutionary puzzle of the eukaryotic glycolytic pathway. *Mol. Biol. Evol.* **17**:989–1000

Hannaert V, Saavedra E, Duffieux F, Szikora JP, Rigden DJ, Michels PA, Opperdoes FR (2003) Plant-like traits associated with metabolism of Trypanosoma parasites. *Proc. Natl. Acad. Sci. USA* **100**:1067–1071

Hasegawa M, Hashimoto T (1993) Ribosomal RNA trees misleading? *Nature* **361**:23

Hashimoto T, Hasegawa M (1996) Origin and early evolution of Eukaryotes inferred from the amino acid sequences of translation elongation factors 1alpha/Tu and 2/G. *Adv. Biophys.* **32**:73–120

Hashimoto T, Sanchez LB, Shirakura T, Muller M, Hasegawa M (1998) Secondary absence of mitochondria in *Giardia lamblia* and *Trichomonas vaginalis* revealed by valyl-tRNA synthetase phylogeny. *Proc. Natl. Acad. Sci. USA* **95**:6860–6865

Hendy M, Penny D (1989) A framework for the quantitative study of evolutionary trees. *Syst. Zool.* **38**:297–309

Hennig W (1966) *Phylogenetic Systematics.* University of Illinois Press, Urbana

Henze K, Horner DS, Suguri S, Moore DV, Sanchez LB, Muller M, Embley TM (2001) Unique phylogenetic relationships of glucokinase and glucosephosphate isomerase of the amitochondriate Eukaryotes *Giardia intestinalis, Spironucleus barkhanus* and *Trichomonas vaginalis. Gene* **281**:123–131

Hickson RE, Simon C, Perrey SW (2000) The Performance of Several Multiple-Sequence Alignment Programs in Relation to Secondary-Structure Features for an rRNA Sequence. *Mol. Biol. Evol.* **17**:530–539

Hillis DM (1996) Inferring complex phylogenies. *Nature* **383**:130–131

Hirt RP, Logsdon JM, Jr., Healy B, Dorey MW, Doolittle WF, Embley TM (1999) Microsporidia are related to fungi: evidence from the largest subunit of RNA polymerase II and other proteins. *Proc. Natl. Acad. Sci. USA* **96**:580–585

Huelsenbeck JP (2002) Testing a covariotide model of DNA substitution. *Mol. Biol. Evol.* **19**:698–707

Imai E, Honda H, Hatori K, Brack A, Matsuno K (1999) Elongation of oligopeptides in a simulated submarine hydrothermal system. *Science* **283**:831–833

Iwabe N, Kuma K, Hasegawa M, Osawa S, Miyata T (1989) Evolutionary relationship of Archaebacteria, Eubacteria, and Eukaryotes inferred from phylogenetic trees of duplicated genes. *Proc. Natl. Acad. Sci. USA* **86**:9355–9359

Jain R, Rivera MC, Lake JA (1999) Horizontal gene transfer among genomes: The complexity hypothesis. *Proc. Natl. Acad. Sci. USA* **96**:3801–3806

Jukes TH, Cantor CR (1969) Evolution of protein molecules. In: Munro HN (ed.) *Mammalian Protein Metabolism.* Academic Press, New York, p 21–132

Katinka MD, Duprat S, Cornillot E, Metenier G, Thomarat F, Prensier G, Barbe V, Peyretaillade E, Brottier P, Wincker P, Delbac F, El Alaoui H, Peyret P, Saurin W, Gouy M, Weissenbach J, Vivares CP (2001) Genome sequence and gene compaction of the Eukaryote parasite *Encephalitozoon cuniculi. Nature* **414**:450–453

Keeling PJ (2001a) Foraminifera and Cercozoa are related in actin phylogeny: two orphans find a home? *Mol. Biol. Evol.* **18**:1551–1557

Keeling PJ (2001b) Parasites go the full monty. *Nature* **414**:401–402

Keeling PJ, Palmer JD (2000) Parabasalian flagellates are ancient Eukaryotes. *Nature* **405**:635–637

Keeling PJ, Palmer JD (2001) Lateral transfer at the gene and subgenic levels in the evolution of eukaryotic enolase. *Proc. Natl. Acad. Sci. USA* **98**:10 745–10 750

Knoll AH (1992) The early evolution of Eukaryotes: a geological perspective. *Science* **256**:622–627

Koonin EV, Makarova KS, Aravind L (2001) Horizontal gene transfer in prokaryotes: quantification and classification. *Annu. Rev. Microbiol.* **55**:709–742

Korbel JO, Snel B, Huynen MA, Bork P (2002) SHOT: a web server for the construction of genome phylogenies. *Trends Genet.* **18**:158–162

Koski LB, Golding GB (2001) The closest BLAST hit is often not the nearest neighbor. *J. Mol. Evol.* **52**:540–542

Koski LB, Morton RA, Golding GB (2001) Codon bias and base composition are poor indicators of horizontally transferred genes. *Mol. Biol. Evol.* **18**:404–412

Kuma K, Nikoh N, Iwabe N, Miyata T (1995) Phylogenetic position of Dictyostelium inferred from multiple protein data sets. *J. Mol. Evol.* **41**:238–246

Lang BF, Gray MW, Burger G (1999) Mitochondrial genome evolution and the origin of Eukaryotes. *Annu. Rev. Genet.* **33**:351–397

Lawrence JG, Ochman H (2002) Reconciling the many faces of lateral gene transfer. *Trends Microbiol.* **10**:1–4

Lawson FS, Charlebois RL, Dillon JA (1996) Phylogenetic analysis of carbamoylphosphate synthetase genes: complex evolutionary history includes an internal duplication within a gene which can root the tree of life. *Mol. Biol. Evol.* **13**:970–977

Lecointre G, Philippe H, Le HLV, Le Guyader H (1993) Species sampling has a major impact on phylogenetic inference. *Mol. Phylogenet. Evol.* **2**:205–224

Lin J, Gerstein M (2000) Whole-genome trees based on the occurrence of folds and orthologs: implications for comparing genomes on different levels. *Genome Res.* **10**:*808–818*

Lockhart P, Steel M, Hendy M, Penny D (1994) Recovering evolutionary trees under a more realistic model of sequence evolution. *Mol. Biol. Evol.* **11**:605–612

Lockhart PJ, Larkum AW, Steel M, Waddell PJ, Penny D (1996) Evolution of chlorophyll and bacteriochlorophyll: the problem of invariant sites in sequence analysis. *Proc. Natl. Acad. Sci. USA* **93**:1930–1934

Lockhart PJ, Steel MA, Barbrook AC, Huson D, Charleston MA, Howe CJ (1998) A covariotide model explains apparent phylogenetic structure of oxygenic photosynthetic lineages. *Mol. Biol. Evol.* **15**:1183–1188

Loomis WF, Smith DW (1990) Molecular phylogeny of *Dictyostelium discoideum* by protein sequence comparison. *Proc. Natl. Acad. Sci. USA* **87**:9093–9097

Lopez P, Casane D, Philippe H (2002) Heterotachy, an important process of protein evolution. *Mol. Biol. Evol.* **19**:1–7

Lopez P, Forterre P, Philippe H (1999) The root of the tree of life in the light of the covarion model. *J. Mol. Evol.* **49**:496–508

Lopez-Garcia P, Rodriguez-Valera F, Pedros-Alio C, Moreira D (2001) Unexpected diversity of small Eukaryotes in deep-sea Antarctic plankton. *Nature* **409**:603–607

Lum JK, Nikaido M, Shimamura M, Shimodaira H, Shedlock AM, Okada N, Hasegawa M (2000) Consistency of SINE insertion topology and flanking sequence tree: quantifying relationships among cetartiodactyls [In Process Citation]. *Mol. Biol. Evol.* **17**:1417–1424

Lwoff A (1943) *L'évolution Physiologique. Etude des Pertes de Fonctions Chez les Microorganismes.* Hermann et Cie, Paris

Lynch M, Conery JS (2000) The evolutionary fate and consequences of duplicate genes. *Science* **290**:1151–1155

Madsen O, Scally M, Douady CJ, Kao DJ, DeBry RW, Adkins R, Amrine HM, Stanhope MJ, de Jong WW, Springer MS (2001) Parallel adaptive radiations in two major clades of placental mammals. *Nature* **409**:610–614

Mai Z, Ghosh S, Frisardi M, Rosenthal B, Rogers R, Samuelson J (1999) Hsp60 is targeted to a cryptic mitochondrion-derived organelle ("crypton") in the microaerophilic protozoan parasite *Entamoeba histolytica*. *Mol. Cell. Biol.* **19**:2198–2205

Margulis L (1996) Archaeal-eubacterial mergers in the origin of Eukarya: phylogenetic classification of life. *Proc. Natl. Acad. Sci. USA* **93**:1071–1076

Martin W (1999) Mosaic bacterial chromosomes: a challenge en route to a tree of genomes. *Bioessays* **21**:99–104

Matte-Tailliez O, Brochier C, Forterre P, Philippe H (2002) Archaeal phylogeny based on ribosomal proteins. *Mol. Biol. Evol.* **19**:631–639

Mayr E (1998) Two empires or three? *Proc. Natl. Acad. Sci. USA* **95**:9720–9723

Moreira D, Le Guyader H, Philippe H (1999) Unusually high evolutionary rate of the elongation factor 1 alpha genes from the Ciliophora and its impact on the phylogeny of Eukaryotes. *Mol. Biol. Evol.* **16**:234–245

Moreira D, Le Guyader H, Philippe H (2000) The origin of red algae: implications for the evolution of chloroplasts. *Nature* **405**:69–72

Murphy WJ, Eizirik E, Johnson WE, Zhang YP, Ryder OA, O'Brien SJ (2001) Molecular phylogenetics and the origins of placental mammals. *Nature* **409**:614–618

Nelson KE, Clayton RA, Gill SR, Gwinn ML, Dodson RJ, Haft DH, Hickey EK, Peterson JD, Nelson WC, Ketchum KA, McDonald L, Utterback TR, Malek JA, Linher KD, Garrett MM, Stewart AM, Cotton MD, Pratt MS, Phillips CA, Richardson D, Heidelberg J, Sutton GG, Fleischmann RD, Eisen JA, Fraser CM et al. (1999) Evidence for lateral gene transfer between Archaea and bacteria from genome sequence of Thermotoga maritima. *Nature* **399**:323–329

Nikoh N, Hayase N, Iwabe N, Kuma K, Miyata T (1994) Phylogenetic relationship of the kingdoms Animalia, Plantae, and Fungi, inferred from 23 different protein species. *Mol. Biol. Evol.* **11**:762–768

Nisbet EG, Sleep NH (2001) The habitat and nature of early life. *Nature* **409**:1083–1091

Ochman H, Lawrence JG, Groisman EA (2000) Lateral gene transfer and the nature of bacterial innovation. *Nature* **405**:299–304

Olsen G (1987) Earliest phylogenetic branching: comparing rRNA-based evolutionary trees inferred with various techniques. *Cold Spring Harb. Symp. Quant. Biol.* **LII**:825–837

Olsen GJ, Woese CR, Overbeek R (1994) The winds of (evolutionary) change: breathing new life into microbiology. *J. Bacteriol.* **176**:1–6

Pace NR (1991) Origin of life – facing up to the physical setting. *Cell* **65**:531–533

Pace NR (1997) A molecular view of microbial diversity and the biosphere. *Science* **276**:734–740

Page RD, Charleston MA (1997) From gene to organismal phylogeny: reconciled trees and the gene tree/species tree problem. *Mol. Phylogenet. Evol.* **7**:231–240

Pawlowski J, Bolivar I, Fahrni JF, de Vargas C, Gouy M, Zaninetti L (1997) Extreme differences in rates of molecular evolution of foraminifera revealed by comparison of ribosomal DNA sequences and the fossil record. *Mol. Biol. Evol.* **14**:498–505

Pennisi E (1998) Genome data shake tree of life. *Science* **280**:672–674

Penny D, McComish BJ, Charleston MA, Hendy MD (2001) Mathematical elegance with biochemical realism: the covarion model of molecular evolution. *J. Mol. Evol.* **53**:711–723

Penny D, Poole A (1999) The nature of the last universal common ancestor. *Curr. Opin. Genet. Dev.* **9**:672–667

Philippe H (2000) Long branch attraction and protist phylogeny. *Protist* **51**:307–316

Philippe H, Adoutte A (1998) The molecular phylogeny of Eukaryota: solid facts and uncertainties. In: Coombs G, Vickerman K, Sleigh M, Warren A (eds) *Evolutionary Relationships among Protozoa*. Kluwer, Dordrecht, p 25–56

Philippe H, Budin K, Moreira D (1999) Horizontal transfers confuse the prokaryotic phylogeny based on the HSP70 protein family. *Molecular Microbiology* **31**:1007–1009

Philippe H, Douzery E (1994) The pitfalls of molecular phylogeny based on four species, as illustrated by the Cetacea/Artiodactyla relationships. *J. Mammal. Evol.* **2**:133–152

Philippe H, Forterre P (1999) The rooting of the universal tree of life is not reliable. *J. Mol. Evol.* **49**:509–523

Philippe H, Germot A (2000) Phylogeny of Eukaryotes based on ribosomal RNA: Long-branch attraction and models of sequence evolution. *Mol. Biol. Evol.* **17**:830–834

Philippe H, Laurent J (1998) How good are deep phylogenetic trees? *Curr. Opin. Genet. Dev.* **8**:616–623

Philippe H, Lopez P, Brinkmann H, Budin K, Germot A, Laurent J, Moreira D, Müller M, Le Guyader H (2000) Early branching or fast evolving Eukaryotes? An answer based on slowly evolving positions. *Philos. Trans. R. Soc. Lond. B Biol. Sci.* **267**:1213–1221

Philippe H, Sörhannus U, Baroin A, Perasso R, Gasse F, Adoutte A (1994) Comparison of molecular and paleontological data in diatoms suggests a major gap in the fossil record. *J. Evolut. Biol.* **7**:247–265

Poe S, Swofford DL (1999) Taxon sampling revisited. *Nature* **398**:299–300

Poole A, Jeffares D, Penny D (1999) Early evolution: prokaryotes, the new kids on the block. *BioEssays* **21**:880–889

Posada D, Crandall KA (2002) The effect of recombination on the accuracy of phylogeny estimation. *J. Mol. Evol.* **54**:396–402.

Ragan MA (2001) On surrogate methods for detecting lateral gene transfer. *FEMS Microbiol. Lett.* **201**:187–191

Roger AJ, Sandblom O, Doolittle WF, Philippe H (1999) An evaluation of elongation factor 1 alpha as a phylogenetic marker for Eukaryotes. *Mol. Biol. Evol.* **16**:218–233

Rokas A, Holland PWH (2000) Rare genomic changes as a tool for phylogenetics. *Trends Ecol. Evol.* **15**:454–459

Ruepp A, Graml W, Santos-Martinez ML, Koretke KK, Volker C, Mewes HW, Frishman D, Stocker S, Lupas AN, Baumeister W (2000) The genome sequence of the thermoacidophilic scavenger *Thermoplasma acidophilum*. *Nature* **407**:508–513

Schopf JW, Kudryavtsev AB, Agresti DG, Wdowiak TJ, Czaja AD (2002) Laser-Raman imagery of Earth's earliest fossils. *Nature* **416**:73–76

Schwartz RM, Dayhoff MO (1978) Origins of prokaryotes, Eukaryotes, mitochondria, and chloroplasts. *Science* **199**:395–403

Sogin ML (1991) Early evolution and the origin of Eukaryotes. *Curr. Opin. Genet. Dev.* **1**:457–463

Stetter KO (1996) Hyperthermophiles in the history of life. *Ciba Found. Symp.* **202**:1–10

Strous M, Fuerst JA, Kramer EH, Logemann S, Muyzer G, van de Pas-Schoonen KT, Webb R, Kuenen JG, Jetten MS (1999) Missing lithotroph identified as new planctomycete. *Nature* **400**:446–449

Swofford DL, Olsen GJ, Waddell PJ, Hillis DM (1996) Phylogenetic inference. In: Hillis DM, Moritz C, Mable BK (eds) *Molecular Systematics*. Sinauer Associates, Sunderland, p 407–514

Tekaia F, Lazcano A, Dujon B (1999) The genomic tree as revealed from whole proteome comparisons. *Genome Res.* **9**:550–557

Tovar J, Fischer A, Clark CG (1999) The mitosome, a novel organelle related to mitochondria in the amitochondrial parasite *Entamoeba histolytica*. *Mol. Microbiol.* **32**:1013–1021

Tovar J, Leon-Avila G, Sanchez LB, Sutak R, Tachezy J, Van Der Giezen M, Hernandez M, Muller M, Lucocq JM (2003) Mitochondrial remnant organelles of Giardia function in iron-sulphur protein maturation. *Nature* **426**:172–176

Van de Peer Y, Ben Ali A, Meyer A (2000) Microsporidia: accumulating molecular evidence that a group of amitochondriate and suspectedly primitive Eukaryotes are just curious fungi. *Gene* **246**:1–8

Venkatesh B, Ning Y, Brenner S (1999) Late changes in spliceosomal introns define clades in vertebrate evolution. *Proc. Natl. Acad. Sci. USA* **96**:10 267–10 271

Vossbrinck CR, Maddox JV, Friedman S, Debrunner-Vossbrinck BA, Woese CR (1987) Ribosomal RNA sequence suggests microsporidia are extremely ancient Eukaryotes. *Nature* **326**:411–414

Waller RF, Keeling PJ, Donald RG, Striepen B, Handman E, Lang-Unnasch N, Cowman AF, Besra GS, Roos DS, McFadden GI (1998) Nuclear-encoded proteins target to the plastid in Toxoplasma gondii and Plasmodium falciparum. *Proc. Natl. Acad. Sci. USA* **95**:12 352–12 357

Wang B (2001) Limitations of compositional approach to identifying horizontally transferred genes. *J. Mol. Evol.* **53**:244–250

Weisburg WG, Giovannoni SJ, Woese CR (1989) The Deinococcus-Thermus phylum and the effect of rRNA composition on phylogenetic tree construction. *Syst. Appl. Microbiol.* **11**:128–134

Williams BA, Hirt RP, Lucocq JM, Embley TM (2002) A mitochondrial remnant in the microsporidian *Trachipleistophora hominis*. *Nature* **418**:865–869

Woese CR (1987) Bacterial evolution. *Microbiol. Rev.* **51**:221–271

Woese CR (2000) Interpreting the universal phylogenetic tree. *Proc. Natl. Acad. Sci. USA* **97**:8392–8396

Woese CR, Achenbach L, Rouviere P, Mandelco L (1991) Archaeal phylogeny: reexamination of the phylogenetic position of Archaeoglobus fulgidus in light of certain composition-induced artifacts. *Syst. Appl. Microbiol.* **14**:364–371

Woese CR, Fox GE (1977) Phylogenetic structure of the prokaryotic domain: the primary kingdoms. *Proc. Natl. Acad. Sci. USA* **74**:5088–5090

Woese CR, Kandler O, Wheelis ML (1990) Towards a natural system of organisms: proposal for the domains Archaea, Bacteria, and Eucarya. *Proc. Natl. Acad. Sci. USA* **87**:4576–4579

Woese CR, Olsen GJ, Ibba M, Soll D (2000) Aminoacyl-tRNA synthetases, the genetic code, and the evolutionary process. *Microbiol. Mol. Biol. Rev.* **64**:202–236

Wolf YI, Rogozin IB, Grishin NV, Tatusov RL, Koonin EV (2001) Genome trees constructed using five different approaches suggest new major bacterial clades. *BMC Evol. Biol.* **1**:8

Yang Z (1996) Among-site rate variation and its impact on phylogenetic analyses. *Trends Ecol. Evol.* **11**:367–370

Zuckerkandl E, Pauling L (1965) Evolutionary divergence and convergence in proteins. In: Bryson V, Vogel HJ (eds) *Evolving Genes and Proteins*. Academic Press, New York, p 97–166

9 Extremophiles

Purificación López-García

9.1 Some Concepts About Extremophiles

Thought sterile for a long time, extreme environments are known today to host a variety of well-adapted organisms so-called "extremophiles" that are their natural inhabitants. Salt-loving organisms were among the first organisms studied from particularly enduring habitats. The first halophilic species was isolated from salted fish by Farlow as early as 1880, although it was properly described some years later (Farlow, 1880). In 1936, Eleazari Volcani isolated a number of microbial strains living at very high salt concentration, up to 30–34% (weight/volume) from the Dead Sea (Wilansky, 1936). He thus demonstrated the inappropriate designation of this water body and became a historical pioneer on extremophile research, since he devoted most of his scientific career to the study of halophilic microbes. However, salt-loving organisms, or halophiles, began to attract much more attention from microbiologists some years later, when it was discovered that most of them belonged to the third domain of life, the archaea, whose recognition dates back to the late 1970s (Woese and Fox, 1977). It was also during this decade that extreme thermophiles and hyperthermophiles growing well above 60 to 80 °C began to be isolated. In 1969, Thomas Brock isolated a thermophilic bacterium living at 60–80 °C, *Thermus aquaticus*, from a hot spring at Yellowstone National Park in the US (Brock and Freeze, 1969). A year later, it was the turn of *Sulfolobus acidocaldarius*, the first isolated hyperthermophilic archaeon, which was able to grow at temperatures up to 85 °C and at low pH (1–5) (Brock et al., 1972). It was a big surprise for biologists who, since Pasteur's times, had the idea in mind that no living being could survive above 80 °C (Madigan et al., 2002). These first discoveries triggered the search for extremophiles that, assisted by important technical developments in microbiology, have had since then an extraordinary impact on our understanding of microbial diversity and evolution in our planet.

The present review aims at summarising the general knowledge about extreme environments and extremophiles, what they are, how we study them, why they are so appealing to us, and what their link is to exo/astrobiology issues. This brief overview is logically very far from exhaustive; references have been kept to minimum numbers and often correspond to general revisions. The reader is therefore invited to consult more specialised work for deeper insights into the different areas.

9.1.1 What is an Extremophile?

An extremophile is an organism that lives optimally in an extreme environment. What is then an extreme environment? Intuitively, we immediately think of a habitat where conditions are harsh and hostile to us. This anthropocentric vision is too subjective, although relatively valid (we are indeed mesophiles). In more neutral terms, an environment can be defined as extreme when the physicochemical parameters characterising it approach the limits within which life is known to occur. This is a strict biological definition. In this sense, we should talk of "bioextremes". Thus, if we take the example of temperature, life expands between approximately $-20\,°C$ (some bacteria) and 110–$120\,°C$ (some archaea). In this interval organisms are usually divided into different categories depending on their minimal and maximal growth temperatures (Fig. 9.1). Any organism grows best at a temperature that is optimal, and growth rate decreases when temperature shifts up or down. Following the above definition, psychrophiles and hyperthermophiles, living close to temperature limits for life, are extremophiles.

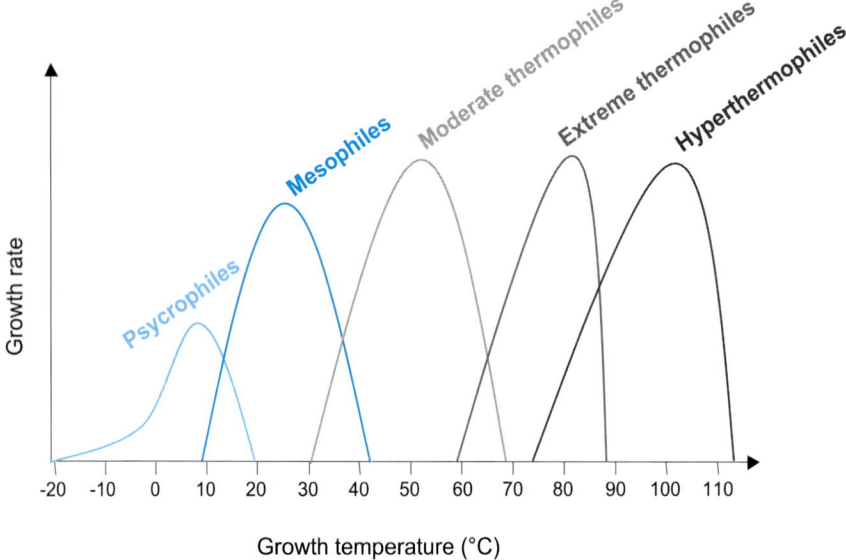

Fig. 9.1. Classification of organisms according to their growth temperature. Psycrophiles and hyperthermophiles are extremophiles

9.1.2 Some Extremophile Features

An extremophilic organism not only tolerates but also requires extreme conditions to live. It is capable of developing its whole life cycle in these conditions. On the contrary, an organism that is extremotolerant can endure extreme values of one or more physicochemical parameters, but its optimal growth occurs under milder conditions. Thus, halotolerant bacteria survive at high salt concentrations but grow better at lower salinity. Similarly, barotolerant organisms stand high pressures but live optimally at lower pressure. We need also to distinguish between life under extreme conditions and under stressing conditions. Stressing environments are often changing biotopes, such as intertidal zones, but they do not display limiting life conditions permanently. Here, organisms are adapted to tolerate change out of the optimal conditions. In fact, extreme environments are most frequently very stable at a biological scale (e.g. many geothermal areas, the Earth crust, the deep ocean, the Antarctic ice).

From a biological point of view, extreme environments are characterised by the predominance of microbial life (bacteria, archaea and, except for very high temperature environments, unicellular eukaryotes). Another important feature is that microbial diversity decreases as one gets closer to life-limiting conditions (increasingly fewer species, although their biomass can be eventually high). This is particularly well illustrated in the case of high temperature. Eukaryotic life is known to exist up to $\sim 60-62\,°C$ (some fungi and protists), but above those values only prokaryotes seem to occur. Thermophilic bacteria and archaea are diverse but, again, hyperthermophily (optimal growth above $80\,°C$, as originally defined by Stetter in 1989) appears limited among bacteria. A few hyperthermophilic bacterial lineages are known, the most hyperthermophilic bacteria being able to grow up to $95\,°C$. Only hyperthermophilic archaea are known to live at temperatures higher than $100\,°C$ (Stetter, 1996).

Extremophiles are normally pluriextremophiles, being adapted to live in habitats where various physicochemical parameters reach extreme values. For instance, many hot springs are acid or alkaline at the same time, and usually rich in metal content; several hypersaline lakes are very alkaline; the deep ocean is generally cold, oligotrophic (very low nutrient content), and exposed to high pressure; pressure and temperature increase progressively in the subsurface, etc. All these are biotopes for specifically adapted microbial communities.

9.1.3 Why Extremophiles are Interesting?

Since they began to be discovered, extremophiles have fascinated and attracted scientists from various disciplines:

- **Molecular biology.** Many different studies have been devoted to understand the molecular adaptations needed to work under conditions that would severely alter, denature or prevent function of nonextremophile's molecules (see below and Table 9.1).
- **Biotechnology.** The potential to use extremophiles or their molecules for industrial application greatly encouraged (and still does) extremophile research. In any organism, enzymes (proteins catalysing transformations of chemical

substrates into products) are logically adapted to work optimally at its natural growth conditions. Enzymes from extremophiles are therefore functional at high temperature, high or low pH, high salt concentration, etc. They are normally very stable and resistant to denaturing agents as well, which may be an additional interest for some industrial processes. Enzymes from organisms adapted to very diverse extreme conditions, also called "extremozymes" by some authors, offer a vast choice to industry due to their functionality in media where traditional enzymes cannot work (Deming, 1998; Horikoshi, 1999; Hough and Danson, 1999; Oren, 1999). Furthermore, many extremophiles display novel metabolic pathways, which increase the panorama of possible reactions for biotechnology.
- **Evolution and phylogeny.** On the one hand, it was partly due to the first studies on extremophiles that the third domain of life, the archaea, was discovered, with fundamental consequences for evolutionary biology (Woese and Fox, 1977; Woese et al., 1990). On the other hand, physicochemical conditions on early Earth could have been similar to those of some contemporary extreme environments. These could be then used as inspiring models to unravel ancient geochemical cycles, the origin of life and its first diversification (Nisbet and Sleep, 2001).
- **Exobiology.** The occurrence of extremophiles living under conditions that are similar to those found in other planets raises more realistic hopes about the possibility to discover extraterrestrial life.

9.2 Microbial Diversity

Methods to study microbial diversity have greatly evolved in the last thirty years. The application of molecular techniques led first to the discovery of archaea, within which we still find today most extremophily records, and later to the detection of an unexpected microbial diversity in very different environments, including extreme biotopes.

Classical microbiology is based on isolation and cultivation of micro-organisms. The objective is to obtain pure cultures (all cells belonging to the same microbial species or strain) that can be assayed for different metabolic or physiological properties in various culture media. However, in most cases, neither morphology nor phenotypic features are informative enough to affiliate micro-organisms to a given phylogenetic (evolutionary-related) family, that is, to achieve a natural or phylogenetic classification. To overcome this problem, we began to use the evolutionary information stored in nucleic acid (DNA and RNA) nucleotide sequences and in protein amino-acid sequences, thus applying the idea that Zuckerkandl and Pauling had published in 1965 (Zuckerkandl and Pauling, 1965). From all biological macromolecules, the RNA of the small ribosomal subunit (16S/18S rRNA) was initially chosen by Woese and colleagues in the 1970s by its properties of universality and reasonable conservation (Woese, 1987; Woese and Fox, 1977). Sequence comparison of rRNAs from many different organisms, including "rare" species coming from extreme environments, led to the recognition of three domains of cellular life on Earth, Archaea, Bacteria and Eukarya. Since then, 16S/18S rRNA has been the most widely used molecular marker, with thousands of sequences from cultivated and noncultivated species being deposited in public databases, and has further confirmed the tripartite division of cellular organisms on Earth (Fig. 9.2).

The application of molecular methods based on 16/18S rRNA gene amplification directly from environmental samples has been revolutionary for microbial diversity studies. This is particularly true for extreme environments for which classical culture methods allow the recovery of usually less than 1% of the actual diversity (Amann et al., 1995). This strategy permits, in principle, the identification of the different rRNA genes present in a given environment (phylotypes). rRNA genes can be amplified by the polymerase chain reaction (PCR) from DNA extracted directly from environmental samples, and then they can be sequenced and the sequences compared with the thousands of rRNA sequences available in databases. In this way, the new environmental sequences can be placed in phylogenetic trees where they represent organisms living in nature even when these have not been or cannot be isolated. The use of this approach has demonstrated that life is much more diversified than we ever thought with the discovery, among many other lineages, of whole organismal groups for which no cultivated member is known (Fig. 9.2) (Hugenholtz et al., 1998; Pace, 1997).

Classical and molecular approaches in microbiology are indeed complementary. Both of them have been, and continue to be, applied to the study of microbes living in extreme habitats leading to the discovery (still very partial) of an astonishing diversity of extremophiles.

9.3 Extreme Environments and Their Inhabitants

9.3.1 Extremophiles and Extremotolerants

It is beyond the scope of this review to decorticate in detail extremophiles and their natural environments. Table 9.1 summarises general features of extremophilic and extremotolerant organisms: their classification as a function of the physicochemical parameter considered, their most outstanding molecular adaptations, as well as some examples of species or groups of extremophiles and the habitats they colonise. Figures 9.3 to 9.9 illustrate different extreme environments. Temperature, pH, salinity (osmotic pressure), pressure, dryness (water availability) and exposure to radiation are parameters whose values usually configure the extreme nature of a particular environment (Table 9.1). Some authors also consider vacuum, gravity, and partial oxygen pressure (Rothschild and Mancinelli, 2001). Nevertheless, no organism is known to live under hypo- or hypergravity and, although many organisms are able to resist vacuum conditions, they do it in an inactive state (dormant or resistance forms), that is, they

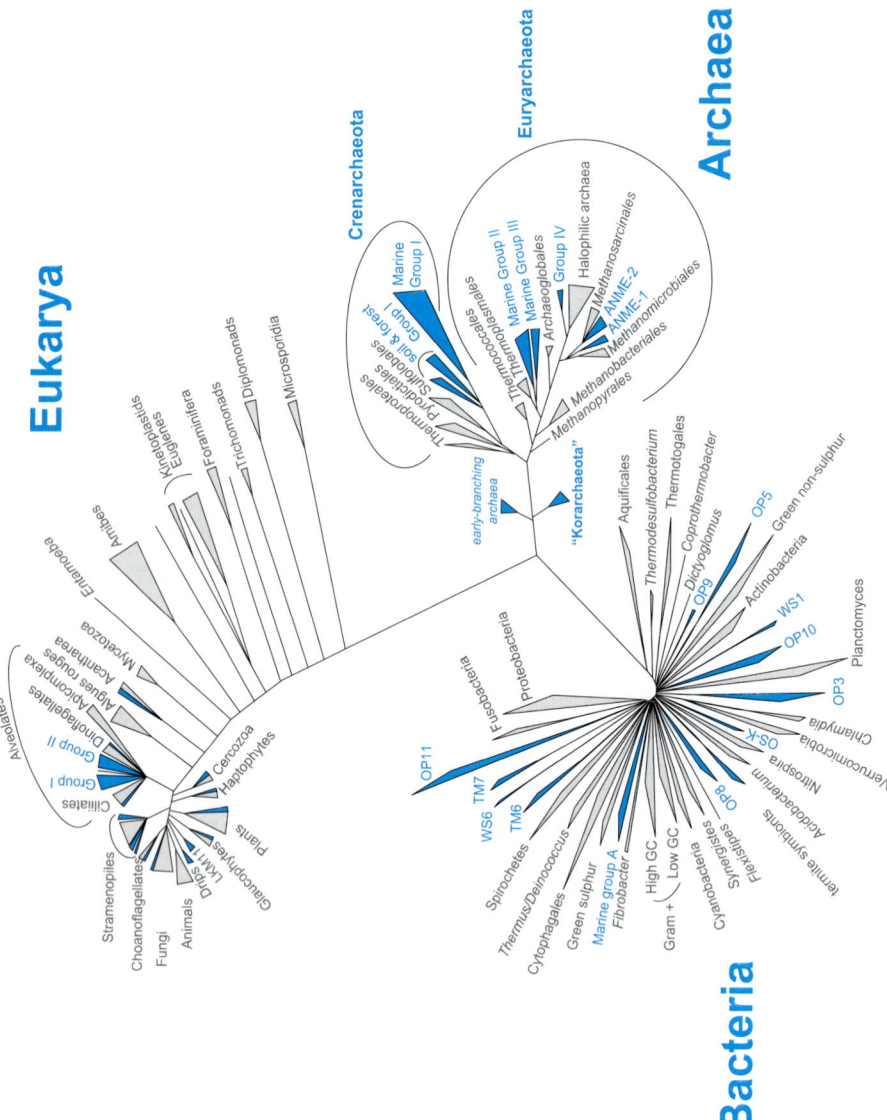

Fig. 9.2. Schematic diagram showing the tree of life based on 16S/18S rRNA sequences. Each *line* represents a sequence (in turn representing the corresponding organism); each *triangle* represents a group of closely related sequences. *Blue triangles* correspond to environmental lineages for which no cultivated members are known

Fig. 9.3. Sampling at a hot mud pool in a hydrothermal field at the Azores archipelago (Photo courtesy of P. Forterre and E. Marguet)

do not depend on those conditions for a living. Oxygen induces the formation of free radicals that can damage cells. Indeed, most organisms have mechanisms to neutralise oxygen-derived deleterious effects. In this sense, aerobic organisms like us could be considered extremophiles. However, a vast diversity of organisms belonging to the three life domains is able to live optimally at all oxygen partial pressures found on Earth: from complete absence to atmospheric levels and even more, for instance in oxygen-producing photosynthetic bacterial mats. This indicates that life has adapted to all possible terrestrial niches with regard to oxygen concentration, without it being a general limiting factor. Therefore, anaerobic organisms cannot be considered extremophiles by this feature.

For some of the parameters mentioned above, we find only extremotolerant organisms. Thus, there are radiotolerant micro-organisms, but none is "radiophile". In addition to the parameters listed in Table 9.1, oligotrophic conditions, or extremely low nutrient concentrations, which very often characterise some extreme environments, constitute a major limiting factor for life.

Table 9.1. Major features of extreme environments and associated biota

Physico-chemical parameter	Type of organism	Definition/ Optimal growth	Molecular adaptations
Temperature	Hyper-thermo-phile	$> 80\,°C$	• DNA stability: existence of a specific enzyme, reverse gyrase, acting on DNA topology • Lipids more saturated, membrane lipids forming a single monolayer more impermeable to protons • Protein stability: increased salt and disulfide interactions, increased hydrophobicity and compaction, less subunits • K^+ or compatible solute-enriched cytoplasm
	Thermo-phile	$60\text{--}80\,°C$	• More saturated lipids • Protein stability: increased hydrophobicity
	Psychro-phile	$< 5\,°C$	• More unsaturated lipids • Proteins: increased flexibility, less interactions among protein domains, high complementarity substrate–enzyme • Antifreezing molecules in cytoplasm
pH	Acidophile	$pH < 2\text{--}3$	• Large internal buffering capacity • External cell surface positively charged • Proton export by specific membrane pumps
	Alkaliphile	$pH > 9\text{--}10$	• Large internal buffering capacity • External cell surface negatively charged • Proton import by antiporters
Salinity	Halophile	High salt concentration ($\sim 2\text{--}5$ M NaCl)	• K^+ or compatible solute-enriched cytoplasm • Proteins enriched in negatively charged amino acids
Pressure	Barophile (Piezophile)	High pressure	• More unsaturated lipids • Specific adaptations in proteins
Desiccation	Xerophile	Conditions of anhydrobiose	• Increased internal osmolarity • DNA stability: protein binding and strong repair mechanisms
Radiation	Radio-tolerant	support high radiation doses (ionising, UV, etc.)	• Proteins stabilising DNA and strong repair systems
Metals	Metallo-tolerant	support high heavy metal concentrations	• Specific detoxification mechanisms • Selective metal accumulation (sequestration) inside or outside cells

Table 9.1. (*continued*)

Physico-chemical parameter	Examples of biotopes	Biodiversity/examples
Tempera-ture	• Continental and deep-sea hydrothermal systems • Geysers • Solfataras • Deep continental and oceanic subsurface regions	• Archaea dominate: *Methanopyrus, Pyrolobus, Thermococcus, Methanothermus, Pyrodictium,* "korarchaeota", etc. • Bacteria: *Aquifex, Thermotoga* • **NO** eukaryotes
	• Hydrothermal systems • Mineral piles • Industrial wastewaters • Deep subsurface	• Archaea: *Thermoplasma, Methanobacterium,* etc. • Bacteria: *Thermus, Bacillus stearothermophilus* • Eukaryotes: some fungi
	• Deep ocean • Arctic and Antarctic ice • Sea ice (seasonal) • Snow and high mountain • Permafrost	• Archaea: *Cenarchaeum symbiosum,* many non-cultivable deep-sea and sea ice lineages • Bacteria: *Flavobacterium, Psychrobacter,* etc. • Eukaryotes: red algae, diatoms, etc.
pH	• Mining areas • Hot acid springs • Acid solfataras	• Archaea: *Picrophilus, Thermoplasma, Sulfolobus,* etc. • Bacteria: *Acidithiobacillus, Thiobacillus, Leptospirillum,* etc. • Eukaryotes: many fungi, the red algae *Cyanidium*
	• Soda lakes • Alkaline hot springs	• Archaea: *Natronobacterium, Natronococcus,* etc. • Bacteria: *Spirulina,* other cyanobacteria • Eukaryotes: several protists
Salinity	• Coastal salterns • Some soda lakes (Natron type) • Deep-sea brines • Evaporites, salt mines	• Archaea: *Halobacterium, Haloferax, Halococcus,* etc. • Bacteria: *Salinibacter, Halomonas,* etc. • Eukaryotes: *Artemia salina, Dunaliella,* some yeast
Pressure	• Deep sea • Deep subsurface	• Archaea: many noncultivable deep-sea lineages • Bacteria: *Shewanella, Colwelia,* etc. • Eukaryotes: fungi, abyssal fauna
Desiccation	• Hot and cold deserts • Many salterns	• Archaea: halophilic archaea and others • Bacteria: *Metallogenium, Pedomicrobium,* etc. • Eukaryotes: fungi, lichens, etc.
Radiation	• Radioactive waste • Natural radioactive mines • Deserts, solar salterns • High mountain	• Archaea: some *Thermococcus* spp., halophiles (UV) • Bacteria: *Deinococcus radiodurans*
Metal	• Mining regions • Metal-contaminated aquifers • Wastewaters	• Archaea: *Acidianus, Thermoplasma,* etc. • Bacteria: *Acidithiobacillus, Leptospirillum* • Eukaryotes: some fungi, algae and plants

Fig. 9.4. Square halophilic archaea as seen by scanning electron microscopy. Several halophilic archaea belonging to the genus *Haloarcula* living close to saturation acquire these unusual morphologies. (Photo courtesy of F. Rodriguez-Valera)

Fig. 9.5. The Rio Tinto (Spain). Its waters are highly acidic (pH 1.9–2.5) and saturated in ferric iron, responsible for the *red colour*. The river bed is densely colonised by acidophilic micro-organisms; filaments of green algae are shown to the *right*. (Photo courtesy of A.I. Lopez-Archilla)

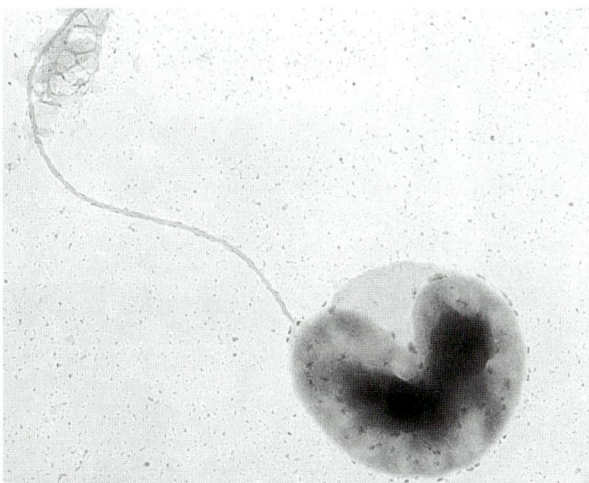

Fig. 9.6. Transmission electron micrograph of an acidophilic chemolitoautotrophic bacterium. (Photo courtesy of A.I. Lopez-Archilla)

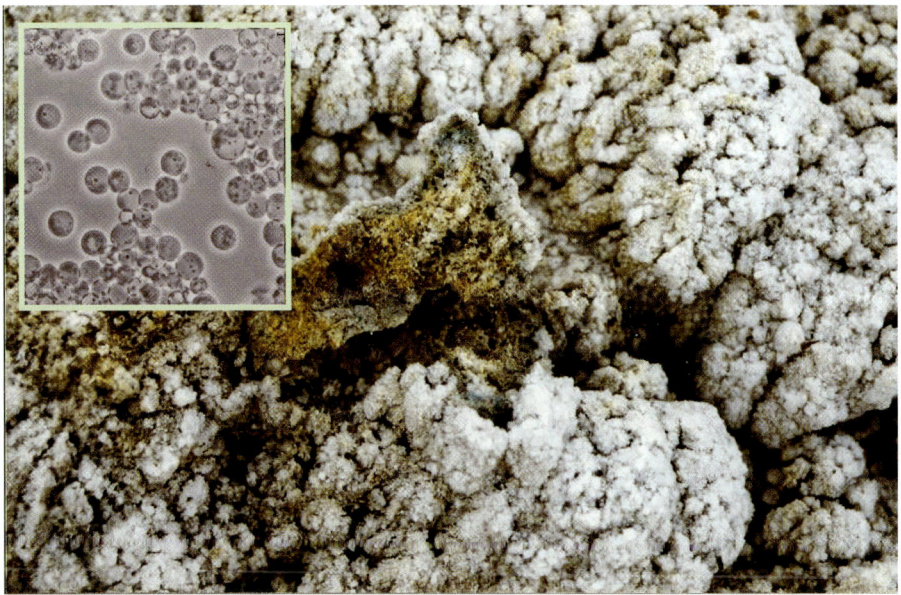

Fig. 9.7. Endolithic community living in gypsum. The green layer is composed of thermophilic algae (50°C), *Galdieria sulfuraria* (shown in *inset*). (Photo courtesy of A.I. Lopez-Archilla)

Fig. 9.8. Metallotolerant organisms living on copper sulfate crystals. (Photo courtesy of A.I. Lopez-Archilla)

Fig. 9.9. Deep subsurface cave (Pozo Alfredo, 800 m deep, Spain). Thermophilic archaea and bacteria have been identified in this subsurface environment. (Photo courtesy of A.I. Lopez-Archilla)

9.3.2 Phylogenetic Groups Best Adapted to Extreme Conditions

Some extremophiles correspond to groups of micro-organisms belonging to specific phylogenetic lineages that are particularly well adapted to specific life-limiting conditions and restricted types of environments. For instance only several archaeal and a few bacterial lineages are hyperthermophiles (Stetter, 1996). Conversely, we also find extremophiles broadly scattered in the phylogenetic tree that are adapted to the same extreme condition. This is the case for different psychrophiles or barophiles, for which we find members dispersed in the three domains of life.

There are also groups of organisms belonging to the same phylogenetic family that, without holding all extremophile records, have adapted to very diverse extreme or moderately extreme conditions. Among bacteria, the best-adapted group to various extreme conditions is possibly that of cyanobacteria. They often form microbial mats with other bacteria, from Antarctic ice to continental hot springs. Having been isolated from Yellowstone living at temperatures of $\sim 70\,°C$ they constitute, together with some green sulfur bacteria[1], the most thermophilic photosynthesisers known (Madigan et al., 2002; Paerl et al., 2000). They can live in both aerobic and anaerobic conditions, and they can also develop in hypersaline and alkaline lakes, they support high metal concentrations and they tolerate xerophilic conditions, forming endolithic communities in desertic regions. However, they do not appear to withstand acid conditions, and they are rarely found at pH values lower than 5–6.

Among eukaryotes, fungi, alone or in symbiosis with cyanobacteria or algae forming lichens, are certainly the most versatile and ecologically successful phylogenetic lineage. With the exception of hyperthermophily, although they represent the most thermophilic eukaryotes, living up to 60–64 °C, they adapt well to the rest of extreme environments. Most are aerobic, but many can be anaerobic. Fungi live in acidic and metal-enriched waters from mining regions, alkaline settings, hypersaline regions such as the Dead Sea, hot and cold deserts, the deep ocean (they have been even isolated from the Mariana Trench at $\sim 11\,000\,m$ deep), etc. (Buchalo et al., 1998; Lopez-Archilla et al., 2001; Sterflinger, 1998; Takami et al., 1997).

Archaea are generally less versatile than bacteria and eukaryotes. In contrast, they are generally very skilful in adapting to different extreme conditions, holding frequently extremophily records. Thus, *Pyrolobus fumarii*, coming from a deep-sea vent from the Mid-Atlantic Ridge and living up to 113 °C, was the most hyperthermophilic organism known until very recently (Blöchl et al., 1997) but, in 2003, another archaeon growing up to 121 °C (strain 121) was isolated from a Pacific deep-sea vent (Kashefi and Lovley, 2003). Similarly, some archaea are among the most halophilic and alkaliphilic micro-organisms

---

[1] Green sulfur bacteria, e.g. *Chlorobium* species, are strictly anaerobic and phototrophic, gaining energy from light (via a bacteriochlorophyll) and using a reduced sulfur species (instead of H_2O) as an electron acceptor.

known (Oren, 1994). In addition, we could mention methanogens as the most versatile among the archaea. They are strict anaerobes that occupy the entire thermal gradient from $\sim 110\,°\text{C}$ (e.g. *Methanopyrus kandlerii*) (Burggraf et al., 1991) to $0\text{--}2\,°\text{C}$, living in cold marine sediments (Lanoil et al., 2001). Many methanogens are also acidophilic, alkaliphilic or halophilic (Madigan et al., 2002).

9.3.3 Resistance Forms and Longevity

How much and how long can micro-organisms resist exceedingly harsh conditions? These are basic questions with regard to the eventual possibility of transporting life forms (e.g. in meteorites) from one planet to another. When the environmental conditions become adverse, most micro-organisms get into latent states that allow them to survive until conditions become favourable again. We can distinguish different cellular states going from normal metabolic activity to death: growing state, resting state (cells do not reproduce), dormancy (absence of metabolism), mummified state (cells undergo already irreversible transformations) (Muliukin et al., 2002) and death. This progression is accompanied by changes at the internal cellular structure. Micro-organisms can survive over more or less long periods of time before death. Many of them, such as Gram positive bacteria or fungi, make spores, special resistance forms that allow them to survive and also to disseminate while assuring an effective protection. Some spores are extraordinarily resistant to extreme conditions and to time. Experiences of exposure to the outer space suggest that *Bacillus subtilis* (a Gram positive bacterium) spores would survive for years if they were just covered by a thin layer protecting them from ultraviolet radiation, such as polysaccharides or a layer of cells. However, they will not last more than a few hours in a single layer. In a similar way, halophilic archaea included in salt crystals resisted the exposure to outer space conditions during the two weeks that the experiment lasted (Rothschild and Mancinelli, 2001). This is a remarkable resistance because halophilic archaea do not form spores. Halophiles are to date the micro-organisms that hold the longevity record (Fish et al., 2002; Grant et al., 1998).

The most efficient ways of biological conservation are cryopreservation and desiccation. All microbiologists use low temperatures to conserve and store most prokaryotic strains, keeping them at $-80\,°\text{C}$ or immersed in liquid nitrogen for years. To get a high survival rate, one has only to add a little glycerol to the culture medium in order to avoid the formation of big water crystals that would damage cells. Cryopreservation occurs also in nature, and reports of species retrieved from permafrost after several thousand years exist in the literature (Kochkina et al., 2001; Shi et al., 1997). Desiccation is also very efficient as it can be readily deduced from the long survival of halophilic archaea in salt crystals. Anecdotally, microbial strains from culture collections are frequently delivered lyophilised.

9.4 Extremophiles and Exobiology

In the above, we have taken a brief look at the different types of extreme environments on Earth, the organisms that occupy them, their adaptations and their long-term survival strategies. As can be realised, life is present on Earth in any place where physicochemical conditions are permissive. If one extends this idea to other planets, the hypothesis of extraterrestrial life becomes more than plausible. The only condition that appears necessary for life, as we know it, is the presence of liquid water (and possibly a carbon-based biochemistry as well).

In our solar system, the two candidates that appear more susceptible to host life or having hosted it in the past are Mars and Europa. Early in its history, the conditions on Mars resembled those of the Earth so that, even if Mars were sterile today, life could have existed on the red planet and therefore left fossil remains. An ocean of liquid water possibly existed that was subsequently lost by evaporation, at least partially, since some water persists frozen at the poles and possibly in the subsurface (Boynton et al., 2002; Forsythe and Zimbelman, 1995; Kerr, 2000). Europa, one of Jupiter's satellites, is fully covered by ice, but some data suggest that an underlying ocean of liquid water might exist and, perhaps, some hydrothermal activity (Carr et al., 1998; Kivelson et al., 2000).

In the following, we will briefly consider the kind of extremophiles and extreme habitats that have raised a special interest among exobiologists.

9.4.1 Hyperthermophiles

Since their discovery, hyperthermophiles have played a key role in generating new hypotheses on the origin of life (on Earth and, by extrapolation, to any other planet exhibiting analogous conditions to those of the early Earth).

Origin-of-life models based on the classical Oparin–Haldane theory propose that life arose at low temperature and was first heterotrophic (Lazcano and Miller, 1996). Life would have been generated in a kind of marine "primitive soup" where simple organic molecules would have been formed, and then accumulated, via prebiotic synthesis helped by ultraviolet radiation and electric discharges. Miller's famous experience in 1953 appeared to be a strong support for this hypothesis, since it demonstrated that amino acids could form in a mixture of reduced gases, supposedly representative of the primitive atmosphere, after exposure to electric discharges (Miller, 1953). However, subsequent experiences showed that amino-acid yield and variety strongly diminished if the starting gas mixture was less reducing and, therefore, in greater agreement with present-day ideas on the early atmosphere composition (Schlesinger and Miller, 1983). From these primordial simple organic ingredients, larger macromolecules would have been formed, RNA molecules, proteins and (somewhat later) DNA, from which primitive cells evolved (Fig. 9.10).

However, during the last decades of the twentieth century, and more particularly after hyperthermophiles were discovered, the idea of a hot origin of life expanded in the scientific community (Pace, 1991). Life would have been initially chemolithoautotrophic (inorganic sources of matter and energy) and would have been born at high temperature in environments similar to contemporary marine hydrothermal vents (Henley, 1996) (Fig. 9.10). The idea of a hot origin of life is sustained in three types of arguments:

- **Geological.** Different geological data suggest that the Earth was hotter than today at the time when life arose about 3.8 Ga ago, with extensive volcanism and hydrothermalism (see Martin's Chap. 4, Part I in this book). Data on oxygen isotopes in archaean sedimentary rocks appear to further sustain this view

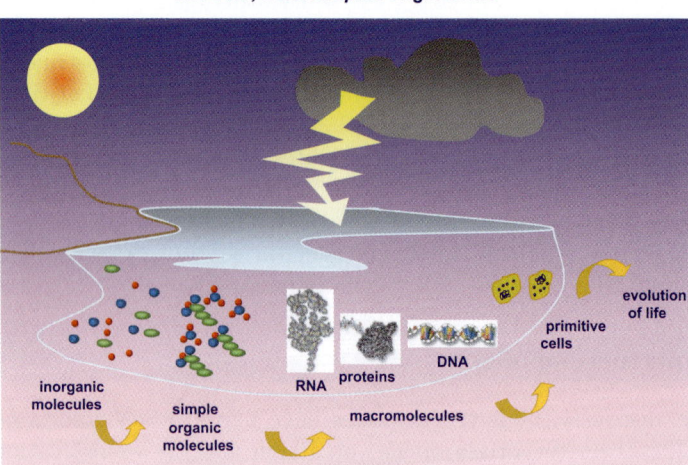

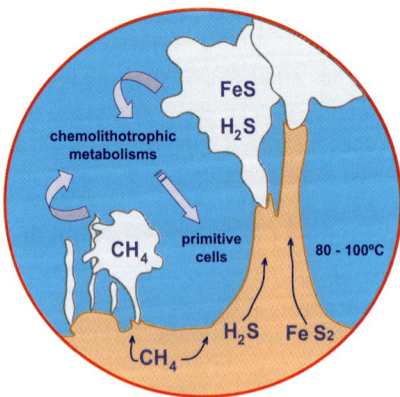

Fig. 9.10. Schematic diagram showing the two main global scenarios for the origin of life

(Knauth, 1992). In addition, the oldest unambiguous microbial fossils known today correspond most likely to thermophilic prokaryotes (Summons, 1999) identified in fossil hydrothermal settings from South Africa. The existence of the earliest putative cyanobacteria fossils in stromatolithe-like structures is very speculative and controversial (Brasier et al., 2002; Schopf et al., 2002). Geological data thus appear compatible with models in which life evolved from a thermophilic ancestor and the subsequent adaptation of microbial communities to increasingly cooler conditions (Nisbet and Sleep, 2001).

- **Phylogenetic.** Several authors have placed the root of the tree of life in the bacterial branch using a strategy based on the comparison of ancient duplicated genes (reviewed in Brown and Doolittle, 1997). If this position is correct, the lineages that appeared to be placed closer to the root were all hyperthermophilic, which supported the idea that the last common universal ancestor was itself a hyperthermophile (Barns et al., 1996; Stetter, 1996). This, of course, led to the idea that life had originated at high temperature, although some authors alternatively suggested that the origin of life took place at low temperature but that the Last Universal Common Ancestor (LUCA) was a hyperthermophile having survived the intensive meteorite bombardments Earth suffered early in its history (e.g. Lazcano and Miller, 1996). In any case, it is difficult to reconstruct ancient phylogenetic relationships among organisms because the evolutionary signal may have been lost with time, or it may be obscured by successive mutations and differences in evolutionary rates (see Brinkmann and Philippe's Chap. 8, Part II, in this book). Indeed, methods of phylogenetic reconstruction today allow the root of the tree of life to be placed safely neither in the bacterial nor in the eukaryotic branch. The phylogenetic argument should thus be taken with caution in this highly controversial debate (Brown et al., 2001; Forterre and Philippe, 1999).
- **Metabolic.** These are based on the fact that many hyperthermophiles are chemolithoautotrophs, displaying a large variety of possible redox reactions to obtain energy (Table 9.2) (Madigan et al., 2002). Furthermore, these reactions frequently involve hydrogen, iron and sulfur. This recalls the model proposed by Wächtershäusser who hypothesized that life appeared on pyrite-like surfaces and that the first organisms gained their energy from FeS reduction by H_2S generating H_2 and FeS_2 (Wächtershäusser, 1988). Most hyperthermophiles are also strict anaerobes (let us recall here that oxygen was not present in significant proportions in the early Earth atmosphere and ocean).

If life arose in a hot environment on Earth, it would be possible that it originated also on Mars, where initial conditions were similar, or in other planets with hydrothermal activity and analogous environmental settings.

9.4.2 Psychrophiles

Psychrophilic micro-organisms living at very low temperatures constitute another focus of research for exobiologists. Extremely cold environments such as

Table 9.2. Chemoautotrophic metabolisms in hyperthermophiles. Bacterial genera are labeled with an asterisk, the remaining genera ascribe to the Archaea (López-García)

Energy-yielding reactions	Representative genera
$H_2 + S^0 \rightarrow H_2S$	*Acidianus, Pyrodictium, Thermoproteus*
$H_2 + NO_3^- \rightarrow NO_2^- + H_2O$ (NO_2, N_2)	*Pyrobaculum, Stygioglobus, Aquifex*, Pyrodictium, Thermoproteus*
$4H_2 + NO_3^- + 2H^+ \rightarrow NH_4 + 3H_2O$	*Pyrolobus*
$2H_2 + O_2 \rightarrow 2H_2O$	*Acidianus, Sulfolobus, Pyrobaculum, Aquifex**
$2S^0 + 3O_2 + 2H_2O \rightarrow 2H_2SO_4$	*Sulfolobus, Acidianus*
$4FeS_2 + 15O_2 + 2H_2O \rightarrow 2Fe_2(SO_4)_3 + 2H_2SO_4$	*Sulfolobus*
$10FeCO_3 + 2NO_3^- + 24H_2O \rightarrow 10Fe(OH)_3 + N_2 + 10HCO_3^- + 8H^+$	*Ferroglobus*
$4H_2 + SO_4^- + 2H^+ \rightarrow 4H_2O + H_2S$	*Archaeoglobus*
$4H_2O + CO_2 \rightarrow CH_4 + 2H_2O$	*Methanopyrus, Methanococcus*

the Antarctic or the permafrost are interesting as analogous environments to cold extraterrestrial regions, for example the hypothetical ocean underlying Europa's ice cover. They could also serve as a model of past fossil life preservation on ice. Subglacial Antarctic lakes, particularly lake Vostok (see J.R. Petit's Chap. 7, Part I, in this book), at a depth of more than 3750 m under the continental ice cover and a pressure of more than 350 bars, may host microbial communities isolated for several million years (Siegert et al., 2001).

9.4.3 Halophiles and Evaporites

Halophilic micro-organisms can survive for long periods within salt crystals in salt mines or in evaporitic rocks (Fish et al., 2002; Grant et al., 1998). They can even be active in fluid inclusions (Rothschild et al., 1994). Evaporites are important in the fossil record since they tend to preserve well microbial structures and algal filaments. Recent Martian exploration revealed the existence of geological structures resembling evaporitic deposits in impact craters showing traces of water flow (Newsom et al., 2001). These are interesting sampling sites due to their potential to have conserved past life signatures.

9.4.4 The Deep Biosphere

The discovery of microbial life in Earth's subsurface has reinforced the idea that life can occur today in other planets of our solar system and even beyond. During the last several years evidence on the existence of a deep microbial biosphere based on chemolithotrophy has been accumulating (Stetter et al., 1993; Pedersen, 2000). Underground biotopes comprise continental and sea floor, deep buried sediments, oil and petroleum reservoirs, deep caves and deep aquifers. The ocean subseafloor close to midoceanic ridges is probably one of the most successfully colonised subterranean habitats, because the mixture of hydrothermal fluid and seawater that penetrates in the substratum is highly enriched in inorganic electron donors and acceptors for chemolithotrophic redox energy reactions (Summit and Baross, 2001). The oceanic crust would thus be the largest habitat on Earth, and its microbial inhabitants could constitute a considerable portion of the total biomass in our planet, even when the microbial density is generally scarce and the metabolic rates low due to overall oligotrophic conditions. The depth reached by biological alterations in subseafloor basaltic glass appears limited by temperatures permissive for life (Fisk et al., 1998; Furnes et al., 1999), which suggests that temperature is a major limiting factor for underground life.

Subterranean microbial communities are characterised by chemolithoautotrophic metabolisms in reducing and strict anaerobic conditions. The role of hydrogen appears to be primordial (Pedersen, 2000). A recent study showed that more than 95% of the population found in a hot deep aquifer were methanogenic archaea utilising H_2 and CO_2 to produce CH_4 (Chapelle et al., 2002). The implications of these studies for exobiology are numerous, since analogous subsurface environments are likely to exist in other planets that still maintain a geothermal activity (Weiss et al., 2000). In addition, these environments are stable and protected from ultraviolet irradiation even when they had a thin atmosphere (e.g. Mars).

9.5 Perspectives

During recent years, the discovery of extremophiles and the limits of life on Earth have allowed questions on the existence of extraterrestrial life to be formulated in a more rigorous way. We need now to continue exploring the micro-organisms colonising extreme environments, particularly those from less-accessible biotopes such as the deep subsurface that still remain very poorly known. The results obtained in this way should allow identification of potential habitats for life (as we know it) on other planets. One of the most important challenges today is the identification of markers of incontestable biogenic origin that could be afterwards detected in other planets, either *in situ* or by direct analysis after an eventual sample return. In the future, exobiology will certainly require a further interdisciplinary approach to answer solidly questions regarding the origin of life, to identify unambiguously biosignatures and, finally, to develop methodologies to detect them on Earth or on other planets.

References

General

Madigan M. T., Martinko J. M. and Parker J. (2002). *Brock Biology of Microorganisms*, 10th edn. New Jersey, Prentice-Hall, Inc.

Specialised

Amann R. I., Ludwig W. and Schleifer K. H. (1995). Phylogenetic identification and in situ detection of individual microbial cells without cultivation. *Microbiol. Rev.* **59**, 143–169.

Barns S. M., Delwiche C. F., Palmer J. D., Dawson S. C., Hershberger K. L. and Pace N. R. (1996). Phylogenetic perspective on microbial life in hydrothermal ecosystems, past and present. *Ciba Found Symp.* **202**, 24–32; discussion 32–39.

Blöchl E., Rachel R., Burgraff S., Hafenbradl D., Jannasch H. W. and Stetter K. O. (1997). *Pyrolobus fumarii*, gen. and sp. nov., represents a novel group of archaea, extending the upper temperature limit for life to 113 °C. *Extremophiles* **1**, 14–21.

Boynton W. V., Feldman W. C., Squyres S. W., Prettyman T. H., Bruckner J., Evans L. G., Reedy R. C., Starr R., Arnold J. R., Drake D. M., Englert P. A., Metzger A. E., Mitrofanov I., Trombka J. I., D'Uston C., Wanke H., Gasnault O., Hamara D. K., Janes D. M., Marcialis R. L., Maurice S., Mikheeva I., Taylor G. J., Tokar R. and Shinohara C. (2002). Distribution of hydrogen in the near surface of Mars: evidence for subsurface ice deposits. *Science* **297**, 81–85.

Brasier M. D., Green O. R., Jephcoat A. P., Kleppe A. K., Van Kranendonk M. J., Lindsay J. F., Steele A. and Grassineau N. V. (2002). Questioning the evidence for Earth's oldest fossils. *Nature* **416**, 76–81.

Brock T. D., Brock K. M., Belly R. T. and Weiss R. L. (1972). *Sulfolobus*: a new genus of sulfur-oxidizing bacteria living at low pH and high temperature. *Arch Mikrobiol.* **84**, 54–68.

Brock T. D. and Freeze H. (1969). *Thermus aquaticus gen. n.* and *sp. n.*, a nonsporulating extreme thermophile. *J. Bacteriol.* **98**, 289–297.

Brown J. R. and Doolittle W. F. (1997). Archaea and the prokaryote-to-eukaryote transition. *Microbiol. Mol. Biol. Rev.* **61**, 456–502.

Brown J. R., Douady C. J., Italia M. J., Marshall W. E. and Stanhope M. J. (2001). Universal trees based on large combined protein sequence data sets. *Nat. Genet.* **28**, 281–285.

Buchalo A. S., Nevo E., Wasser S. P., Oren A. and Molitoris H. P. (1998). Fungal life in the extremely hypersaline water of the Dead Sea: first records. *Proc. R. Soc. Lond. B Biol. Sci.* **265**, 1461–1465.

Burggraf S., Stetter K. O., Rouviere P. and Woese C. R. (1991). *Methanopyrus kandleri*: an archaeal methanogen unrelated to all other known methanogens. *Syst. Appl. Microbiol.* **14**, 346–351.

Carr M. H., Belton M. J., Chapman C. R., Davies M. E., Geissler P., Greenberg R., McEwen A. S., Tufts B. R., Greeley R., Sullivan R., Head J. W., Pappalardo R. T., Klaasen K. P., Johnson T. V., Kaufman J., Senske D., Moore J., Neukum G., Schubert G., Burns J. A., Thomas P. and Veverka J. (1998). Evidence for a subsurface ocean on Europa. *Nature* **391**, 363–365.

Chapelle F. H., K, O. N., Bradley P. M., Methe B. A., Ciufo S. A., Knobel L. L. and Lovley D. R. (2002). A hydrogen-based subsurface microbial community dominated by methanogens. *Nature* **415**, 312–315.

Deming J. W. (1998). Deep ocean environmental biotechnology. *Curr. Opin. Biotechnol.* **9**, 283–287.

Farlow, W. G. (1880). *On the nature of the peculiar reddening of salted codfish during the summer season.* U.S. Commission of Fish and Fisheries, pp. 969–974.

Fish S. A., Shepherd T. J., McGenity T. J. and Grant W. D. (2002). Recovery of 16S ribosomal RNA gene fragments from ancient halite. *Nature* **417**, 432–436.

Fisk M. R., Giovannoni S. J. and Thorseth I. H. (1998). Alteration of oceanic volcanic glass: textural evidence of microbial activity. *Science* **281**, 978–980.

Forsythe R. D. and Zimbelman J. R. (1995). A case for ancient evaporite basins on Mars. *J. Geophys. Res.* **100**, 5553–5563.

Forterre P. and Philippe H. (1999). Where is the root of the universal tree of life? *Bioessays* **21**, 871–879.

Furnes H., Muehlenbachs K., Tumyr O., Torsvik T. and Thorseth I. H. (1999). Depth of active bio-alteration in the ocean crust: Costa Rica Rift (Hole 504B). *Terra Nova* **11**, 228–233.

Grant W. D., Gemmell R. T. and McGenity T. J. (1998). Halobacteria: the evidence for longevity. *Extremophiles* **2**, 279–287.

Henley R. W. (1996). Chemical and physical context for life in terrestrial hydrothermal systems: chemical reactors for the early development of life and hydrothermal ecosystems. *Ciba Found. Symp.* **202**, 61–76; discussion 76–82.

Horikoshi K. (1999). Alkaliphiles: some applications of their products for biotechnology. *Microbiol Mol Biol Rev* **63**, 735–750, table of contents.

Hough D. W. and Danson M. J. (1999). Extremozymes. *Curr. Opin. Chem. Biol.* **3**, 39–46.

Hugenholtz P., Pitulle C., Hershberger K. L. and Pace N. R. (1998). Novel division level bacterial diversity in a Yellowstone hot spring. *J. Bacteriol.* **180**, 366–376.

Kashefi K. and Lovley D. R. (2003). Extending the upper temperature limit for life. *Science* **301**, 934.

Kerr R. A. (2000). Planetary science. Buried channels may have fed Mars ocean. *Science* **287**, 1727–1728.

Kivelson M. G., Khurana K. K., Russell C. T., Volwerk M., Walker R. J. and Zimmer C. (2000). Galileo magnetometer measurements: a stronger case for a subsurface ocean at Europa. *Science* **289**, 1340–1343.

Knauth L. P. (1992). Origin and diagenesis of cherts: An isotopic perspective. In *Isotopic signatures and sedimentary records*, pp. 123–152. Edited by N. Clauer and S. Chanduri. Berlin: Springer-Verlag.

Kochkina G. A., Ivanushkina N. E., Karasev S. G., Gavrish E., Gurina L. V., Evtushenko L. I., Spirina E. V., Vorob'eva E. A., Gilichinskii D. A. and Ozerskaia S. M. (2001). Micromycetes and actinobacteria under conditions of many years of natural cryopreservation. *Mikrobiologiia* **70**, 412–420.

Lanoil B. D., Sassen R., La Duc M. T., Sweet S. T. and Nealson K. H. (2001). Bacteria and archaea physically associated with gulf of Mexico gas hydrates. *Appl. Environ. Microbiol.* **67**, 5143–5153.

Lazcano A. and Miller S. L. (1996). The origin and early evolution of life: prebiotic chemistry, the pre-RNA world, and time. *Cell* **85**, 793–798.

Lopez-Archilla A. I., Marin I. and Amils R. (2001). Microbial community composition and ecology of an acidic aquatic environment: the tinto river, Spain. *Microb. Ecol.* **41**, 20–35.

Madigan M. T., Martinko J. M. and Parker J. (2002). *Brock Biology of Microorganisms*, 10th edn. New Jersey: Prentice-Hall, Inc.

Miller S. (1953). A production of amino acids undert possible primitive Earth conditions. *Science* **117**, 528–529.

Muliukin A. L., Sorokin V. V., Vorob'eva E. A., Suzina N. E., Duda V. I., Gal'chenko V. F. and El'-Registan G. I. (2002). Detection of microorganisms in the environment and the preliminary appraisal of their physiological state by X-ray microanalysis. *Mikrobiologiia* **71**, 836–848.

Newsom H. E., Hagerty J. J. and Thorsos I. E. (2001). Location and sampling of aqueous and hydrothermal deposits in Martian impact craters. *Astrobiology* **1**, 71–88.

Nisbet E. G. and Sleep N. H. (2001). The habitat and nature of early life. *Nature* **409**, 1083–1091.

Oren A. (1994). The ecology of extremely halophilic archaea. *FEMS Microbiol. Rev.* **13**, 415–440.

Oren A. (1999). Bioenergetic aspects of halophilism. *Microbiol. Mol. Biol. Rev.* **63**, 334–348.

Pace N. R. (1991). Origin of life–facing up to the physical setting. *Cell* **65**, 531–533.

Pace N. R. (1997). A molecular view of microbial diversity and the biosphere. *Science* **276**, 734–740.

Paerl H. W., Pinckney J. L. and Steppe T. F. (2000). Cyanobacterial-bacterial mat consortia: examining the functional unit of microbial survival and growth in extreme environments. *Environ. Microbiol.* **2**, 11–26.

Pedersen K. (2000). Exploration of deep intraterrestrial microbial life: current perspectives. *FEMS Microbiol. Lett.* **185**, 9–16.

Rothschild L. J., Giver L. J., White M. R. and Mancinelli R. L. (1994). Metabolic activity of microorganisms in gypsum-halite crusts. *J. Phycol.* **30**, 431–438.

Rothschild L. J. and Mancinelli R. L. (2001). Life in extreme environments. *Nature* **409**, 1092–1101.

Schlesinger G. and Miller S. (1983). Prebiotic synthesis in atmospheres containing CH_4, CO and CO_2. I. Amino acids. *J. Mol. Evol.* **19**, 376–382.

Schopf J. W., Kudryavtsev A. B., Agresti D. G., Wdowiak T. J. and Czaja A. D. (2002). Laser–Raman imagery of Earth's earliest fossils. *Nature* **416**, 73–76.

Shi T., Reeves R. H., Gilichinsky D. A. and Friedmann E. I. (1997). Characterization of viable bacteria from Siberian permafrost by 16S rDNA sequencing. *Microb. Ecol.* **33**, 169–179.

Siegert M. J., Ellis-Evans J. C., Tranter M., Mayer C., Petit J. R., Salamatin A. and Priscu J. C. (2001). Physical, chemical and biological processes in Lake Vostok and other Antarctic subglacial lakes. *Nature* **414**, 603–609.

Sterflinger K. (1998). Temperature and NaCl-tolerance of rock-inhabiting meristematic fungi. *Antonie Van Leeuwenhoek* **74**, 271–281.

Stetter K.O. (1989) Extremely thermophilic chemolithoautotrophic Archaebacteria. *In* H.G. Schlegel and B. Bowien (eds.), *Extremely Thermophilic Chemolithoautotrophic Archaebacteria*. Springer-Verlag, Berlin, pp. 167–176.

Stetter K. O., Huber R., Blöchl E., Kurr M., Eden R. D., Fiedler M., Cash H. and Vance I. (1993). Hyperthermophilic archaea are thriving in deep North Sea and Alaskan oil reservoirs. *Nature* **365**, 743–745.

Stetter K. O. (1996). Hyperthermophilic prokaryotes. *FEMS Microbiol. Rev.* **18**, 149–158.

Summit M. and Baross J. A. (2001). A novel microbial habitat in the mid-ocean ridge subseafloor. *Proc. Natl. Acad. Sci. USA* **98**, 2158–2163.

Summons R. (1999). Molecular probing of deep secrets. *Nature* **398**, 752–753.

Takami H., Inoue A., Fuji F. and Horikoshi K. (1997). Microbial flora in the deepest sea mud of the Mariana Trench. *FEMS Microbiol. Lett.* **152**, 279–285.

Wächtershäusser G. (1988). Pyrite formation, the first energy source for life: a hypothesis. *Syst. Appl. Microbiol.* **10**, 207–210.

Weiss B. P., Yung Y. L. and Nealson K. H. (2000). Atmospheric energy for subsurface life on Mars? *Proc. Natl. Acad. Sci. USA* **97**, 1395–1399.

Wilansky B. (1936). Life in the Dead Sea. *Nature* **138**, 467.

Woese C. R. (1987). Bacterial evolution. *Microbiol. Rev.* **51**, 221–271.

Woese C. R. and Fox G. E. (1977). Phylogenetic structure of the prokaryotic domain: the primary kingdoms. *Proc. Natl. Acad. Sci. USA* **74**, 5088–5090.

Woese C. R., Kandler O. and Wheelis M. L. (1990). Towards a natural system of organisms: proposal for the domains Archaea, Bacteria, and Eucarya. *Proc. Natl. Acad. Sci. USA* **87**, 4576–4579.

Zuckerkandl E. and Pauling L. (1965). Molecules as documents of evolutionary history. *J Theor. Biol.* **8**, 357–366.

Appendices

1 Earth Structure and Plate Tectonics: Basic Knowledge

Hervé Martin

1.1 Earth Internal Structure

The Zapoliarny borehole in Russia only reaches a depth of 12.5 km; samples of deeper parts of our planet were brought up by volcanoes, but their original depth never exceeds a few hundred kilometres. Consequently, most of our knowledge of the Earth's internal composition and structure is indirect and mostly due to studies of propagation rates of seismic waves, although gravimetric data have also been used. The structure of the Earth can be schematically described as consisting of concentric shells that are from the centre to the surface (Fig. 1.1).

1.1.1 Inner Core (from 6378 to 5155 km Depth)

It is solid and mainly consists of iron with smaller amounts of nickel; its density is about 12 and its temperature at least 6000 K.

Photo 1.1. TTG sampled at 11 047.9 m depth in the Zapoliarny borehole in Kola Peninsula (Russia). This is the deepest borehole in the world until today (Photo H. Martin)

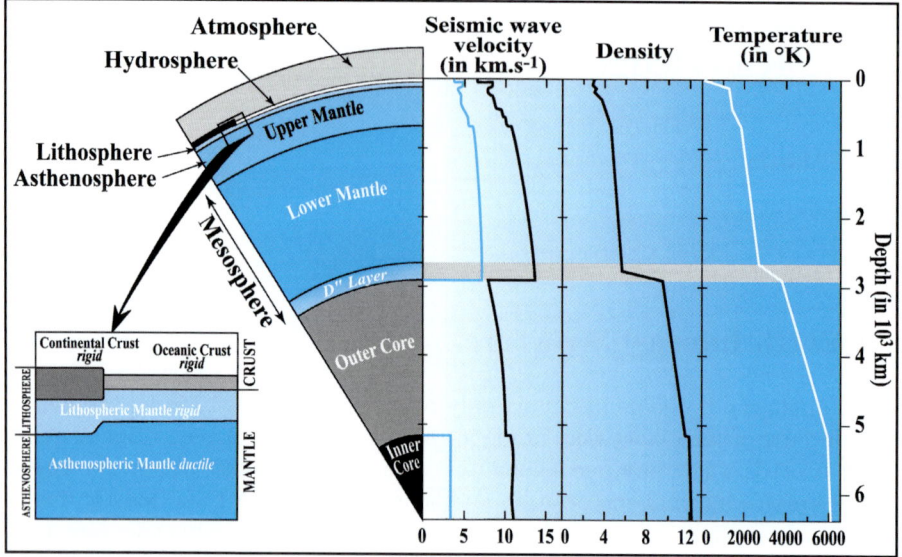

Fig. 1.1. Schematic cross section of the Earth showing its internal structure. On the right side are shown seismic wave velocity, density and temperature variations vs. depth. Shear seismic waves are in blue and compression waves in black. It must be noted that shear waves do not travel through the outer core, pointing to its liquid state. The inset on the left side illustrates the distinction that is made between mantle/crusts and asthenosphere/lithosphere pairs

1.1.2 Outer Core (from 5155 to 2891 km Depth)

It is separated from the inner core by the Lehman discontinuity at 5155 km.

Shear seismic waves (S waves) do not travel through the outer core (Fig. 1.1), which indicates that it is liquid. As for the inner core, it is composed of iron and nickel but also contains lighter element such as S, Si, O, and possibly K. From top to bottom its density ranges between 9.5 and 11.5 and its temperature between 3800 and 6000 K. As the Earth cooled it progressively crystallized at high pressure thus creating the solid inner core. During crystallization, light elements (Si, O, S, K) are rejected and remain in the liquid core.

The relative motions of the inner and outer cores as well as the convection in the outer core result in the creation of electromagnetic currents that are responsible for the Earth's magnetic field.

1.1.3 Lower Mantle (from 2891 to 670 km Depth)

It is separated from the outer core by the Gutenberg discontinuity at 2891 km.

Sometimes called the mesosphere, it consists of a silicate-rich solid, made up of perovskite $(Fe,Mg)SiO_3$ and magnesiowurstite $(Fe,Mg)O$. From top to bottom its density ranges between 4.5 and 5.5 and its temperature between 1900 and 2500 K. Velocities of shear seismic waves show that the lower mantle is solid, but on the geological timescale ($\sim 10^8$ years) it behaves as a fluid and is affected by convection.

The transition between the mantle and core consists of a 100–200 km thick heterogeneous layer called the D″ layer; through this zone temperature increases from 2500 to 3800 K. Through this thermal boundary layer, heat is only transferred by conduction and not by convection, consequently, this interface is the place where energy exchanges can take place, however, only a little matter exchange is suspected. Instabilities in this zone are considered as responsible for hot-spot magmatism. It has also been proposed that cold avalanches due to accumulation of subducted oceanic crust could create these instabilities.

1.1.4 Upper Mantle (from 670 km to 7 km Depth Under Oceans and 30 km Depth Under Continents)

It mainly consists of peridotite: a rock made up of olivine $(Fe,Mg)_2SiO_4$, orthopyroxene $(Fe,Mg)_2Si_2O_6$, clinopyroxene $Ca(Fe,Mg)Si_2O_6$ and an aluminium-rich mineral that is plagioclase $CaAl_2Si_2O_8$ (depth $< 30-40$ km), spinel $MgAl_2O_4$ ($30-40$ km $<$ depth < 70 km) or garnet $Mg_3Al_2(SiO_4)_3$ (depth > 70 km). Its density ranges from 3.3 to 4.5 and its temperature increases from 900–1000 K to 1900 K. It is also the place of active convection. Until now, scientific research has tried to determine if the whole mantle is affected by a single convection system or if upper and lower mantles convert independently.

The transition between lower and upper mantle is due to the transformation of olivine at low pressure into perovskite + magnesiowurstite at high pressure.

Between 100 and 200 km depth, seismic waves show a diminution of their propagation rate, this zone is called the low-velocity zone (LVZ). It is interpreted as being due to the diminution of peridotite rigidity but this zone remains solid as it allows transmission of shear seismic waves. (It must be noted that the notion of rigidity is relative and strongly depends on the timescale. For instance, the upper mantle can have an immediate rigid behaviour but is ductile when million years period is taken into account.)

1.1.5 Crusts (from 7 km Depth Under Oceans and 30 km Depth Under Continents to Surface)

It is separated from the upper mantle by the Mohorovicic (Moho) discontinuity.
There exist two different types of crusts:

Oceanic Crust
Its average composition is that of a basalt, it is thin (~ 7 km), has a density of 3.1 and its age never exceeds 180 Ma. It is generated in oceanic ridge systems by partial melting of the upper mantle and recycled into the mantle in subduction zones.

Continental Crust
It has an average thickness of ~ 30 km but under mountains, it can be 70 km thick. It is highly heterogeneous with an average composition of granodiorite. When compared with mantle it has a relatively low density (2.75) such that it has high buoyancy and is almost not recycled into the mantle. This is why continental crust as old as 4.04 Ga can be found. Most of the new continental crust is generated in subduction-zone environments.

1.1.6 Hydrosphere

It consists of surface water: ocean, lakes, rivers, ice caps, clouds, etc. It strongly interacts with surface rocks.

1.1.7 Atmosphere

This is the gaseous envelope of the Earth, it can extend up to 500 km in altitude. This interface between the surface of the Earth and the extraterrestrial domain plays the role of energy filter.

Both the hydrosphere and atmosphere are the places where life developed.

1.1.8 Lithosphere and Asthenosphere

The distinction between core, mantle and crust is based on compositional differences. However, for the same chemical and mineralogical composition, depending, for instance, on temperature, rheological properties can be completely different. For instance, viscosity is strongly temperature dependent, it decreases when temperature increases. Consequently the uppermost part of the mantle that is cold has high viscosity and a rigid behaviour even at long timescale, whereas the deeper and hotter parts have a lower viscosity resulting in a ductile behaviour on the geological timescale (~ 100 Ma). Consequently, in the mantle–crusts system two rheological shells are distinguished:

Lithosphere
It consists of the crust plus the rigid part of the upper mantle, its average thickness is about 100 km under oceans and 150–200 km under continents. It has a rigid behaviour and is not affected by internal convection; it corresponds to a domain where heat is transferred by conduction.

Asthenosphere
It corresponds to the deeper and hotter parts of the mantle where peridotite has a ductile behaviour and where heat is dissipated by convection. It is generally considered that the lithosphere–asthenosphere limit corresponds to the $800\,°C$ isotherm; it roughly begins with the LVZ. As it cools the asthenosphere is transformed into lithosphere.

1.2 Plate Tectonics

Since the 1960s, the concept of sea-floor spreading has been established. As the surface of the Earth can be considered as constant, the oceanic crust, which is progressively generated in rift systems, must return and be recycled into the mantle (subduction zone). Then plate tectonics was born and this concept progressively evolved and refined.

1.2.1 Plates on the Surface of the Earth

The surface of the Earth is divided into lithospheric plates made up of crust (oceanic or continental) and with the rigid part of the upper mantle; they move over the convecting asthenospheric mantle. Today 12 major plates are recognized on the surface of the Earth but several small (micro-plates) also exist (Table 1.1; Fig. 1.2).

Table 1.1. Name and surface of the main lithospheric plates

Plate name	Area (10^6 km^2)	Plate name	Area (10^6 km^2)
Pacific	103.28	South-American	43.62
African	78.00	Nazca	15.63
North American	75.89	Philippine	5.45
Eurasian	67.81	Arabian	5.01
Antarctic	60.92	Caribbean	3.32
Indo-Australian	59.07	Cocos	2.86

1.2.2 Margin Definitions

Some of these plates contain both oceanic and continental crust (i.e. Eurasian or African), others are exclusively oceanic (i.e. Pacific or Nazca). Most of the seismic and volcanic activity of our planet is concentrated at plate margins. Two kinds of continent–ocean limit must be distinguished:

- ***Passive margin:*** both oceanic and continental crust belong to the same plate (limit between Europe and East North Atlantic), there is no seismic or volcanic activity;
- ***Active margin:*** oceanic and continental crusts do not belong to the same plate; generally the oceanic crust disappears under continental crust giving rise to an important seismic and volcanic activity.

There exist 3 main kinds of boundary between lithospheric plates:

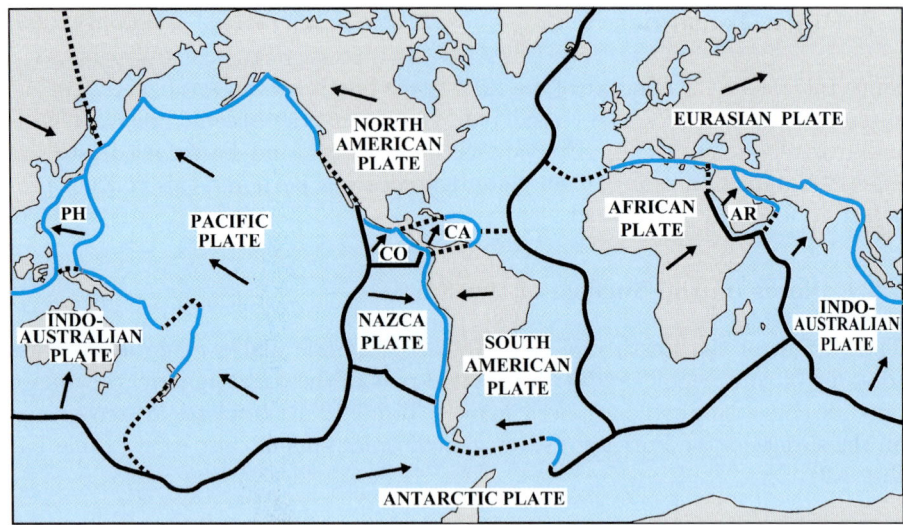

Fig. 1.2. Schematic map of the surface of the Earth showing the 12 main lithospheric plates. *Black heavy line* = divergent margin; *blue heavy line* = convergent margin; *heavy dotted line* = transform margin. Arrows show the main direction of plate motion. CO = Coco plate; CA = Caribbean plate; PH = Philippine plate; AR = Arabian plate

- **Divergent margin:** also called constructive margin, this is where new material resulting from upper-mantle melting is added to the oceanic lithosphere; the two plates pull away from each other. They correspond to the midoceanic ridges that form a 80 000 km long submarine relief.
- **Convergent margin:** also called destructive margin. This is where the oceanic crust plunges under another plate and sinks into the mantle (subduction). Subduction can occur under either continental or oceanic lithosphere. Because of its low density and of its correlated high buoyancy, continental crust is almost never subducted deep into the mantle. Consequently when both plates contain a continent, the two continents collide and form a mountain chain as, for instance, the subduction of the Indian plate under the Eurasian plate led to the collision of continents and gave rise to the Himalaya Mountains.
- **Transform margin:** also called conservative margin, there, crust is neither produced nor destroyed as the plates slide horizontally past each other. The San Andreas Fault perfectly illustrates this type of contact between Pacific and North-American plates.

1.2.3 Divergent Margin

Midoceanic ridge
The rate of oceanic spreading ranges between 2 and $17 \, \text{cm} \, \text{yr}^{-1}$.

Divergent margins are located over the ascending branch of a convection cell (Fig. 1.3). Due to the low thermal diffusion rates, solid upper mantle peridotite ascents adiabatically. Figure 1.4 shows that during adiabatic ascent a dry ($H_2O \sim 0.1\%$) peridotite will crosscut its dry solidus temperature (temperature where dry peridotite begins to melt) at low pressure (in the domain of stability of spinel or even of plagioclase) thus giving rise to a basaltic magma whose crystallization generates new oceanic crust.

As this magmatism is generally submarine, ocean water penetrates into the fractures of the newly formed oceanic crust.

During its descent into the crust water progressively warms up, thus creating a convective water circulation into the oceanic crust. This water will not only exchange energy with basalts but it will also chemically interact with it. This is well exemplified by the black or white smokers that correspond to the escape of warm ($> 300\,°C$) hydrothermal water rich in sulfurs dissolved from the basalts. Another important effect of water is that a basalt initially composed of anhydrous minerals (olivine $(Fe,Mg)_2SiO_4$, orthopyroxene $(Fe,Mg)_2Si_2O_6$, clinopyroxene $Ca(Fe,Mg)Si_2O_6$ and plagioclase $CaAl_2Si_2O_8$) will be transformed into a hydrous mineral-bearing rock (talc $(Fe,Mg)_3Si_4O_{10}(OH)_2$;

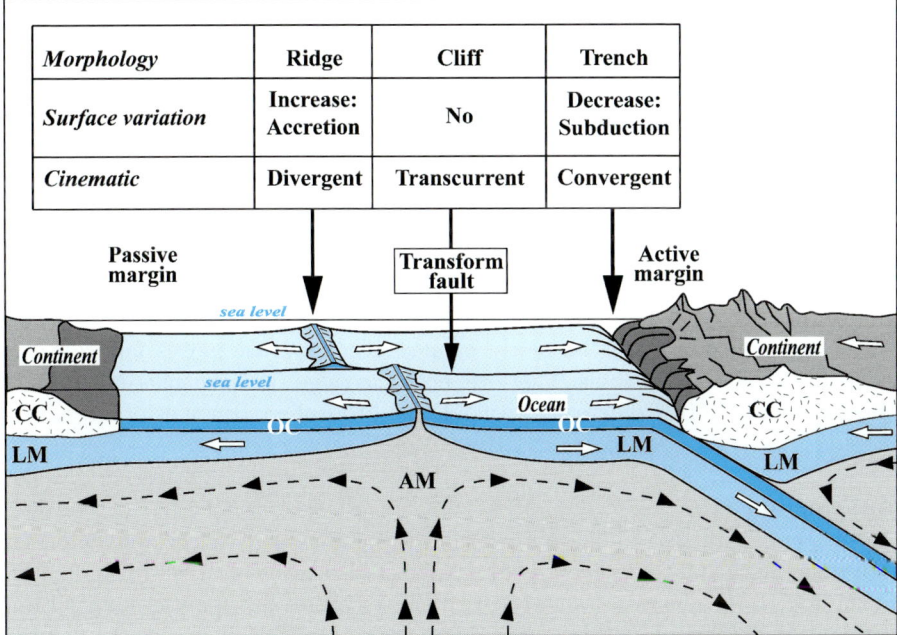

Fig. 1.3. Schematic block diagram showing different possible relationships between plates as well as the terminology generally used. CC = continental crust; OC = oceanic crust; LM = lithospheric mantle; AM = asthenospheric mantle. *Black arrows* and *dotted lines* show convective displacement into the upper mantle

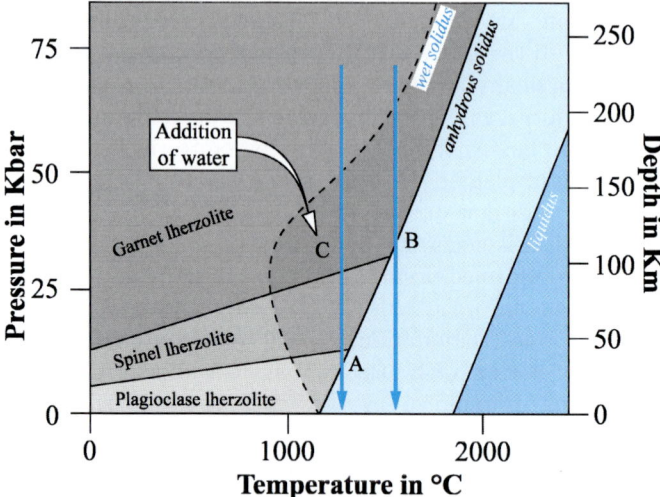

Fig. 1.4. Pressure vs. temperature diagram showing the domains of stability of garnet, spinel and plagioclase lherzolite of the lithospheric mantle. *Heavy lines* represent solidus and liquidus of dry mantle, *dotted line* is the solidus of hydrous mantle lherzolite. In ridge environments mantle melts by adiabatic decompression, the dry *solidus curve* is cross-cut at shallow depth (**A**). Melting in hot-spot environments is also due do adiabatic decompression but as mantle has a deeper origin it is hotter and melting takes place at greater depth (**B**). In subduction zones, mantle melts by addition of water due to dehydration of the subducted oceanic crust, water significantly lowers the solidus temperature (**C**)

serpentine $(Fe,Mg)_6Si_4O_{10}(OH)_8$, amphibole $Ca_2(Fe,Mg)_5Si_8O_{22}(OH)_2$, chlorite $Fe_3Mg_3AlSi_3O_{10}(OH)_8$). The oceanic crust that contained about 0.3% water when it emplaced will contain between 1.5 and 5% water after hydrothermal alteration.

In the course of time the newly formed oceanic crust moves far from the ridge and progressively cools. Therfore, near the ridge the lithosphere was warm and consequently thin whereas as it cools the lithosphere becomes progressively thicker.

Continental rift

A continental rift is a zone of continental lithosphere where extensional deformation takes place and that generally evolves towards continent break-up and ocean birth (Fig. 1.5). The process starts with the ascent of hot mantle material under continental crust. As this material is hot and buoyant, it causes the doming of the overlying plate. The warmed lithosphere transforms into asthenosphere. Doming of continental crust provokes its stretch and thinning due to its split into large blocks limited by faults, thus resulting in a collapsed valley called rift. As thinning proceeds, crust begins to be pulled apart, to extend and as in oceanic ridge, adiabatic decompression of mantle peridotite leads to its melting and to the genesis of basaltic magma. Progressive and successive injection of basalt in the central part of the rift contributes to pushing plates apart, the opened space being immediately filled with magma. This mechanism results in the genesis of oceanic crust, whose surface increases as continents separate (Fig. 1.5).

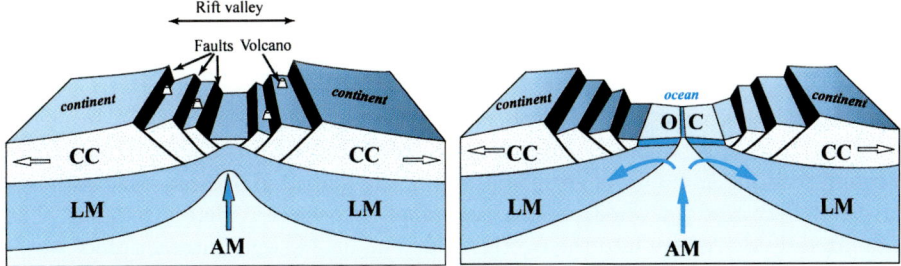

Fig. 1.5. Schematic diagrams showing how continental crust breaks (*left*) and how ocean birth occurs (*right*). CC = continental crust; OC = oceanic crust; LM = lithospheric mantle; AM = asthenospheric mantle

1.2.4 Convergent Margin

Subduction zone

To compensate for new material formed in ridges, old oceanic lithosphere has to be destroyed which is realised in subduction zones (Fig. 1.2). As it cools, the density of oceanic lithosphere increases and consequently, its buoyancy decreases. When its density becomes greater than that of the underlying asthenosphere, then oceanic lithosphere can spontaneously sink into the mantle. In this case as it is old, oceanic lithosphere is thick and subduction angle is high (Fig. 1.6).

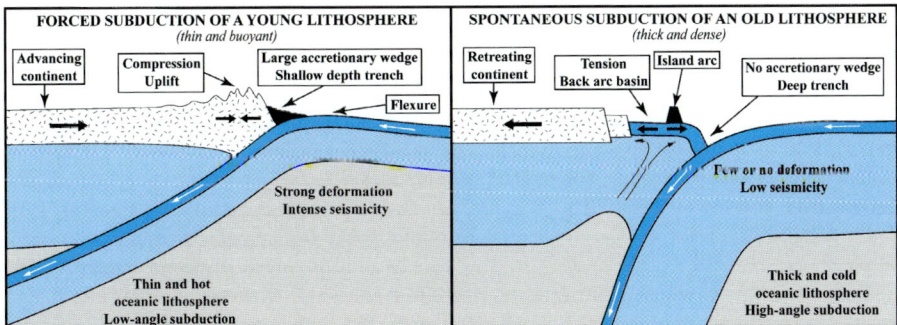

Fig. 1.6. Schematic cross sections showing the contrasted characteristics of both spontaneous and forced subduction

This geometry induces a deep trench and generally the associated seismicity is moderate (i.e. Marianna, Japan).

When strain constraints are too high young oceanic lithosphere can be forced to subduct, even if its density is lower than the asthenosphere one. In this situation (Fig. 1.5) the lithosphere is thin and angle of subduction is low (it can even be almost flat). As oceanic lithosphere resists subduction, seismic activity is generally very important. All sediments emplaced on the ocean floor do not enter into subduction and form an accretionary wedge in a relatively shallow trench. Thus, the continental crust is compressed and can form mountains (i.e. Andes).

Contrary to ridge systems, subduction corresponds to descending branches of convection cells (Fig. 1.2); consequently dry mantle cannot melt by adiabatic decompression. As it enters into subduction, the hydrated oceanic crust progressively dehydrates (metamorphism) such that it becomes unable to melt. The liberated fluids rise up into the overlying dry mantle wedge. Interaction of water with dry peridotite significantly lowers its solidus temperature (Fig. 1.6) such that it can melt. Fluids released by subducted oceanic crust not only lower mantle solidus temperature, but they also transport dissolved elements (U, K, Rb, etc.) that modify the mantle peridotite composition (metasomatism). The generated magmas are andesitic or granodioritic in composition, they form new (juvenile) continental crust. The descending residual oceanic crust accumulates at

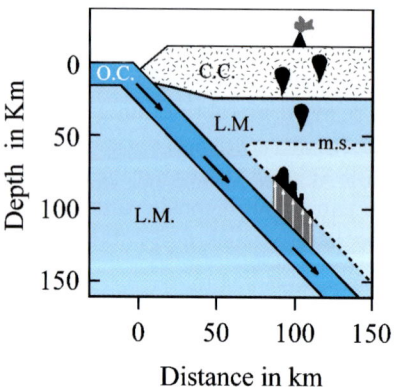

Fig. 1.7. Schematic cross section of a subduction zone showing the process of magma generation: as it enters into the mantle, the subducted oceanic crust dehydrates. Liberated fluids rise-up through the overlying mantle wedge, which becomes progressively rehydrated, this lowers its solidus temperature and induces its partial melting (Fig. 1.3). O.C. = oceanic crust; C.C. = continental crust; L.M. = lithospheric mantle; m.s. = lithospheric mantle hydrous solidus (the *pale blue part* of the mantle wedge corresponds to the part whose temperature is greater than the hydrous solidus one: it can melt if water is available); *black domains* correspond to place where magma is generated and emplaced; whereas *grey area* with *white arrows* represents place where fluids pass through

the 660-km seismic discontinuity. When stored oceanic crust exceeds a threshold it suddenly sinks into the mantle as a cold avalanche that reaches the mantle–core boundary (D″ layer). In some exceptional situations generally related to subduction of very young oceanic lithosphere, oceanic crust can melt thus generating magmas called adakites that also will contribute to form juvenile continental crust.

Collision zone

When the plate implied in subduction carries a continent, this latter, due to its high buoyancy and contrarily to oceanic lithosphere, does not significantly sink into the mantle; which results in continental collision (Fig. 1.8). The overstacking of continental crust segments significantly thickens continental lithosphere and generates mountain chains. Part of the continental crust transported to depth, can melt and generates granitic magmas. As these latter are produced by melting of the continental crust they do not constitute new addition to this crust whose volume remains unchanged: this mechanism is called recycling.

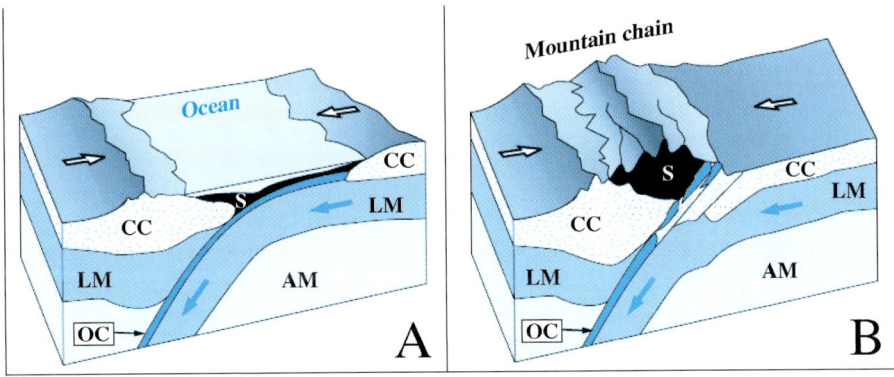

Fig. 1.8. Schematic cross sections showing how subduction and closure of an ocean (**A**) can evolve towards the collision of two continental crusts. This process, still active in the Alps and the Himalayas, generates mountain chains and thickens continental crust (**B**). O.C. = oceanic crust; S. = sediments; C.C. = continental crust; L.M. = lithospheric mantle; A.M. asthenospheric mantle

1.2.5 Hot Spots

In some places volcanism occurs in the middle of plates (Fig. 1.9) it is called hot-spot magmatism (Hawaii, La Réunion, etc.). These magmas are considered as being generated by the ascent of hot lower mantle (mesosphere) diapir, possibly coming from the core mantle boundary (D″ layer). The source of these diapirs appears as immobile on the scale of several tens of millions of years. Lithospheric

294   1 Earth Structure and Plate Tectonics: Basic Knowledge

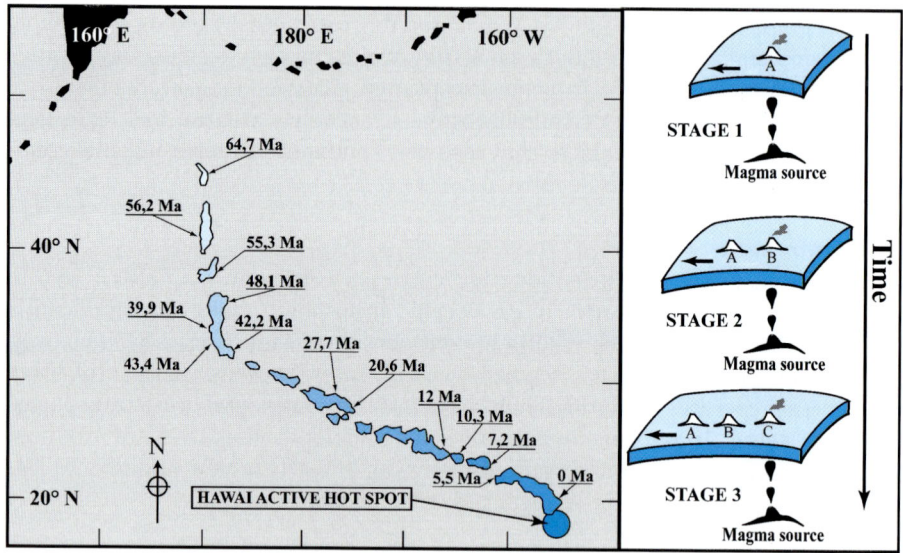

Fig. 1.9. Map of Hawaii-Emperor chain showing that the age of volcanoes regularly decreases towards Hawaii that remains the unique active edifice. *Inset*, on the *right*, schematically illustrates hot-spot island genesis. Magma source is immobile and deep seated, the lithospheric plate is perforated by ascending magma and its motion generates the volcanic chain

plates move over these fixed spots and are perforated by ascending magma, thus this motion generates chains of volcanic island (i.e. Emperor Chain, Hawaii; Fig. 1.9). Melting of the mantle is due to its adiabatic ascent, but as diapir is hotter than in the case of the midocean ridge, melting occurs at greater depth in asthenosphere between 100 and 200 km (Fig. 1.3).

Hot-spot activity can also generate large igneous provinces (LIP), which corresponds to the emplacement of huge amounts of basaltic magma either in continental (i.e. Deccan trapps) or oceanic environments (i.e. Ontong-Java, Kerguelen oceanic plateaus). Their formation requires that enormous volumes of hot mantle rise up until the base of the lithosphere. They are generally attributed to melting in the head of large starting mantle plumes (diapirs).

Recently, three great mesosphere domains (Hawaiian, Tristan and Icelandic) were recognized based on their hot-spot characteristics. They are called "mesoplates", their upper surface corresponds to the asthenosphere-mesosphere frontier and they are limited by the deep subduction zones that extend below 200 km.

1.2.6 Wilson Cycle

In 1968 Wilson developed the concept of cyclicity of oceanic spreading: Ocean begins its life by breaking a continent, it extends through a ridge system and disappears in subduction or collision; this evolution is known as the Wilson cycle. On the other hand, collision between two independent continents assembles them and leads to the formation of a single and greater new continent. Succession of subductions and continental collisions gives rise to supercontinents. The supercontinent cycle involves formation of a supercontinent from smaller continental blocks followed by fragmentation, new ocean opening and again assembly of a new supercontinent. Such a cycle lasts about 500 Ma. For instance, about 245 Ma ago, almost all present-day continents were joined and formed a supercontinent called Pangea. At 245 Ma the Atlantic Ocean began to open and India started to migrate from Madagascar towards Asia: a new supercycle started. Continent breaking continues today, for example in the Red Sea and in the East African rift.

The breakup of an existing supercontinent is due to hot-spot activity and also to the thermal shield effect due to continental crust that acts as a screen that makes internal heat release more difficult. In both cases, heat accumulates under continental crust and provokes its thickening that leads to its breakup.

1.2.7 Energy for Plate Tectonics

All the energy necessary to allow plate tectonics to work is the internal energy of the Earth. Today, this energy has been evaluated at 42 TW (42×10^{12} W). Its main source is the decay of long-lived radiogenic elements (^{235}U, ^{238}U, ^{232}Th and ^{40}K). They are mostly concentrated in the continental crust that produces about 20% of the heat of the Earth. They are less concentrated in the mantle, but due to its greater volume, it contributes about 55% of the heat budget. The core does not contain significant amounts of radiogenic elements. However, crystallization of the outer core produces latent heat ($\sim 10\%$). The last 15% of released heat corresponds to fossil accretion energy. When it formed, due to accretion energy, the Earth was hot, since 4.55 Ga it slowly cooled (200 to 250°C since 4.0 Ga).

References

Condie, K.C., 1997. *Plate Tectonics and Crustal Evolution*. Pergamon, Oxford, 476 pp.
Fowler, C.M.R., 2004. *The Solid Earth: An Introduction to Global Geophysics*. Cambridge University Press, Cambridge, 500 pp.
Turcotte, D.L. and Schubert, G., 2002. 1.2 *Geodynamics*. Wiley, John and Sons, 472 pp.
Wilson, J.T. (ed.), 1976. *Continents Adrift and Continents Aground*. W.H. Freeman and Co., San Francisco, 230 pp.
Wilson, M., 1991. *Igneous Petrogenesis: a Global Tectonic Approach*. Harper Collins Academic, 466 pp.

2 Useful Astrobiological Data

2.1 Physical and Chemical Data

1. International System Units

	Name	Symbol	
Basic units			
Length	Metre	m	
Mass	Kilogram	kg	
Time	Second	s	
Electric current	Ampere	A	
Thermodynamic temperature	Kelvin	K	
Amount of substance	Mole	mol	
Luminous intensity	Candela	cd	
Additional units			
Plane angle	Radian	rad	
Solid angle	Steradian	sr	
Derived units			
Energy	Joule	J	$m^2\,kg\,s^{-2} \equiv N\,m$
Force	Newton	N	$m\,kg\,s^{-2} \equiv J\,m^{-1}$
Pressure	Pascal	Pa	$m^{-1}\,kg\,s^{-2} \equiv N\,m^{-2}$
Electrical charge	Coulomb	C	$s\,A$
Magnetic flux	Weber	Wb	$m^2\,kg\,s^{-2}\,A^{-1}$
Magnetic flux density	Tesla	T	$kg\,s^{-2}\,A^{-1} \equiv V\,s\,m^{-2}$
Magnetic field strength			$A\,m^{-1}$
Inductance	Henry	H	$m^2\,kg\,s^{-2}\,A^{-2}$
Power	Watt	W	$m^2\,kg\,s^{-2} \equiv J\,s^{-1}$
Electric potential difference	Volt	V	$m^2\,kg\,s^{-3}\,A \equiv J\,C^{-1}$
Frequency	Hertz	Hz	s^{-1}

2. Other Units

	Name	Symbol	
Length	Fermi	fm	10^{-15} m
	Angstrom	Å	10^{-10} m
	Micron	µm	10^{-6} m
Time	Minute	min, mn	60 s
	Hour	h	3600 s
	Day	d	86 400 s
	Year	a, y	3.156×10^7 s
Volume	Litre	l	10^{-3} m$^3 \equiv 1$ dm^3
Mass	Ton	t	10^3 kg
Temperature	Degree Celsius	°C	Temperature comparison:
			0°C = 273.15 K = 32°F
			100°C = 373.15 K = 212°F
			Temperature scale:
			K = °C = 1.8°F
Force	Dyne (cgs system)	dyn	10^{-5} N
Energy	Erg (cgs system)	erg	10^{-7} J
	Calorie	cal	4.184 J
	Electron-volt	eV	1.6×10^{-19} J
	Kilowatt-hour		3600×10^3 J =
			8.6042×10^5 cal
Pressure	Bar	bar	10^5 Pa
	Atmosphere	atm	1.01325×10^5 Pa
	mercury mm	torr	1.3332×10^2 Pa
Magnetic flux density	Gauss	gauss	10^{-4} T
Concentration	Molarity	M	mol l^{-1}
Molar mass			g/mol
Angle	Degree	°	$1° = \pi/180$
			$= 1.74 \times 10^{-2}$ rad
	Minute	′	$1' = \pi/10\,800$
			$= 2.91 \times 10^{-4}$ rad
	Second	″	$1'' = \pi/648\,000$
			$= 4.85 \times 10^{-6}$ rad
			1 Radian = 57.296°
			$= 3.44 \times 10^3\,'$
			$= 2.06 \times 10^5\,''$

3. International System Prefixes

10^{-1}	deci	d		10^{1}	deca	da
10^{-2}	centi	c		10^{2}	hecto	h
10^{-3}	milli	m		10^{3}	kilo	k
10^{-6}	micro	µ		10^{6}	mega	M
10^{-9}	nano	n		10^{9}	giga	G
10^{-12}	pico	p		10^{12}	tera	T
10^{-15}	femto	f		10^{15}	peta	P
10^{-18}	atto	a		10^{18}	exa	E
10^{-21}	zepto	z		10^{21}	zetta	Z
10^{-24}	yocto	y		10^{24}	yotta	Y

4. Fundamental Physical Constants

Newtonian constant of gravitation	G	$6.67259(85) \times 10^{-11}\,\mathrm{N\,m^2\,kg^{-2}}$
Acceleration of free fall on Earth	g	$9.80665\,\mathrm{m\,s^{-1}}$
Speed of light in vacuum	c	$2.99792458 \times 10^{8}\,\mathrm{m\,s^{-1}}$
Planck constant	h	$6.6260755(40) \times 10^{-34}\,\mathrm{J\,s}$
Planck mass	$(hc/2\pi G)^{1/2}$	$2.17671(14) \times 10^{-8}\,\mathrm{kg}$
Planck length	$(hG/2\pi c^3)^{1/2}$	$1.61605(10) \times 10^{-35}\,\mathrm{m}$
Planck time	$(hG/2\pi c^5)^{1/2}$	$5.39056(34) \times 10^{-44}\,\mathrm{s}$
Boltzmann constant	k	$1.380658(12) \times 10^{-23}\,\mathrm{J\,K^{-1}}$
Black body constant	a	$7.564 \times 10^{-16}\,\mathrm{J\,m^{-3}\,K^{-4}}$
Stefan Boltzmann constant	σ	$5.67051(19) \times 10^{-8}\,\mathrm{W\,m^{-2}\,K^{-4}}$
Perfect gas constant	R	$8.314510(70)\,\mathrm{J\,K^{-1}\,mol^{-1}}$
Elementary charge	e	$1.602176462(63) \times 10^{-19}\,\mathrm{C}$
Electron rest mass	m_e	$9.1093897(54) \times 10^{-31}\,\mathrm{kg}$
Proton rest mass	m_p	$1.6726231(10) \times 10^{-27}\,\mathrm{kg}$
Neutron rest mass	m_n	$1.6749286(10) \times 10^{-27}\,\mathrm{kg}$
Proton/electron mass ratio		$1836.152701(37)$
Hydrogen atomic mass		$1.6735344 \times 10^{-27}\,\mathrm{kg}$
Electron rest mass energy equivalent (eV)	$m_e c^2$	$0.511 \times 10^{6}\,\mathrm{eV}$

(continued)

Atomic mass unit	u	$1.6605402(10) \times 10^{-27}\,\text{kg} \equiv$ $1/12$ of ^{12}C mass
Avogadro constant	N	$6.0221367(36) \times 10^{23}\,\text{mol}^{-1}$
Vacuum permittivity constant	$1/(\mu_0 c^2)$	$8.854187817 \times 10^{-12}\,\text{m}^{-3}\,\text{kg}^{-1}\,\text{s}^4\,\text{A}^2$
Vacuum permeability constant	μ_0	$4\pi 10^{-7}\,\text{m}\,\text{kg}\,\text{s}^{-2}\,\text{A}^{-2}$
Ice melting temperature ($P = 1\,\text{atm}$)	$0\,^\circ\text{C}$	$273.150\,\text{K}$
Water triple point (H_2O)		$T = 273.160\,\text{K}$, $P = 611.73\,\text{Pa}$ (6.11 mbar)
Bohr radius	a_0	$0.529177249(24) \times 10^{-10}\,\text{m}$
Electron volt-energy relationship		$1\,\text{eV} = 1.60217733(49) \times 10^{-19}\,\text{J}$
Electron volt-wavelength relationship		$1\,\text{eV} = 12\,398.428 \times 10^{-10}\,\text{m}$
Electron volt-frequency relationship		$1\,\text{eV} = 2.41798836(72) \times 10^{14}\,\text{s}^{-1}$
Electron volt-temperature relationship		$1\,\text{eV} = 11\,604.45\,\text{K}$

5. Other Physical, Chemical and Astronomical Symbols and Abbreviations

c	Mass concentration
	Light speed
E_a	Activation energy
k	Reaction rate
R	"Rectus" configuration (chirality)
S	"Sinister" configuration (chirality)
L,D	Old nomenclature still used to define the absolute configuration of oses and amino acids
δ	Chemical shift (RMN)
	Fractional charge
λ	Wavelength
μ	Dipolar moment
	Micron
	Reduced mass
ν	Frequency
[ ]	Concentration $(\mathrm{mol\,l^{-1}})$
$\rightleftarrows$	Chemical equilibrium
$\leftrightarrow$	Mesomery
UV	Ultraviolet
IR	Infrared
NMR	Nuclear Magnetic Resonance
a	Ellipse semi great axis
e	Ellipse eccentricity
i	Inclination
PSR	Pulsar
QSO	Quasar
γ	Photon
v	Neutrino
☉	Sun
☿	Mercury
♀	Venus
⊕	Earth
♂	Mars
♃	Jupiter
♄	Saturn
♅	Uranus
♆	Neptune
♇	Pluto

References: *Allen's: Astrophysical quantities*, Arthur N. Cox Editor, (2000); *Introduction aux éphémérides astronomiques*, EDP Sciences, (1998); *Handbook of Chemistry and Physics*, D.R. Lide, editor-in-Chief 82nd Edition (2001–2002) CRC Press.

6. Periodic Table of Elements

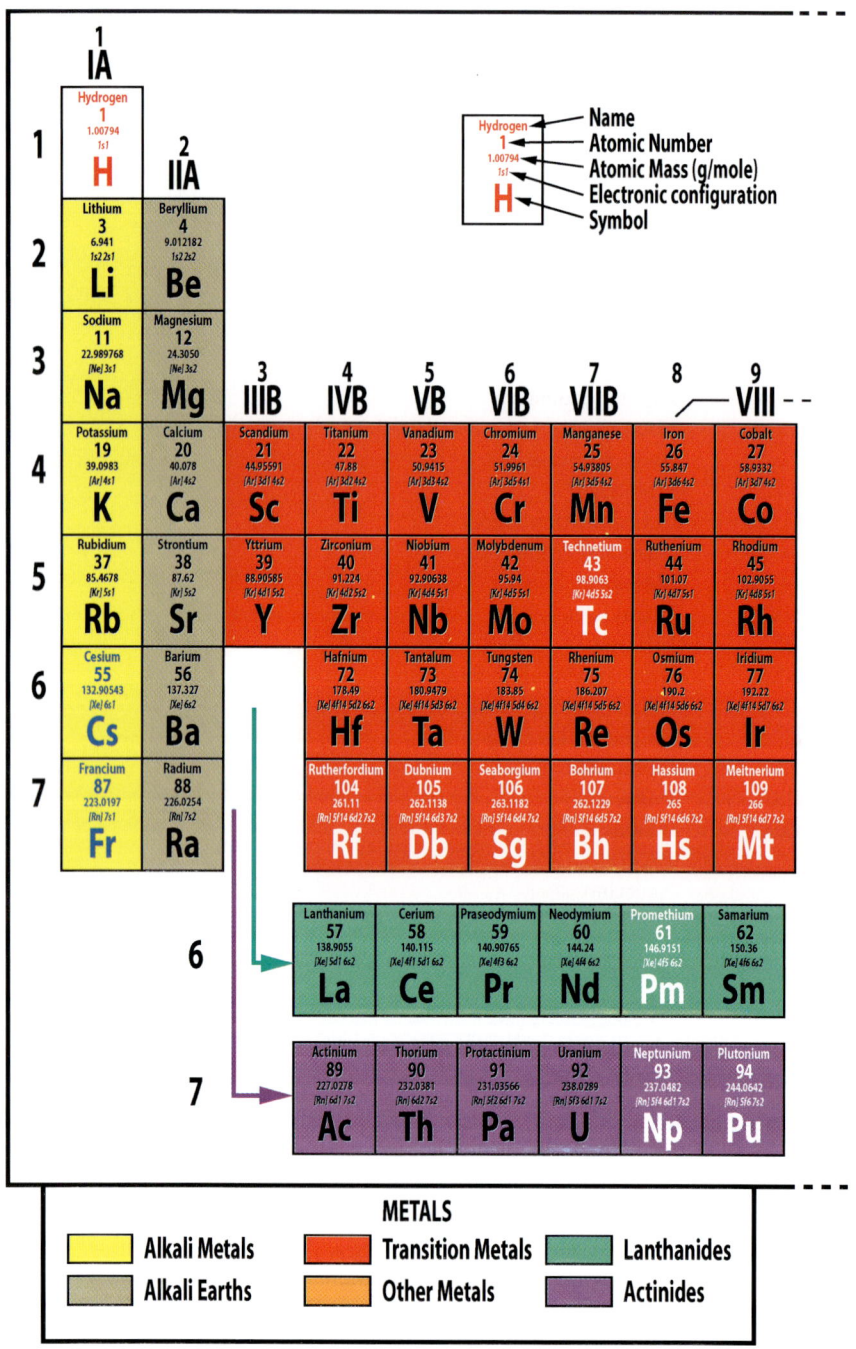

2.1 Physical and Chemical Data

13 IIIA	14 IVA	15 VA	16 VIA	17 VIIA	18 VIIIA
					Helium 2 4.002602 1s2 **He**
Boron 5 10.811 1s22s22p1 **B**	Carbon 6 12.011 1s22s22p2 **C**	Nitrogen 7 14.00674 1s22s22p3 **N**	Oxygen 8 15.9994 1s22s22p4 **O**	Fluorine 9 18.9984032 1s22s22p5 **F**	Neon 10 20.1797 1s22s22p6 **Ne**
Aluminium 13 26.981539 [Ne]3s23p1 **Al**	Silicon 14 28.0855 [Ne]3s23p2 **Si**	Phosphorus 15 30.973762 [Ne]3s23p3 **P**	Sulfur 16 32.066 [Ne]3s23p4 **S**	Chlorine 17 35.4527 [Ne]3s23p5 **Cl**	Argon 18 39.948 [Ne]3s23p6 **Ar**

10	11 IB	12 IIB
Nickel 28 58.6934 [Ar]3d84s2 **Ni**	Copper 29 63.546 [Ar]3d104s1 **Cu**	Zinc 30 65.39 [Ar]3d104s2 **Zn**
Palladium 46 106.42 [Kr]4d10 **Pd**	Silver 47 107.8682 [Kr]4d10 5s1 **Ag**	Cadmium 48 112.411 [Kr]4d10 5s2 **Cd**
Platinum 78 195.08 [Xe]4f14 5d9 6s1 **Pt**	Gold 79 196.96654 [Xe]4f14 5d10 6s1 **Au**	Mercury 80 200.59 [Xe]4f14 5d10 6s2 **Hg**

(Row continuations for groups 13–18, periods 4–6:)

Gallium 31 69.723 [Ar]3d104s24p1 **Ga**	Germanium 32 72.61 [Ar]3d104s24p2 **Ge**	Arsenic 33 74.92159 [Ar]3d104s24p3 **As**	Selenium 34 78.96 [Ar]3d104s24p4 **Se**	Bromine 35 79.904 [Ar]3d104s24p5 **Br**	Krypton 36 83.80 [Ar]3d104s24p6 **Kr**
Indium 49 114.82 [Kr]4d10 5s2 5p1 **In**	Tin 50 118.71 [Kr]4d10 5s2 5p2 **Sn**	Antimony 51 121.757 [Kr]4d10 5s2 5p3 **Sb**	Tellurium 52 127.60 [Kr]4d10 5s2 5p4 **Te**	Iodine 53 126.90447 [Kr]4d10 5s2 5p5 **I**	Xenon 54 131.29 [Kr]4d10 5s2 5p6 **Xe**
Thallium 81 204.3833 [Xe]4f14 5d10 6s2 6p1 **Tl**	Lead 82 207.2 [Xe]4f14 5d10 6s2 6p2 **Pb**	Bismuth 83 208.98037 [Xe]4f14 5d10 6s2 6p3 **Bi**	Polonium 84 208.9824 [Xe]4f14 5d10 6s2 6p4 **Po**	Astatine 85 209.9871 [Xe]4f14 5d10 6s2 6p5 **At**	Radon 86 222.0176 [Xe]4f14 5d10 6s2 6p6 **Rn**

| Europium 63 151.965 [Xe]4f7 6s2 **Eu** | Gadolinium 64 157.25 [Xe]4f7 5d1 6s2 **Gd** | Terbium 65 158.92534 [Xe]4f9 6s2 **Tb** | Dysprosium 66 162.50 [Xe]4f10 6s2 **Dy** | Holmium 67 164.93032 [Xe]4f11 6s2 **Ho** | Erbium 68 167.26 [Xe]4f12 6s2 **Er** | Thulium 69 168.93421 [Xe]4f13 6s2 **Tm** | Ytterbium 70 173.04 [Xe]4f14 6s2 **Yb** | Lutetium 71 174.967 [Xe]4f14 5d1 6s2 **Lu** |

| Americium 95 243.0614 [Rn]5f7 7s2 **Am** | Curium 96 247.0703 [Rn]5f7 6d1 7s2 **Cm** | Berkelium 97 247.0703 [Rn]5f9 7s2 **Bk** | Californium 98 251.0796 [Rn]5f10 7s2 **Cf** | Einsteinium 99 252.0829 [Rn]5f11 7s2 **Es** | Fermium 100 257.0951 [Rn]5f12 7s2 **Fm** | Mendelevium 101 258.0986 [Rn]5f13 7s2 **Md** | Nobelium 102 259.1009 [Rn]5f14 7s2 **No** | Lawrencium 103 260.1053 [Rn]5f14 6d1 7s2 **Lr** |

NON-METALS
- Hydrogen
- Other Non-Metals
- Halogens
- Noble Gases

STANDARD CONDITIONS (P = 100 KPa; T = 298 K)

Gaz — Liquid — Solid — Synthetic

2.2 Astrophysical Data

1. Units and General Data

Distance units

Astronomical unit (AU)	$1.4959787066 \times 10^{11}$ m
1 parsec (pc)	206 264.8 au
	3.0856776×10^{16} m
	3.2615638 light-years
1 light-year	9.460730×10^{15} m
	6.324×10^{4} AU

Time units

1 sideral day	23 h 56 mn 04.098 s (average time)
1 tropic year	365.2422 average solar days
1 sideral year	365.2564 average solar days

Known universe

Number of nucleons	$\sim 10^{80}$
Universe radius	$\sim 10^{26}$ m
Number of galaxies	$\sim 10^{11}$
Nebula recession speed	$\sim 3 \times 10^{-18}\,\mathrm{s}^{-1}\,(\sim 100\,(\mathrm{km/s})/\mathrm{Mpc})$

Our galaxy

Number of stars	$\sim 1.6 \times 10^{11}$
Diameter	$\sim 10^{21}$ m
Mass	$\sim 8 \times 10^{41}$ kg

Sun

Radius	6.95508×10^{8} m
Mass	1.9891×10^{30} kg
Sun luminosity	$3.845(8) \times 10^{26}$ W
Core temperature	1.557×10^{7} K
Core pressure	2.334×10^{16} Pa

(continued)

Photosphere temperature	5780 K
Corona temperature	2 to 3×10^6 K
Average mass density	1.41×10^3 kg m^{-3}
Earth/Moon	
Earth's mass	5.9742×10^{24} kg
Moon's mass	7.34×10^{22} kg
Earth's average radius	6371.23×10^3 m
Earth's equatorial radius	6378.136×10^3 m
Earth–Moon distance (semi-major axis)	3.844×10^8 m
Earth–Sun distance (semi-major axis)	1.496×10^{11} m

2. Compared Planetology

Object		Semi major axis of the orbit[1]		Revolution period (d: day, yr: year, terrestrial)	Eccentricity (e)
		AU	Million km		
Mercury	☿	0.3871	57.9092	87.969 d	0.2056
Venus	♀	0.7233	108.2089	224.701 d	0.0068
Earth	⊕	1.0000	149.65979	365.2564 d	0.0167
Moon		2.570×10^{-3}	0.3844	27.3217 d	0.0549
Mars	♂	1.5234	227.9364	686.98 d	0.0934
Jupiter	♃	5.2034	778.4120	11.862 yr	0.0484
Europe		4.4851×10^{-3}	0.671	3.551181 d	0.009
Saturn	♄	9.5371	1426.7254	29.457 yr	0.0542
Titan		8.167×10^{-3}	1.2218	15.945 d	0.0291
Uranus	♅	19.1913	2870.9722	84.019 yr	0.0472
Neptune	♆	30.0690	4498.2529	164.767 yr	0.0086
Pluto	♇	39.4817	5906.3762	247.688 yr	0.2488

[1] around the Sun or, in the case of satellites, around their planet

2 Useful Astrobiological Data

Object		Equatorial radius km	Mass $\oplus = 1$	Average density g cm^{-3}	Surface gravity[1] m s^{-2}	Surface escape velocity[1] km s^{-1}
Sun	☉	695 508	332 946.0	1.41	274.0	617.5
Mercury	☿	2439.7	0.0553	5.43	3.72	4.25
Venus	♀	6051.8	0.8150	5.24	8.87	10.36
Earth	⊕	6378.136	1.000	5.515	9.81	11.18
Moon		1737.4	0.012300	3.34	1.62	2.38
Mars	♂	3397	0.1074	3.94	3.71	5.02
Jupiter	♃	71 492	317.82	1.33	23.12	59.54
Europe		1565	0.008026	3.04	1.8	2.0
Saturn	♄	60 268	95.161	0.70	8.96	35.49
Titan		2575	0.02226	1.90	0.14	2.7
Uranus	♅	25 559	14.371	1.30	8.69	21.29
Neptune	♆	24 764	17.147	1.76	11.0	23.71
Pluto	♇	1737.4	0.0022	1.1	0.81	1.27

[1] or at $P = 1$ bar for gaseous planets

Object		Mean obliquity (degree)	Visual geometric Albedo	Sideral rotation peroid (d:day, h; hour)
Sun	☉		–	25.7 d
Mercury	☿	0.0	0.11	58.6462 d
Venus	♀	177.3	0.65	243.0185 d [1]
Earth	⊕	23.45	0.37	23.9345 h
Moon		6.683	0.12	27.3217 d
Mars	♂	25.19	0.15	24.6230 h
Jupiter	♃	3.12	0.52	9.9249 h
Europe		–	0.67	2.5512 d
Saturn	♄	26.73	0.47	10.6562 h
Titan		–	0.22	15.9454 h
Uranus	♅	97.86	0.51	17.2400 h [1]
Neptune	♆	29.58	0.51	16.1100 h
Pluto	♇	119.61	0.3	153.29 h [1]

[1] retrograde rotation

3. Electromagnetic Spectrum

Wavelength	Frequency	Wave number	Equivalent energy	Equivalent temperature		Spectral Region	
1000 km	300 Hz				ULF		
100 km	3 kHz						
1 km	300 kHz						Radio
10 m	30 MHz						
1 m	300 MHz				VHF		
					UHF		
10 cm	3 GHz					centimetric	
1 cm	30 GHz	1 cm^{-1}	1.2 meV	3 K			Microwave
						millimetric	
1 mm	300 GHz	10 cm^{-1}					
100 μm	3 THz	100 cm^{-1}	0.012 eV	30 K		submillimetric	
40 μm					FIR		
10 μm	30 THz	1000 cm^{-1}	0.12 eV	300 K			Infrared
5 μm					MIR		

(continued)

Wavelength	Frequency	Wave number	Equivalent energy	Equivalent temperature	Spectral Region
1 μm	3×10^{14} Hz	10 000 cm^{-1}	1.24 eV	3000 K	NIR
0.7 μm					
					Visible
0.4 μm					
0.1 μm	3×10^{15} Hz		12.4 eV	3×10^4 K	Ultraviolet
10 nm	3×10^{16} Hz		124 eV	3×10^5 K	
1 nm	3×10^{17} Hz		1.24 keV	3×10^6 K	
1 Å	3×10^{18} Hz		12.4 keV	3×10^7 K	X-rays
10^{-11} m	3×10^{19} Hz		124 keV		High Energies
10^{-12} m	3×10^{20} Hz		1.24 MeV		Gamma-rays

Remarks:
- Wave numbers, energies and temperatures can be calculated for all existing wavelengths (here, they have been intentionally limited to the domains where they are generally used).
- Spectrum domain limits are no more than indicative and correspond to astronomical use.
- ULF: Ultra Low Frequency, VHF: Very High Frequency, UHF: Ultra High Frequency, FIR: Far Infrared, MIR: Mid Infrared, NIR: Near Infrared, UV: Ultraviolet
- Formula: $E = h\nu = hc/\lambda = eV = kT$

2.3 Geological Data

1. General Information

Whole Earth	
Earth's Age	4.55 Ga
Age of the oldest known terrestrial material (zircon from Jack Hills; Australia)	4.404 Ga
Age of the oldest known terrestrial rock (Acasta gneiss; Canada)	4.03 Ga
Equatorial radius	6378.136 km
Polar radius	6356.753 km
Ellipticity	0.00335281
Total surface	$510 \times 10^6 \text{ km}^2$
Ocean surface	$357 \times 10^6 \text{ km}^2$
Continent surface	$153 \times 10^6 \text{ km}^2$
Mass	$5.9742 \times 10^{24} \text{ kg}$
Mean density	5.515
Deeper oceanic trench (Marianas)	11 034 m
Higher mountain (Everest)	8863 m
Present day oceanic ridge length	~80 000 km
Solid inner core	
Depth	5155 km
Thickness	~1223 km
Average density	12
Volume	$8.45 \times 10^{10} \text{ km}^3$
Mass	$1.18 \times 10^{23} \text{ kg}$
Liquid outer core	
Depth	2891 km
Thickness	~2264 km
Average density	10
Volume	$1.65 \times 10^{11} \text{ km}^3$
Mass	$1.65 \times 10^{24} \text{ kg}$
Lower mantle (Mesosphere)	
Depth	660 km
Thickness	~2231 km
Average density	5
Volume	$5.87 \times 10^{11} \text{ km}^3$
Mass	$2.93 \times 10^{24} \text{ kg}$

(continued)

Upper mantle	
Depth	30 km
Thickness	~630 km
Average density	3.2
Volume	3.08×10^{11} km^3
Mass	0.986×10^{24} kg
Oceanic crust	
Average thickness	5 km
Average density	3.1
Age of the oldest known oceanic crust	0.165 Ga
Volume	2.0×10^9 km^3
Mass	6.2×10^{21} kg
Continental crust	
Average thickness	30 km
Average density	2.75
Age of the oldest known continental crust	4.01 Ga
Volume	7×10^9 km^3
Mass	1.9×10^{22} kg
Average surface magnetic flux density	5×10^{-5} T
Heat flux	42×10^{12} W
Range of lithospheric plate motion rate	1 to 10 cm yr^{-1}

2. Decay Constants (λ)

Parent nuclide	Daughter nuclide	Decay constant
^{228}Th	^{224}Ra	3.63×10^{-1} yr^{-1}
^{210}Pb	^{210}Bi	3.11×10^{-2} yr^{-1}
^{32}Si	^{32}P	2.1×10^{-3} yr^{-1}
^{226}Ra	^{222}Rn	4.33×10^{-4} yr^{-1}
^{14}C	^{14}N	1.245×10^{-4} yr^{-1}

(continued)

Parent nuclide	Daughter nuclide	Decay constant
^{231}Pa	^{227}Ac	$2.11 \times 10^{-5}\,\mathrm{yr}^{-1}$
^{230}Th	^{226}Ra	$9.21 \times 10^{-6}\,\mathrm{yr}^{-1}$
^{59}Ni	^{59}Co	$9.12 \times 10^{-6}\,\mathrm{yr}^{-1}$
^{41}Ca	^{41}K	$6.93 \times 10^{-6}\,\mathrm{yr}^{-1}$
^{81}Kr	^{81}Br	$3.03 \times 10^{-6}\,\mathrm{yr}^{-1}$
^{234}U	^{230}Th	$2.83 \times 10^{-6}\,\mathrm{yr}^{-1}$
^{36}Cl	^{36}Ar	$2.30 \times 10^{-6}\,\mathrm{yr}^{-1}$
^{26}Al	^{26}Mg	$9.80 \times 10^{-7}\,\mathrm{yr}^{-1}$
^{107}Pd	^{107}Ag	$6.5 \times 10^{-7}\,\mathrm{yr}^{-1}$
^{60}Fe	^{60}Ni	$4.62 \times 10^{-7}\,\mathrm{yr}^{-1}$
^{10}Be	^{10}B	$4.59 \times 10^{-7}\,\mathrm{yr}^{-1}$
^{182}Hf	^{182}W	$7.7 \times 10^{-8}\,\mathrm{yr}^{-1}$
^{129}I	^{129}Xe	$4.3 \times 10^{-8}\,\mathrm{yr}^{-1}$
^{92}Nb	^{92}Zr	$1.93 \times 10^{-8}\,\mathrm{yr}^{-1}$
^{53}Mn	^{53}Cr	$1.87 \times 10^{-8}\,\mathrm{yr}^{-1}$
^{244}Pu	$^{131-136}$Xe	$8.66 \times 10^{-9}\,\mathrm{yr}^{-1}$
^{235}U	^{207}Pb	$9.849 \times 10^{-10}\,\mathrm{yr}^{-1}$
^{146}Sm	^{142}Nd	$6.73 \times 10^{-10}\,\mathrm{yr}^{-1}$
^{40}K	^{40}Ar	$5.50 \times 10^{-10}\,\mathrm{yr}^{-1}$
^{40}K	^{40}Ca	$4.96 \times 10^{-10}\,\mathrm{yr}^{-1}$
^{238}U	^{206}Pb	$1.551 \times 10^{-10}\,\mathrm{yr}^{-1}$
^{130}Te	^{130}Xe	$8.66 \times 10^{-22}\,\mathrm{yr}^{-1}$
^{87}Rb	^{87}Sr	$1.42 \times 10^{-11}\,\mathrm{yr}^{-1}$
^{187}Re	^{187}Os	$1.64 \times 10^{-11}\,\mathrm{yr}^{-1}$
^{176}Lu	^{176}Hf	$1.93 \times 10^{-11}\,\mathrm{yr}^{-1}$
^{232}Th	^{208}Pb	$4.95 \times 10^{-11}\,\mathrm{yr}^{-1}$
^{40}K	^{40}Ar	$5.81 \times 10^{-11}\,\mathrm{yr}^{-1}$
^{147}Sm	^{143}Nd	$6.54 \times 10^{-12}\,\mathrm{yr}^{-1}$
^{138}La	^{138}Ce	$2.24 \times 10^{-12}\,\mathrm{yr}^{-1}$
^{130}Te	^{130}Xe	$8.66 \times 10^{-22}\,\mathrm{yr}^{-1}$

3. Temperature Range of Magma Emplacement

Granitic magma	700–800 °C
Basaltic magma	1250–1350 °C
Komatiitic magma	1650 °C

4. Average Compositions of Earth Shells

	N = Atomic number	M = Atomic mass (g/mol)	Unit	C1 Chondrite	Bulk Solid Earth
H	1	1.00794	%	2.02	0.0257
He	2	4.002602		ppm	56 (nL/g)
Li	3	6.941	ppm	1.49	1.08
Be	4	9.012182	ppm	0.0249	0.046
B	5	10.811	ppm	0.87	0.203
C	6	12.011	%	3.5	0.73
N	7	14.00674	ppm	3180	55
O	8	15.9994	%	46.4	30.5
F	9	18.998403	ppm	58.2	17.1
Ne	10	20.1797	ppm	203 (pL/g)	
Na	11	22.989768	%	0.51	0.18
Mg	12	24.3050	%	9.65	15.39
Al	13	26.981539	%	0.86	1.59
Si	14	28.0855	%	10.65	16.5
P	15	30.973762	ppm	1080	1101
S	16	32.066	%	5.4	0.634
Cl	17	35.4527	ppm	680	74.5
Ar	18	39.948	ppm	751 (pL/g)	
K	19	39.0983	%	0.0550	0.0164
Ca	20	40.078	%	0.925	1.71
Sc	21	44.95591	ppm	5.92	10.93
Ti	22	47.88	%	0.044	0.0813
V	23	50.9415	ppm	56	94.4
Cr	24	51.9961	ppm	2650	3720
Mn	25	54.93805	ppm	1920	1974
Fe	26	55.847	%	18.1	30.3
Co	27	58.9332	ppm	500	838
Ni	28	58.6934	ppm	10500	17 690
Cu	29	63.546	ppm	120	60.9
Zn	30	65.39	ppm	310	37.1
Ga	31	69.723	ppm	9.2	2.74

Table showing the compositions of the different Earth shells. Data are from Anders and Grevesse (1989) Taylor and Mc Lennan, Mc Donough et al. (1992), Mc Donough and Sun (1995), Rudnick and Fountain (1995), McDonough (1998), Rudnick and Gao (2003). Several other data are coming from the Geochemical Earth Reference Model (GERM) Internet site 'http://www.earthref.org'.

(continued)

Core	Primitive Mantle	Upper Mantle	Oceanic Crust	Bulk Continental Crust	Ocean	Atmosphere
0.06	0.01	0.00025		0.14	10.8	0.000053
					0.0000069	5.2
0	1.6	1.57	10	16	0.18	
0	0.068	0.0442	0.5	1.9	0.000225	
0	0.3	0.07	4	11	4.39	
0.2	0.0120	0.0018		0.02	0.0028	0.0337
170	2	0.05		56	8.5	780900
0	44	44		45	85.7	20.95
0	25	9.8		553	1.3	
				0.00007	0.04	18
0	0.267	0.214	2.08	2.36	1.05	
0	22.8	22.8	4.64	3.20	0.135	
0	2.35	2.23	8.47	8.41	0.00000016	
6.4	21	21	23.1	26.77	0.0003	
3200	90	54	873	2000	0.07	
1.9	0.025	0.024		0.0404	0.0885	0.0001
200	17	2.55		244	19000	0.001
				1.2	0.6	9300
0	0.024	0.0024	0.125	0.99	0.038	
0	2.53	2.4	8.08	5.29	0.04	
0	16.2	15.4	38	21.9	0.0000006	
0	0.12	0.0928	0.9	0.54	0.0000001	
120	82	82	250	138	0.0019	
9000	2625	2625	270	135	0.0003	
4000	1045	1045	1000	852	0.0002	
85	6.26	6.26	8.16	5.6	0.00000056	
2500	105	105	47	26.6	0.00039	
52 000	1960	1960	135	59	0.0054	
125	30	29.1	86	27	0.0002	
0	55	55	85	72	0.0037	
0	4	3.8	17	16	0.00003	

(continued)

	N = Atomic number	M = Atomic mass (g/mol)	Unit	C1 Chondrite	Bulk Solid Earth
Ge	32	72.61	ppm	31	7.24
As	33	74.92159	ppm	1.85	1.67
Se	34	78.96	ppb	21 000	2560
Br	35	79.904	ppb	3570	300
Kr	36	83.80	ppb	8.7 *(pL/g)*	
Rb	37	85.4678	ppm	2.3	0.405
Sr	38	87.62	ppm	7.25	13.4
Y	39	88.90585	ppm	1.57	2.95
Zr	40	91.224	ppm	3.82	7.1
Nb	41	92.90638	ppm	0.240	0.45
Mo	42	95.94	ppm	0.9	1.66
Tc	43	98.9063	ppm		
Ru	44	101.07	ppb	710	1310
Rh	45	102.9055	ppb	130	242
Pd	46	106.42	ppb	550	1010
Ag	47	107.8682	ppb	200	54.4
Cd	48	112.411	ppb	710	76
In	49	114.82	ppb	80	7.5
Sn	50	118.71	ppb	1650	251
Sb	51	121.757	ppb	140	46.2
Te	52	127.60	ppb	2330	285.7
I	53	126.90447	ppb	450	50
Xe	54	131.29	ppb	8.6 *(pL/g)*	
Cs	55	132.90543	ppb	190	35
Ba	56	137.327	ppm	2.41	4.46
La	57	138.9055	ppm	0.237	0.437
Ce	58	140.115	ppm	0.613	1.148
Pr	59	140.90765	ppm	0.0928	0.171
Nd	60	144.24	ppm	0.457	0.844
Pm	61	146.9151	ppm		
Sm	62	150.36	ppm	0.148	0.274
Eu	63	151.965	ppm	0.0563	0.104
Gd	64	157.25	ppm	0.199	0.367
Tb	65	158.92534	ppm	0.0361	0.068
Dy	66	162.50	ppm	0.246	0.455
Ho	67	164.93032	ppm	0.546	0.102

2.3 Geological Data

(continued)

Core	Primitive Mantle	Upper Mantle	Oceanic Crust	Bulk Continental Crust	Ocean	Atmosphere
20	1.1	1.1	1.5	1.3	0.000051	
5	0.05	0.01	1	2.5	0.0026	
8000	75	71	160	130	0.00012	
700	50	5		880	65 000	
					0.29	1140
0	0.6	0.041	2.2	49	0.124	
0	19.9	12.93	130	320	8.1	
0	4.3	3.65	32	19	0.000013	
0	10.5	6.19	80	132	0.000026	
0	0.658	0.11	2.2	8	0.000015	
5	0.05	0.03	1	0.8	0.01	
4000	4.9	4.9	1	0.57	0.0007	
740	0.91	0.91	0.2	0.06		
3100	3.9	3.9	0.2	1.5		
150	8	7.8	26	56	0.28	
150	40	39.2	130	80	0.00011	
0	11	8.5	72	52	0.000115	
500	130	97.5	1400	1700	0.81	
130	5.5	1.1	17	200	0.33	
850	12	12	3	5		
130	10	1		710	64	
					0.066	86
65	21	504	30	2600	0.5	
0	6.6	0.45	25	456	0.021	
0	0.648	0.08	3.7	20	0.0000029	
0	1.675	0.538	11.5	43	0.0000028	
0	0.254	0.114	1.8	4.9	0.00000064	
0	1.25	0.738	10	20	0.0000028	
0	0.406	0.305	3.3	3.9	0.00000045	
0	0.154	0.119	1.3	1.1	0.0000013	
0	0.544	0.430	4.6	3.7	0.0000007	
0	0.099	0.080	0.87	0.6	0.00000014	
0	0.674	0.559	5.7	3.6	0.00000091	
0	0.149	0.127	1.3	0.77	0.00000022	

(continued)

	N = Atomic number	M = Atomic mass (g/mol)	Unit	C1 Chondrite	Bulk Solid Earth
Er	68	167.26	ppm	0.16	0.296
Tm	69	168.93421	ppm	0.0247	0.046
Yb	70	173.04	ppm	0.161	0.297
Lu	71	174.967	ppm	0.0246	0.046
Hf	72	178.49	ppm	0.103	0.191
Ta	73	180.9479	ppm	0.0136	0.025
W	74	183.85	ppm	0.093	0.170
Re	75	186.207	ppb	40	75.3
Os	76	190.2	ppb	490	900
Ir	77	192.22	ppb	455	835
Pt	78	195.08	ppb	1010	1866
Au	79	196.96654	ppb	140	164
Hg	80	200.59	ppb	300	23.1
Tl	81	204.3833	ppb	140	12.2
Pb	82	207.2	ppb	2470	232
Bi	83	208.98037	ppb	110	9.85
Po	84	208.9824			
At	85	209.9871			
Rn	86	222.0176			
Fr	87	223.0197			
Ra	88	226.0254			
Ac	89	227.0278			
Th	90	232.0381	ppm	0.029	0.054
Pa	91	231.03588			
U	92	238.0289	ppm	0.0074	0.0139

References

Anders, E. and Grevesse, N., 1989. Abundances of the elements: Meteoritic and solar. Geochimica Cosmochimica Acta, 53: 197–214.

McDonough, W.F., 1998. Earth's core. In: C.P. Marshall and R.W. Fairbridge (Editors), Encyclopedia of geochemistry. Kluwer Academic Publishers, Dordrecht, pp. 151–156.

McDonough, W.F. and Sun, S.-S., 1995. Composition of the Earth. Chemical Geology, 120: 223–253.

(continued)

Core	Primitive Mantle	Upper Mantle	Oceanic Crust	Bulk Continental Crust	Ocean	Atmosphere
0	0.438	0.381	3.7	2.1	0.00000087	
0	0.068	0.060	0.54	0.28	0.00000017	
0	0.441	0.392	5.1	1.9	0.00000082	
0	0.0675	0.061	0.56	0.30	0.00000015	
0	0.283	0.167	2.5	3.7	0.0000016	
0	0.037	0.006	0.3	0.7	0.0000025	
0.47	0.029	0.002	0.5	1	0.000001	
230	0.28	0.27	0.9	0.188	0.0084	
2750	3.43	3.43	0.004	0.041	0.0000017	
2550	3.18	3.18	0.02	0.037	0.00000115	
5700	7.07	7.07	2.3	1.5	0.000117	
500	0.98	0.96	0.23	1.3	0.011	
50	10	9.8	20	30	0.15	
30	3.5	3.5	12	500	0.0123	
4000	150	20	800	11 000	0.03	
25	2.5	0.5	7	18	0.017	
0	0.0795	0.06	0.22	5.6	0.0000004	
0	0.0203	0.002	0.1	1.3	0.0033	

McDonough, W.F., Sun, S.-S., Ringwood, A.E., Jagoutz, E. and Hofmann, A.W., 1992. Potassium, rubidium, and cesium in the Earth and Moon and the evolution of the mantle of the Earth. Geochimica et Cosmochimica Acta, 56(3): 1001–1012.

Rudnick, R.L. and Fountain, D.M., 1995. Nature and composition of the continental crust: a lower crustal perspective. Reviews in Geophysics, 33: 267–309.

Rudnick, R.L. and Gao, S., 2003. The Composition of the Continental Crust. In: R.L. Rudnick (Editor), The Crust. Treatise on Geochemistry. Elsevier-Pergamon, Oxford, pp. 1–64.

Taylor, S.R. and McLennan, S.M., 1985. The continental crust: its composition and evolution. Blackwell scientific Publications, Oxford, 312 pp.

5. Geological Time Scale and Live Evolution

Aeon	Era		System	Epoch	Ma	Life evolution (Ma)
PHANEROZOIC	Cenozoic	Quaternary	Quaternary	Holocene	Present–0,01	Modern human beings (0,12)
				Pleistocene	1,75	Human evolution (2)
		Tertiary	Neogene	Pliocene	5,3	Australopithecus (4)
				Miocene	23,5	
			Paleogene	Oligocene	33,7	
				Eocene	53	Mammal's expansion (50)
				Paleocene	65	1st Primates (58)
	Mesozoic	Secondary	Cretaceous			Dinosaur's extinction (65)
					135	
			Jurassic			1st Mammals (180)
					203	
			Triassic			1st Dinosaurs (230)
					250	
	Paleozoic	Primary	Permian			
					295	
			Carboniferous			
					355	
			Devonian			1st Reptiles (360)
					410	
			Silurian			
					435	
			Ordovician			1st Amphibians (440)
					500	1st Aerial plants (450)
			Cambrian			1st Vertebrates (fishes) (500)
					540	
PRECAMBRIAN	PROTEROZOIC					1st Shelled Invertebrates (560)
						1st Pluricellular organisms (1000)
						1st Eukariotes (2000-2100)
					2500	
	ARCHAEAN					1st Cyanobacterias (2600-2700)
						1st Bacterias (3500)
					4000	
	HADEAN				4550	

Simplified geological time scale for Earth; main stages in life development are also shown. The International Stratigraphic Chart is available on UNESCO-IUGS Internet site: ftp://ftp.iugs.org/pub/iugs/iugs_intstratchart.pdf

2.4 Biochemical Data

1. The Main Prebiotic Precursors of Biomolecules (in brackets, the IUPAC nomenclature recommendation)

water (hydrogen oxyde)	H—O—H
acetaldehyde (ethanal)	H_3C-CHO
acetronitrile (methyl cyanide)	$H_3C-C\equiv N$
acetylene (ethyne)	$H-C\equiv C-H$
ammonia	NH_3
cyanhydric acid (hydrogen cyanide)	$H-C\equiv N$
cyanoacetylene (2-Propynenitrile)	$H-C\equiv C-C\equiv N$
ethylene (ethene)	$H_2C=CH_2$
formaldehyde (methanal)	$H-CHO$
formic acid (methanoic acid)	$H-COOH$
hydrogen sulfide	H—S—H
imidazole	(imidazole ring)
methane	CH_4
methanethiol	H_3C-S-H

2. Nucleic Bases

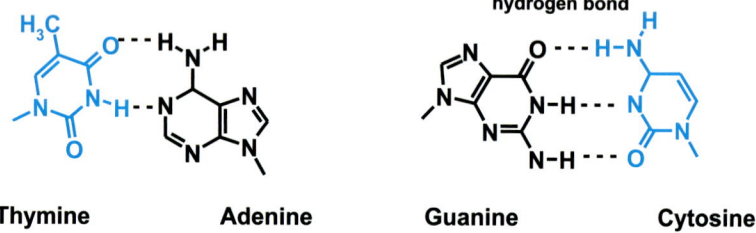

The five major nucleic bases (with atom numerotation)

Watson-Crick base pairs

3. Nucleosides and Nucleotides

Nucleosides

Adenosine, 2'-deoxyadenosine
Guanosine, 2'-deoxyguanosine
Thymidine, 2'-deoxythymidine
Uridine
2'-deoxythymidine

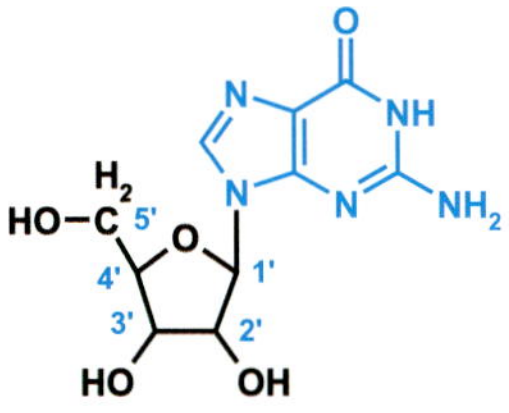

Guanosine (RNA)

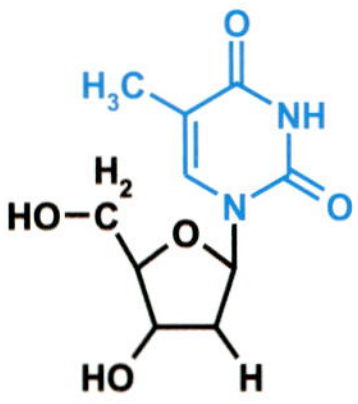

2'-deoxythymidine (DNA)

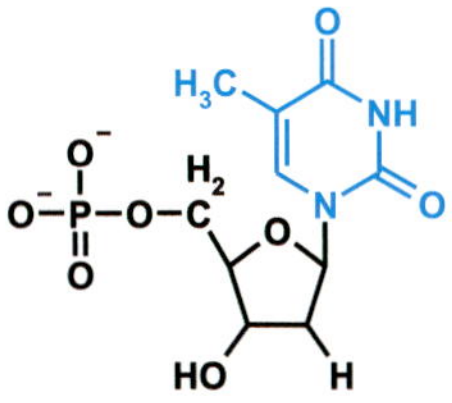

2'-deoxythymidine-5'monophosphate (DNA)

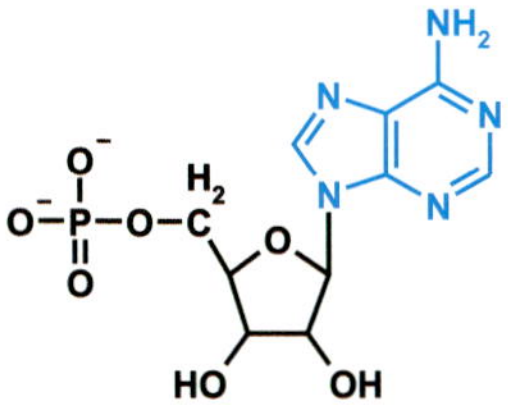

Adenosine-5'-monophosphate (AMP) (RNA)

Some nucleotides

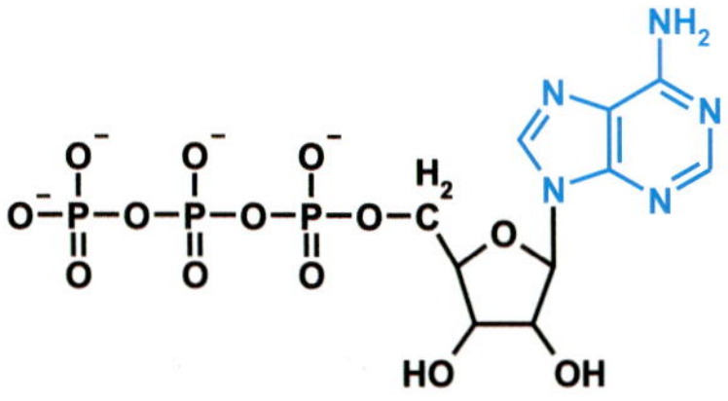

Adenosine-5'-triphosphate (ATP)

4. Amino Acids Genetically Coded

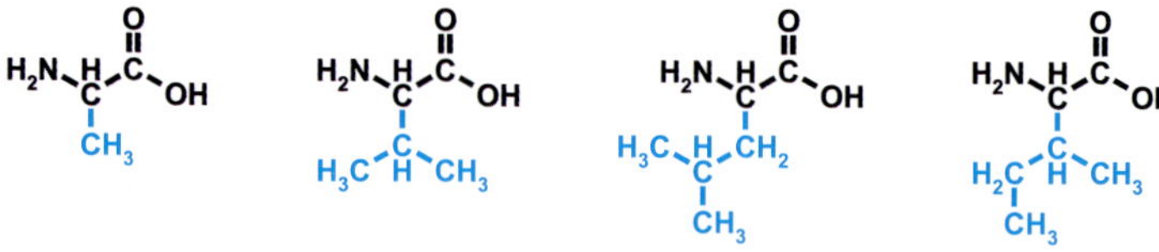

Alanine (Ala, A) **Valine (Val, V)** **Leucine (Leu, L)** **Isoleucine (Ile, I)**

Hydrophobic aliphatic amino acids

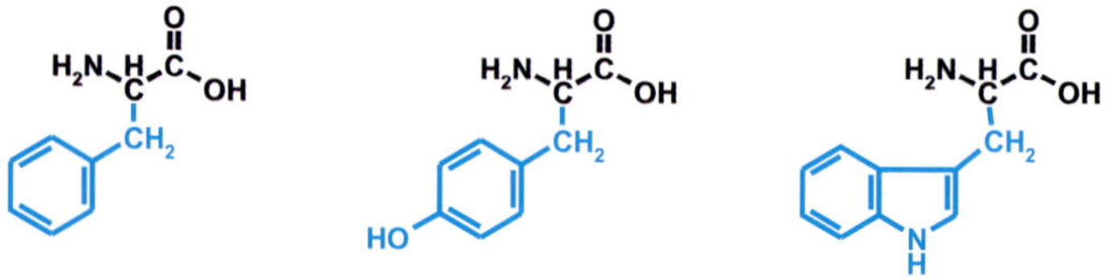

Phenylalanine (Phe, F) **Tyrosine (Tyr, Y)** **Tryptophane (Trp, W)**

Aromatic amino acids

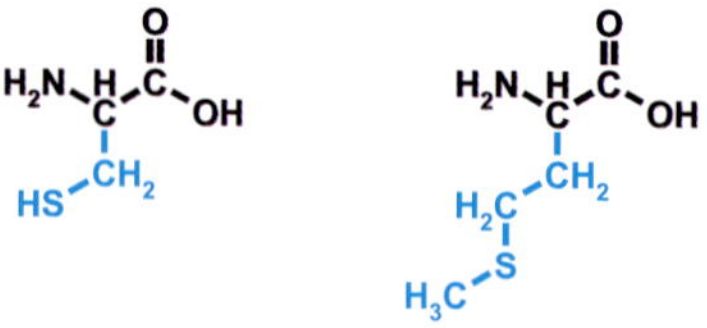

Sulfur containing amino acids

Cysteine (Cys, C) **Methionine (Met, M)**

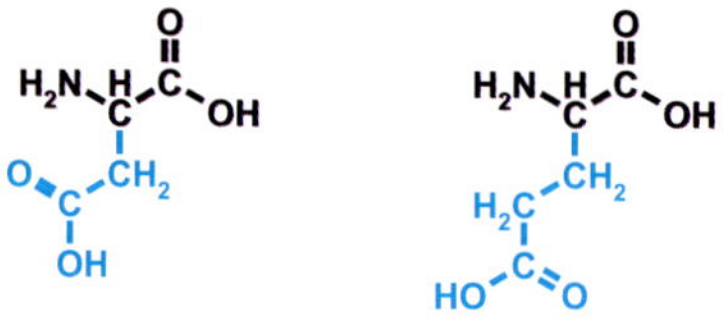

Acidic amino acids

Aspartic acid (Asp, D) **Glutamic acid (Glu, E)**

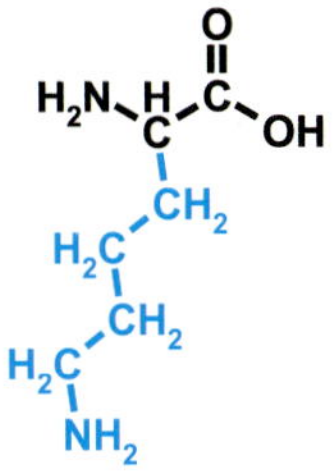

2.4 Biochemical Data

Lysine (Lys, K)

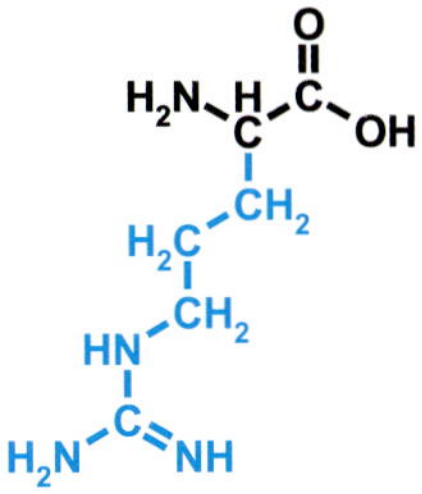

Arginine (Arg, R)

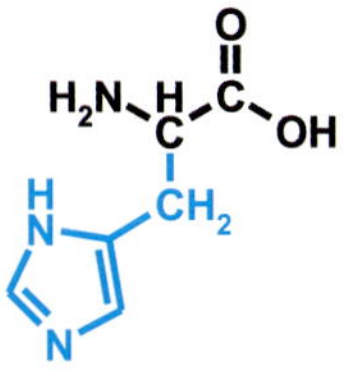

Histidine (His, H)

Basic amino acids

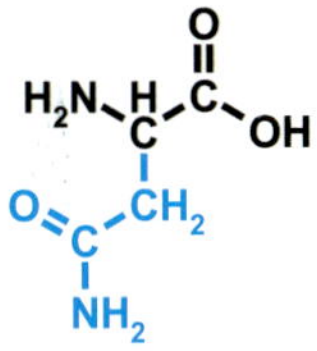

Asparagine (Asn, N)

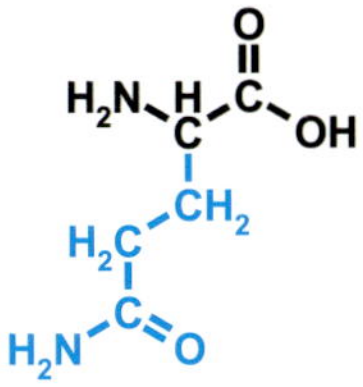

Glutamine (Gln, Q)

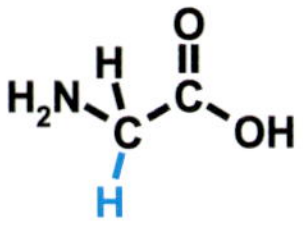

Glycine (Gly, G)

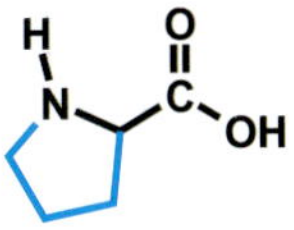

Proline (Pro, P)

N-alkyl amino acid

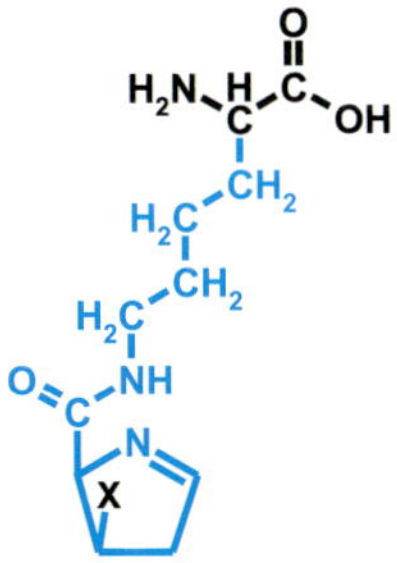

Pyrrolysine

X = NH_2, CH_3 or OH

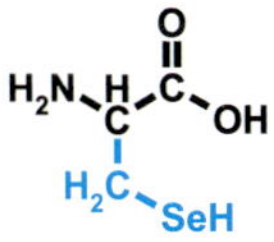

Selenocysteine

Rare amino acids

5. Peptide Bond Formation

leucine + aspartic acid ⟶ peptide + water

$$H_2N-\underset{\underset{CH_3}{\overset{}{CH_2}}}{\overset{H}{\underset{|}{C}}}-\overset{O}{\overset{\|}{C}}-OH \quad H-\underset{H}{\overset{H}{N}}-\underset{\underset{COOH}{\overset{}{CH_2}}}{\overset{}{C}}-COOH$$

Leucine   **Aspartic acid**

$$H_2N-\underset{\underset{CH_3}{\overset{}{CH_2}}}{\overset{H}{\underset{|}{C}}}-\overset{O}{\overset{\|}{C}}-\underset{H}{\overset{H}{N}}-\underset{\underset{COOH}{\overset{}{CH_2}}}{\overset{H}{\underset{|}{C}}}-COOH$$

Dipeptide Leu-Asp

3 Glossary

As editors, we would like to thank Doctor D. Despois who initiated this glossary, for the book entitled "L'environnement de la Terre Primitive" published in 2001 by the Presses Universitaires de Bordeaux. Since this initial version, new terms have been added and today, the glossary contains more than 700 items. Considering the diversity of scientific fields that are of direct interest in astrobiology, any glossary devoted to astrobiology is necessarily incomplete. Nevertheless, the editors hope that the following glossary, even far from perfect, will help specialists in one field to obtain some pertinent information concerning other scientific fields without being obliged, at least in a first step, to refer to original papers or specialized books.

A

a (*astronomy*): Orbital parameter, semimajor axis of elliptical orbit.

Abiotic: In the absence of life.

Ablation (*astronomy*): Loss of matter by fusion or vaporization during the entrance of an object into the atmosphere.

Absolute magnitude: Magnitude of a stellar object when seen from a distance of 10 parsecs (32.6 light-years) from Earth.

Acasta: Region of Northern Territories (Canada) where the oldest relics of continental crust were discovered. These gneisses were dated at 4.03 Ga.

Accretion (*astronomy*): Matter aggregation leading to the formation of larger objects (stars, planets, comets, asteroids). Traces of accretion of their parent body (asteroid) can be observed in some meteorites.

Accretion disk: Disk of matter around a star (or around a black hole) such that matter is attracted by the central object and contributes to its growth. Disks around protostars, newborn stars and T-Tauri stars are accretion disks.

Accretion rate (*astronomy*): Mass accreted per time unit (typically 10^{-5} to 10^{-8} solar mass per year).

Acetaldehyde: CH_3CHO, ethanal.

Acetic acid: CH_3COOH, ethanoic acid.

Acetonitrile: CH_3CN, cyanomethane, methyl cyanide.

Achiral: Not chiral.

Acidophile: Organism that "likes" acidic media, which needs an acidic medium.

Actinolite: Mineral. Inosilicate (double-chain silicate). $[Ca_2(Fe,Mg)_5\ Si_8O_{22}(OH,F)_2]$. It belongs to the amphibole group.

Activation energy (E_a): Empirical parameter that permits, via an Arrhenius law, the temperature dependence of a reaction rate to be described. In the case of a reaction $A + B \rightarrow C$, E_a can be described as the energy difference between the reactants $A + B$ and the activated complex (AB^*) on the reaction pathway to C. It corresponds to an energy barrier.

Active site (*of an enzyme*): Part of an enzyme where substrate is specifically fixed, in a covalent or noncovalent way, and where catalysis takes place.

Activity (*solar or stellar*): All physical phenomena that are time dependent and that are related to star life (like stellar wind, solar prominence, sunspots). Their origin is mainly magnetic, they correspond to emissions of electromagnetic waves at different frequencies (from radiowaves to X-rays) and also to emissions of charged particles (protons, alpha particles and heavier particles). In the case of the Sun, the solar activity is cyclic. The shorter cycle is an eleven-year cycle.

Activity: In nonideal solution, activity plays the same role as mole fraction for ideal solution.

Adakite: Volcanic felsic rock generated in subduction zone, by partial melting of the subducted basaltic oceanic crust.

Adenine: Purine derivative that plays an important role in the living world as a component of nucleotides.

Adenosine: Nucleoside that results from condensation of adenine with ribose or desoxyribose.

ADP: Adenosine diphosphate (see also ATP).

Aerobian: Organism whose life requires free oxygen in its environment.

Aerobic respiration: Ensemble of reactions providing energy to a cell, oxygen being the ultimate oxidant of organic or inorganic electrons donors.

Aerosols: Liquid or solid submillimetric particles in suspension in a gas. Aerosols play an important role in atmospheric physics and chemistry.

AIB: See aminoisobutyric acid.

Akilia: Region of Greenland near Isua where are exposed sediments and volcanic rocks similar in both composition and age (3.865 Ga) to the Isua gneisses.

Alanine (*Ala*): Proteinic amino acid containing three carbon atoms.

Albedo: Fraction of the incident light that is reflected by a reflecting medium (i.e. atmosphere) or surface (i.e. ice cap). A total reflection corresponds to an albedo of 1.

Albite: Mineral. Tectosilicate (3D silicate). Sodic plagioclase feldspar $NaAlSi_3O_8$.

Alcohol: R–OH.

Aldehyde: R–CHO.

Aldose: Any monosaccharide that contains an aldehyde group (–CHO).

ALH84001: Martian meteorite found, in 1984, in the Alan Hills region (Antarctica). In 1996, the claim that it contains traces of metabolic activity and even, possibly, microfossils was the starting point of strong debates.

Allende: Large carbonaceous chondrite (meteorite) of 2 tons, of the C3/CV type and found in Mexico in 1969.

Allochtonous sediment: Sediment that formed in a place different from its present-day location (transported).

Alpha helix: A type of secondary helical structure frequently found in proteins. One helix step contains approximately 3.6 amino acid residues.

Alteration (*Weathering*): Modification of physical and chemical properties of rocks and minerals by atmospheric agents.

Amide: $R-CO-NH_2$, R–CO–NH–R', R–CO–NR'R" depending if the amide is primary, secondary or tertiary. The bond between the CO group and the N atom is generally called the amide bond but when it links two amino acid residues in a polypeptide or in a protein, it is called the peptide bond.

Amine: Derivatives of ammonia NH_3 in which one, two or three H atoms are substituted by an R group (a group containing only C and H atoms) to give primary, secondary or tertiary amines.

Amino acid (*AA*): Organic molecule containing a carboxylic acid function (COOH) and an amino function (generally, but not always, a NH_2 group). If the two functions are linked to the same carbon atom, the AA is an alpha amino acid. All proteinic AA are alpha amino acids.

Amino acids (*biological*): AA directly extracted from organisms, fossils, sedimentary rocks or obtained, after hydrolysis, from polypeptides or proteins found in the same sources.

Amino acids (*proteinic*): AA found as building blocks of proteins. All proteinic AA are homochiral and characterized by an L absolute configuration.

Amino isobutyric acid (*AIB*): Nonproteinic alpha amino, alpha methyl AA. Detected in chondrites.

Amino nitrile: Molecule containing a CN group and an amine function. It could have played a role in the prebiotic synthesis of amino acids.

Amitsôq: Region of Greenland where the oldest huge outcrops ($\sim 3000\,km^2$) of continental crust, dated at 3.822 Ga (gneiss) are exposed.

Amorphous: Solid state characterized by a lack of order at large distances.

Amphibole: Mineral family. Inosilicate (water-bearing double chain silicate), including actinolite, hornblende, glaucophane, etc.

Amphibolite: Rock generated by metamorphism of basalt. It mainly consists of amphibole and plagioclase feldspar crystals, sometimes associated with garnet.

Amphiphile: Molecule with a hydrophilic part and a lipophylic part.

Amplification (*of DNA*): Production, in relatively large quantity, of fragments of DNA by in vitro replication, starting from a very small initial sample (see: PCR).

Amu (atomic mass unit): Atomic mass unit such that the atomic mass of the ^{12}C (carbon isotope) has exactly a mass equal to 12.0000.

Anabolism: General term to designate a group of biochemical reactions involved in the biosynthesis of different components of living organisms.

Anaerobian: An organism that does not need free oxygen for his metabolism. In some cases, free oxygen is a poison for anaerobian organisms.

Anaerobic respiration: Ensemble of reactions providing energy to a cell, the ultimate oxidant being an inorganic molecule other than oxygen.

Anatexis: High-degree metamorphism where rock begins to undergo partial melting.

Andesite: Effusive mafic magmatic rock (volcanic); it mainly consists of sodic plagioclase feldspar + amphibole ± pyroxene crystals. These magmas are abundant in subduction zones. Diorite is its plutonic equivalent.

Anorthite: Mineral. Tectosilicate (3D silicate). Calcic plagioclase feldspar $CaAl_2Si_2O_8$.

Anticodon: Triplet of nucleotides of tRNA able to selectively recognize a triplet of nucleotides of the mRNA (codon).

Antisense: Strand of DNA that is transcribed into a mRNA (messenger RNA).

Aphelia: In the case of an object in elliptical motion around a star, the point that corresponds to the largest distance with respect to the star.

Apollo: Ensemble of asteroids whose orbits intersect Earth orbit.

Archaea: One of the three main domains of the living world. All the organisms of this domain are prokaryotes. Initially, these organisms were considered as the most primitive form of life but this is no longer accepted by most biologists. Most extremophiles such as hyperthermophiles and hyperacidophiles belong to the Archaea domain.

Archaean (*Aeon*): Period of time (Aeon) ranging from 4.0 to 2.5 Ga. Archaean aeon belongs to Precambrian. Unicellular life existed and possibly was already aerobe.

Archebacteria: See Archaea.

Arginine: Proteinic alpha amino acid containing six carbon atoms and a guanido group in the side chain.

Aspartic acid (*Asp*): Proteinic AA with an acidic side chain.

Asteroid belt: Ring-shaped belt between Mars and Jupiter where the majority of the asteroids of the Solar System are located.

Asteroid: Small object of the Solar System with a diameter less than 1000 km. Many of them are orbiting around the Sun, between Mars and Jupiter (asteroid belt).

Asthenosphere: Layer of the Earth mantle, located under the lithosphere and having a ductile behaviour. It is affected by convective movements. Depending on geothermal gradient, its upper limit varies between 0 km under midoceanic ridges and 250 km under continents.

Asymmetric: An atom is called asymmetric when it is surrounded by four ligands that are oriented in 3D space in such a way that they define an irregular tetrahedron. Such an atomic arrangement is chiral and can exist as two enantiomeric forms. These two enantiomers are called D or L depending on their absolute configuration. The D/L nomenclature is now replaced by the R/S nomenclature of Cahn, Ingold and Prelog but the D/L system is still accepted for amino acids and oses. The asymmetric carbon atom is a particular (but very important) example of an asymmetric atom.

Atmosphere: Gaseous envelope around a star, a planet or a satellite. In the absence of a rigid crust or ocean, the atmosphere is defined as the most external part of the object. *The primary atmosphere* of a young planet corresponds to the first gaseous envelope directly generated from protostellar nebula. *The primitive atmosphere* of the Earth corresponds to the atmosphere in which prebiotic chemistry may have occurred, when free oxygen was at very low concentrations. In the second part of the past century, this primitive atmosphere was considered as being highly reductive. Today, it is considered as mainly consisting of carbon dioxide, nitrogen and water vapour.

ATP synthetase: Enzyme involved in ATP synthesis.

ATP: Adenosine triphosphate. Molecule that plays an important role in the living world for energy transfers. Its hydrolysis in ADP and in inorganic phosphate is an exergonic reaction.

AU (*astronomical unit*): Average Earth–Sun distance corresponding to 149.6×10^6 km (or approximately 8 light-minutes or 100 Sun diameters).

Authigenic minerals: In a sedimentary environment, minerals generated by direct local precipitation of dissolved ions.

Autocatalysis: Chemical reaction such that a reaction product acts as a catalyst for its own synthesis.

Autochthonous sediment: Sediment that formed in the place where it is now.

Automaton (*chemical automaton*): As defined by A. Brack, chemical system able to promote its own synthesis.

Autotroph: Organism that is able to synthesize its own constituents from simple molecules like water and carbon dioxide.

B

Bacteria: One of the three domains of the living world. Organisms of this domain are all prokaryotes. They are the more abundant micro-organisms on Earth. One of their characteristics consists of the presence of a cell-wall containing muramic acid. Some of these organisms are extremophiles like *Aquifex* and *Thermoga*.

Bacteriochlorophylls: Pigments found in micro-organisms and containing a tetrapyrole acting as a ligand for an Mg cation (similarly to chlorophylls of green plants).

Barophile: Micro-organism that lives optimally (or can only grow) in high-pressure conditions such as deep-sea environments.

Basalt: Effusive mafic magmatic rock (volcanic); it mainly consists of plagioclase feldspar + pyroxenes ± olivine crystals. It results from 20 to 25% melting of the mantle. Gabbro is its plutonic equivalent.

Bases in nucleotides: See nucleic bases, purine bases and pyrimidine bases.

Benioff plane: In a subduction zone, interface between the subducted slab and the overlying mantle wedge.

Beta sheet: A secondary pleated structure frequently observed in proteins.

BIF (*banded iron formation*): Sedimentary rocks widespread in Archaean terrains and no longer generated today. They consist of alternation of black iron-rich (magnetite) and white amorphous silica-rich layers.

Bifurcation (thermodynamics): This term describes the behaviour of a system that, submitted to a very small variation in exchange conditions with its surroundings, jumps suddenly from one stationary state to another one. The characteristics of the new stationary state cannot be predicted on the basis of a complete knowledge of the initial stationary state.

Biofilm: Tiny film (few µm thick) consisting of colonies of micro-organisms fixed on an inorganic or organic surface.

Biogenic sediment: Sediment formed by precipitation of ions dissolved in water by the means of living beings (e.g., calcium ion in shells).

BIOPAN: Experimental module made by ESA to be fixed on a Russian satellite of the Photon type, the aim of the system is to expose samples or dosimeters to space conditions, such as vacuum, microgravity or radiation.

Biosignature: Observable considered as a piece of evidence for the presence of life.

Biosphere: Ensemble of species living on Earth.

Biosynthesis: Production of a chemical compound by a living organism, equivalent to anabolism.

Biotite: Mineral. Phyllosilicate (water-bearing sheet silicate). [K(Fe,Mg)$_3$Si$_3$AlO$_{10}$ (OH)$_2$]. It belongs to the mica group and is also called black mica.

Biotope: Smallest unit of habitat where all environment conditions and all type of organisms found within it are the same throughout.

Bipolar flow (*astronomy*): Flow of matter in two opposite directions, perpendicularly to the circumstellar disk associated to a newborn star. The components of the bipolar flow are molecules, atoms, ions and dust particles.

Birthline: Locus, in the Hertzsprung–Russell diagram where young stars become optically visible.

Black body (*radiation*): Radiation emitted by a body at a temperature T and such that the coupling between matter and radiation is perfect. Such a body is black. The total power emitted by unit area and the power emitted at a well-defined frequency depends only on the temperature (the T^4 dependence is given by the empirical Stefan law and by the theoretical Planck law). As a first approximation and on the basis of their emission properties, stars and planets can be described as black bodies.

Black smoker: Also called hydrothermal vent. Structure observed on the oceans floor generally associated with midocean ridges. There, hot hydrothermal fluid, rich in base metal sulfides, enters in contact with cold oceanic water. Polymetallic sulfides and calcium sulfate precipitate progressively building a columnar chimney around the vent.

Blast: Blast (basic local alignment search tool) represents a powerful and rapid way to compare new sequences to an already existing database, which may either contain nucleotides (Blastn) or proteins (Blastp). Since the BLAST algorithm establishes local as well as global alignments, regions of similarity embedded in otherwise unrelated proteins could be detected. The Blast program gives a value for each of the high-scoring results, together with the probability to find an identical score in a database of the same size by chance.

Blue algae: See cyanobacteria. Old name for some procaryotic unicellular that are not algae.

Bolometric light (*or bolometric magnitude*): Total radiation output by time unit of a stellar object.

Bootstrap: Statistical method with resampling, commonly used to measure the robustness and the reliability of phylogenic trees.

Branching ratio: In the case of a chemical or of a nuclear reaction, such that different reaction paths exist, the branching ratio is the relative rate constants or probabilities of occurrence per unit of time for each different path.

Breccias (geology): Sedimentary or magmatic rocks consisting of an accumulation of angular fragments in a sedimentary or magmatic matrix.

Bremsstrahlung: German word also used in French and English to describe the electromagnetic radiation emitted by high-speed particle (electron, proton) when deviated by a magnetic field.

Brown Dwarf: Space body born as a star, but whose mass is too small (< 80 Jupiter mass) to allow nuclear reaction: both core temperature and core pressure are insufficient to initiate hydrogen fusion.

C

C (*alpha*): Carbon atom linked to a chemical function we are interested in. More specifically, the carbon atom directly linked to the carboxylic function in amino acids. In alpha amino acids, the amino group is linked to the alpha carbon. The Greek letters alpha, beta, gamma … are used to describe carbon atoms separated from the function by one, two, three … other carbon atoms.

CAI (*Ca–Al Inclusions*): Ca-and Al-rich inclusions abundant in some chondrite meteorites.

Caldera: Circular km-sized structure due to the collapse of superficial formation induced by the emptying of an underlying magma chamber.

Carbonaceous chondrite: Chondrite with high carbon content. The famous Murchison meteorite is a carbonaceous chondrite.

Carbonate: $(CO_3)^{2-}$-bearing mineral e.g., Calcite = $CaCO_3$; Dolomite = $MgCa(CO_3)_2$.

Carbonation: In Ca-, Mg-, K-, Na- and Fe-bearing minerals, chemical reaction of alteration resulting in the formation of carbonates.

Carbonyl: –(CO)–, this chemical function is found in carboxylic acids, ketones, aldehydes, amides in the peptide bond and many other organic molecules.

Carboxylic acid: Organic molecule containing a COOH group.

C-asteroid: Asteroid containing carbon. C-asteroids are the parent-bodies of carbonaceous chondrites.

Catalysis: Chemical process such that a substance (catalyst) increases the reaction rate by changing reaction pathway but without being chemically modified during reaction. Enzymes are very efficient catalysts able to increase reaction rates by several orders of magnitude and also able to limit the number of secondary products and therefore to increase reaction selectivity.

CCD (*charge-coupled device*): Silicon photoelectronic imaging device containing numerous photosensors (often at least 1000×1000). The most used astronomic detector in the visible wavelength domain.

Cell: Complex system surrounded by a semipermeable membrane that can be considered as the basis unit of all living organisms.

Cenozoic or Caenozoic(*Era*): Period of time (Era) ranging from 65 Ma to today, it is also called Tertiary Era but in addition it also includes the Quaternary Era.

Chemical sediment: Sediment formed by direct precipitation of ions dissolved in water.

Chemolithoautotroph: Chemotroph that uses CO_2 as only source of carbon.

Chemolitotroph: Chemotroph that takes its energy from the oxidation of inorganic molecules. On Earth, the first living organisms could have been chemolitotrophs.

Chemoorganotroph: Chemotroph that takes its energy from oxidation of organic molecules.

Chemotroph: Organism that takes its free energy from the oxidation of chemicals.

Chicxulub: Large (180 km in diameter) impact crater, located in the Gulf of Mexico. It is assumed to be the result of the collision of a big (10 km in diameter) meteorite, 65 Ma ago. This impact is considered as the cause of the important biological crisis at the Cretaceous–Tertiary boundary that led to mass extinction of thousands of living species, such as ammonites, dinosaurs, etc.

Chirality: Property of an object (and therefore of a molecule) to be different from its mirror image in a plane mirror. A hand is an example of a chiral object (in Greek: *cheir* means hand). Any object is chiral or achiral and if it is chiral, it can exist as two enantiomorphous forms (called enantiomers for molecules).

Chlorophylls: Pigments of major importance in the oxygenic photosynthesis. The chemical structure of all chlorophylls is based on a porphyrin ring system chelating an Mg^{2+} cation. Chlorophylls are found in higher plants, algae and some micro-organisms.

Chloroplast: Subcellular structure that plays a fundamental role in photosynthesis in all photosynthetic eukaryotes. Chloroplasts have more probably an endosymbiotic origin.

Chondre: Small spherical aggregate of radiated silicate minerals (typically 1 mm in diameter) that is frequent in stony meteorites and especially in chondrites. Olivine is the main component of chondres.

Chondrite: Undifferentiated stony meteorite unmelted and frequently considered as a very primitive object. Chondrites have the same composition as the Sun except for volatile elements.

Chromatography: Preparative or analytical chromatography: experimental method based on the properties of all molecules to be absorbed more or less selectively by a solid phase (the stationary phase) and therefore to migrate at different rates when they are "pushed away" by a mobile phase that can be a gas (gas chromatography or GC) or a liquid (LC or HPLC for high-performance LC).

Chromosome: Subcellular structure containing most of the genetic material of the cell. Prokaryotic cells generally contain only one chromosome made of a circular DNA molecule, while eukaryotic cells generally have several chromosomes, each of them containing a linear DNA molecule.

CIP (*for Cahn–Ingold–Prelog*): General nomenclature used in chemistry to describe chiral molecules and more generally stereoisomers. Following the International Union of Pure and Applied Chemistry (IUPAC), CIP nomenclature must replace all other nomenclatures including D, L Fisher nomenclature, except for amino acids and sugars for which the D/L nomenclature is still accepted.

Circular dichroism: The absorption coefficients of right and left circularly polarized light are different if the absorbing medium is chiral. By plotting the difference between the absorption coefficients as a function of the light wavelength, the curve obtained corresponds to the circular dichroism curve of the medium.

Circumstellar disk: Disk of gas and dust particles around a star.

Class [0, I, II, III] (*astronomy*): Classification of young stellar objects based on their electromagnetic emission in the microwave and infrared domains. It consists of an evolution sequence from the protostars (0 and I) to the T-Tauri stars (III).

Clast: Fragment of mineral or rock included in another rock.

Clay: Mineral family. Phyllosilicate (water-bearing sheet silicate), e.g., kaolinite, illite, smectite, montmorillonite.

CM matter: Pristine material of the Solar System, analogue to constitutive matter of CM carbonaceous meteorites (M = Mighei) and very abundant in micrometeorites.

Coacervat (*droplet*): Protein and polysaccharides containing emulsion. According to Oparin, model of protocells.

Codon: Triplet of nucleotides in a mRNA molecule that corresponds to a specific amino acid of a protein synthesized in a ribosome or that corresponds to a punctuation signal in protein synthesis.

Coenzyme: Small molecule that binds with an enzyme and that is necessary to its activity. ATP, CoA, NADH, biotine are examples of coenzymes. Coenzymes are often derived from vitamins.

Collapse (*astronomy*): Process that describes the formation of stars from dense cores. The process seems to be fast: less than 10^5 years.

Coma: Broadly spherical cloud of dust particles and gaseous molecules, atoms and ions surrounding cometary nucleus. It appears when a nucleus becomes active, generally when approaching the Sun.

Combustion: Used to describe any exothermal chemical reaction involving dioxygen and organic reactants. Sometimes used in astrophysics to describe the thermonuclear fusion reactions taking place in the stars and that, obviously, are also exothermal.

Comet: Small body of the Solar System with an average size of 1 to 100 km, travelling generally on a strongly elliptic orbit around the Sun. Comets are constituted of ice and dust and are considered as the most primitive objects of the Solar System.

Cometary nucleus: Solid part of a comet (1–100 km diameter), made of ice (H_2O, CH_3OH, CO...) and of dust.

Cometary tail: Part of an active comet; three different cometary tails are known depending on their composition (dust particles and molecules, neutral sodium atoms, ions).

Complex molecule (*astronomy*): Molecule containing more than 3 atoms.

Complexation: Chemical term used to describe the noncovalent interaction of a molecule or, more frequently, an inorganic ion with other molecules (called ligands) to give a supramolecular system described as a complex.

Condensation: Chemical reaction involving two molecules and leading to the formation of a new chemical bond between the two subunits but also to the elimination of a small molecule (generally a water molecule). The formation of bonds between the subunits of many biochemical polymers are condensation reactions (examples: polynucleotides, polypeptides, polysaccharides).

Configuration: Term used in stereochemistry. Stereoisomers (except if they are conformers) have different configurations. As an example, butene can be cis or trans and it corresponds to two different configurations. In the case of chiral molecules like the D- and L-valine, the stereoisomers are enantiomers and they have "opposite" configurations. It is important to make a clear distinction between the relative configurations of two enantiomers and the absolute configuration of each of them.

Conglomerate (*chemistry*): In the restricted case of crystallized chiral molecules, a crystalline state such that all molecules of the same chirality (homochiral) crystallize together giving a mixture of crystal that, themselves, are of opposite chirality.

Conglomerate (*geology*): Detrital sedimentary rock consisting of an accumulation of rounded fragments in a sedimentary matrix.

Continental margin: Submarine part of a continent making the boundary with the oceanic crust.

Continents: Emerged and associated shallow depth (< 300 m) parts of the Earth's surface. Their average composition is that of a granitoid.

Continuum (*astronomy*): in emission (or absorption) spectroscopy, emission (or absorption) background of a spectrum extending over a large frequency domain. Frequently, lines are superimposed on the continuous spectrum. Blackbody radiation corresponds to a continuous spectrum.

Cool Early Earth: Period of time between Earth accretion (4.55 Ga) and the late heavy bombardment (4.0–3.8 Ga). This model considers that period of early Earth as quiescent with respect to meteoritic impacts, thus being potentially favourable for life development.

Core (*geology*): Central shell of a planet. On Earth the core mainly consists of iron with minor amounts of nickel and some traces of sulfur; it represents 16% of the volume but 33% of the mass of the planet. It is subdivided in a solid inner core from 5155 to 6378 km depth and a liquid outer core from 2891 to 5155 km depth.

COROT: French project for the search of extrasolar planets based on a 25-cm telescope and able to detect planets having a diameter equal to twice the Earth diameter and located at 0.5 AU from its star.

Cosmic rays: Highly accelerated ions coming from the Sun (solar wind, essentially protons) or coming from other and extrasolar sources (galactic cosmic rays).

Covalent bonds: Interatomic bonds such that two atoms share one, two or three electron pairs leading to the formation of a single, a double or a triple bond. A covalent bond is described as a polarized bond if the two bonded atoms have different electronegativities.

CPT (*theorem*): general theorem of physics that assumes that physical laws are unchanged if, simultaneously, space is reversed (parity operation), time is reversed (the sense of motion is reversed) and matter is replaced by antimatter. In many cases, but not all, the CP theorem alone is valid.

Craton: Huge block of old (often Precambrian in age) and very stable continental crust (see also shield).

Crust (*geology*): The more superficial shell of the solid Earth. Its lower limit with the mantle is called Morohovicic (Moho) discontinuity. Together with the rigid part of the upper mantle it forms the lithosphere. Two main crusts exist 1) oceanic crust, basaltic in composition and about 7 km thick, it constitutes the ocean floor; 2) continental crust is granitic in composition, with a thickness ranging between 30 and 80 km, it constitutes the continents.

Cryosphere: Part of the Earth surface made of ice.

Cumulate: Igneous rock generated by accumulation of crystals extracted from magma.

Cyanamid: NH_2–CO–CN.

Cyanhydric acid: HCN, hydrogen cyanide. Triatomic molecule that during prebiotic period could have played the role of starting material for purine synthesis.

Cyanoacetylene or better cyanoethyne: H–CC–CN.

Cyanobacteria: Micro-organism belonging to the Bacteria domain and able to perform oxygenic photosynthesis. In the past, these micro-organisms were improperly called "blue algae". The cyanobacteria could be the ancestors of chloroplasts.

Cyanogen: C_2N_2.

Cysteine (*Cys*): Protinic amino acid containing three C atoms and a –SH group in its side chain. In a proteinic chain, two cysteine residues can be linked together by a –S-S- bond (disulfide bond) often used to stabilize protein conformation. Two cysteines linked together by a disulfide bond is a cystine molecule.

Cytidine: The ribonucleoside of cytosine. The corresponding deoxyribonucleoside is called deoxycytidine.

Cytochrome c: One particular example of the large cytochrome family. Cytochrome c is a protein involved in the electron transfers associated to aerobic respiration. In the eukaryotic cells, cytochrome c is localized in the mitochondrias.

Cytoplasm: Whole content of a cell (protoplasm) except the nucleus whose content is called nucleoplasm.

Cytoplasmic membrane: also called plasmatic membrane or cell membrane.

Cytosine (*C*): One of the nucleic bases of pyrimidine type.

D

D/H ratio: D = deuterium; H = hydrogen. Due to their mass difference, molecules containing either H or D are able to fractionate. For instance, during vaporization, a molecule containing the light isotope (H) is more efficiently vaporized than its heavy equivalent (D). Chemical reactions, including biochemical ones are also able to fractionate these isotopes. Consequently, the D/H ratio can provide information on the biotic or abiotic origin of some organic molecules. The D/H ratio can be very high in chondritic organic matter due to D-enrichment during reactions taking place in the interstellar clouds.

Dalton (*Da*): Molecular mass unit equal to the sum of the atomic masses given in amu (atomic mass unit).

DAP: Abbreviation frequently used to design two different molecules, diamino propionic acid and diamino-pimelic acid.

Darwin: Research programme from the European Space Agency devoted to the search of extrasolar planets and the study of their atmosphere composition by spatial interferometry, in the infrared spectral region. Five independent telescopes of 1.5 m each constitute the basis of this very sophisticated system.

Daughter molecule (*in comet*): In the cometary coma, any molecule produced by photodissociation of a parent molecule coming from the nucleus.

Deamination: Reaction associated to the elimination of an amine group (NH_2, NHR or NRR′).

Decarboxylation: Reaction associated to the elimination of a CO_2 molecule.

Deccan (*Trapps*): Voluminous stacking of basaltic flows emplaced in Northwest India at the end of Cretaceous period. It constitutes evidence of an extremely important volcanic event contemporaneous with the extinction of dinosaurs.

Delta (*isotopic*) (δ): Difference between the isotopic ratio of a sample (R_e) and that of a standard (R_s). $\delta = 1000 \times (R_e - R_s)/R_s$, e.g., $\delta^{18}O$.

Denaturation: Change of the native conformation of a biopolymer. More specifically and in the case of proteins, denaturation can be induced by increasing temperature and/or pressure or by adding a chemical reagent (like urea). Denaturation can be reversible or irreversible. It is generally associated with a loss of enzymatic properties.

Dense clouds: See interstellar clouds.

Dense core (*astronomy*): Gravitationally bound substructure located inside a molecular cloud and surrounded by protoplanetary disks. Dense core collapsing leads to the formation of one or several stars.

Deoxyribose: Ose or monosaccharide having a structure identical to ribose except that the OH group in position 2' is replaced by an H atom. In living organisms, deoxyribose comes from ribose via a reduction reaction.

Depletion: Impoverishment of chemical element abundance when compared to a standard of reference composition.

Diagenesis: Chemical and/or biochemical transformation of sediment after its deposition. This process, which generally consists of cementation and compaction, transforms a running sediment into a compact rock.

Diapirism: Gravity driven magma or rock ascent in the Earth. Generally, low-density materials rise up into denser rocks.

Diastereoisomers (*or diastereomers*): Stereoisomers that are not enantiomers. Diastereisomers are characterized by physical and chemical properties that can be as different as observed for isomers having different connectivity (constitutional isomers).

Differentiation (*Earth*): Separation from a homogeneous body of several components whose physical and chemical properties are contrasted. In the case of Earth these components are core, mantle, crusts, hydrosphere and atmosphere.

Diffuse clouds: See interstellar clouds.

Dinitrogen: N_2, also frequently called nitrogen.

Dioxygen: O_2, also frequently called oxygen.

Disk (of second generation): Disk around a star resulting from the breaking off of solid bodies previously formed like planetesimals, asteroids or comets. The beta Pictoris disk is probably a secondary disk.

Dismutation: Reaction that from a single reactant gives two products one of them being more oxidized than the reactant and the other one more reduced.

A typical example is the transformation of an aldehyde into an alcohol and an acid.

Disulfide bond: Covalent bond between two S atoms. When the –SH groups of two cysteines react on each other in the presence of an oxidant, it gives a cystine molecule, i.e. a cysteine dimer linked by a –S-S-bond. Disulfide bridges stabilize the ternary and quaternary structures of proteins. Some irreversible denaturazing of proteins can be associated to the formation of disulfide bonds.

DNA: Desoxyribonucleic acid, long-chain polymer of desoxynucleotides. Support of the genetic code in most living cells. Frequently observed as a double helix made by two complementary strands.

Drake (*equation of*): Empirical formula (containing several adjustable parameters) that gives a rough estimation of the number of "intelligent civilizations" in the galaxy.

Dust (*interstellar, cometary*): Small solid particles (0.1–1 mm), generally made of silicates, metal ions and/or carbonaceous matter.

E

e (*astronomy*): Orbital parameter that measures the eccentricity of an elliptical orbit.

Eccentricity: Parameter (e) characterizing the shape of an orbit. e is equal to 0 for a circle, equal to 1 for a parabola, higher than 1 for a hyperbola and between 0 and 1 for an ellipse.

Ecliptic: Geometric plane of the Earth orbit. More precisely, average planar of Earth–Moon barycentre orbit.

Eclogite: High-degree metamorphosed basalt. It consists of an anhydrous rock made up of pyroxene and garnet.

Ecosystem: Community of organisms and their natural environment. Ecosystem = community + biotope.

Eddington: ESA project devoted to asteroseismology and search for extrasolar planets by the transits method. Kepler: similar project for exoplanetary science.

Electrophoresis: Analytical method used in chemistry and based on the difference between diffusion rates of ions when placed in an electric field. Initially the ions are adsorbed on a support or immersed into a viscous medium. Capillary electrophoresis is a technique adapted to the analysis of small samples.

Enantiomeric excess (*i.e. in per cent*): In the case of a mixture of two enantiomers whose respective concentrations are D and L with D greater than L, i.e. $= (D - L/D + L) \times 100$.

Enantiomers: Two stereoisomers that differ only due to their opposite chirality (like an idealized left hand and an idealized right hand).

Enantiomorph: Two objects that differ only due to their opposite chirality (like an idealized left hand and an idealized right hand).

Enantioselectivity: A reaction leading to products that are chiral is said to be enantioselective if it gives an excess of one enantiomer. Enantioselectivity is generally induced by a chiral reactant or a chiral catalyst (like an enzyme). Enantioselectivity is also used to describe a reaction such that starting from a mixture of enantiomers, one of them reacts faster than the other. In this case, the reactant must be chiral.

Enantiotopic (*chemistry*): A planar molecule like H–CO–CH$_3$ can be seen as a scalene triangle with summits of different colours. In the 2D space, such an object is chiral and the two faces are said to be enantiotopic. A chiral reagent is able to differentiate enantiotopic faces.

Endergonic: A chemical reaction or a physical change is endergonic if it requires a supply of free energy from its surroundings to succeed.

Endolithic (*biology*): Micro-organisms living in rocks.

Endosymbiosis: Process by which a eukaryotic cell lives in symbiosis with another cell (generally a prokaryotic cell) located in its cytoplasm. Chloroplasts and mitochondria are considered as vestiges of endosymbiotic prokaryotes.

Endothermic: A chemical reaction or a physical change is endothermic if it needs a supply of energy from its surroundings to succeed.

Enstatite: Mineral. Inosilicate (simple chain silicate). Mg-rich Pyroxene [MgSiO$_3$].

Envelope (*circumstellar*): In astronomy, cloud of dust particles and gas surrounding a newborn star (protostar).

Enzyme: In biochemistry, a molecule that catalyses a reaction. Most enzymes are proteins but some are polynucleotides (ribozymes).

Equator (*celestial*): Plane perpendicular to the rotation axis of the Earth and corresponding to an extension in space of the terrestrial equator.

Escherichia Coli: Bacteria found in the intestine and commonly used in experimental bacteriology. Its size is of the order of one micrometre. Its genome codes, approximately, for 3000 proteins.

Ester: Molecule that can be described as the result of condensation of an acid and an alcohol associated with elimination of a water molecule. (–O–CO–) is the ester bond.

Eubacteria: Sometimes used instead of bacteria, in order to point out the difference between Bacteria and Archaea. Archaeas themselves are sometimes called Archebacterias.

Eukaria: One of the three domains of life (together with bacteria and archaea). Eukaryotic cells are members of the Eukaria domain.

Eukaryote: Any organism from the Eukarya domain and characterized by a nucleus (containing the genetic material) separated from the cytoplasm by a membrane. Eukaryotes can be unicellular or pluricellular.

Europa: Satellite of Jupiter with a diameter of about 3100 km (same size as the Moon). An ocean could exist beyond its icy surface.

Evaporite: Sedimentary rock generated by evaporation of huge volumes of water; chemical elements dissolved in water precipitate leading to the deposition of minerals such as halite (NaCl) or gypsum $= (CaSO_4, 2H_2O)$.

Exergonic: A chemical reaction or a physical change is exergonic when it provides free energy to its surroundings.

EXOCAM: Special reactor used to experimentally simulate exobiological processes.

Exon: Sequence of transcripted nucleotides that is present in natural RNA and that corresponds to a DNA sequence. In DNA, exons are separated by introns (intervening sequences). Exons and introns are respectively coding and noncoding sequences for proteins.

Exoplanet: Planet orbiting around a star other than the Sun. The number of already discovered exoplanets is greater than 100. (More information on this fast developing field is available on www.obspm.fr/planets.)

Exothermic: A chemical reaction or a physical change is exothermic if it provides energy to its surroundings.

Extinction (*interstellar, atmospheric*): Decrease of the light intensity of a star due to light diffusion or absorption by a medium (planetary atmosphere, interstellar cloud). In an interstellar cloud, visible magnitude can be reduced by a factor of 1, while the reduction can be as large as 100 in a protostar dense nucleus.

Extremophile: Micro-organism that optimally lives (or optimally grows) in "extreme" physicochemical environments (P, T, pH...).

F

Fatty acid: Carboxylic acid R–COOH where R is a long chain containing only C and H atoms.

Feldspar: Mineral family. Tectosilicate (3D silicate). Feldspars are subdivided into two main chemical groups: 1) Alkali feldspars ($NaAlSi_3O_8$ = Albite and $KAlSi_3O_8$ = Orthoclase); 2) Plagioclase feldspars ($NaAlSi_3O_8$ = Albite and $CaAl_2Si_2O_8$ = Anorthite). These minerals constitute 52% of the continental and oceanic crusts.

Fermentation: Biochemical process such that complex organic molecules (i.e. glucose) are transformed into low molecular mass molecules (i.e. ethanol) by cells, in anaerobic conditions. Fermentation corresponds to an oxidation process but the final electron acceptor is an organic molecule instead of oxygen. During fermentation as during respiration, ATP is produced but less efficiently.

Ferrihydrite: Iron hydroxide, $5Fe_2O_3 9H_2O$.

Fisher: German chemist who was the first to introduce the D/L nomenclature to differentiate enantiomers and to characterize their absolute configurations.

Fisher–Tropsch (*reaction of*): Reaction that gives hydrocarbons from a mixture of H_2 and CO. The FT reaction had and still has a great industrial importance but it could also have been important in prebiotic chemistry. The FT reaction requires metallic catalysts.

Flint: Rock mainly made of amorphous silica and having a biogenic origin. It frequently appears as nodules in chalk or limestone.

Fluid inclusion: 1 to 100 µm-sized cavities in minerals that contain fluids trapped during mineral crystallization.

Formaldehyde or methanal: The simplest aldehyde (H–CO–H).

Formation (*geology*): Group of terrains or rocks having the same characteristics.

Formic acid: HCOOH, methanoic acid.

Formose (*reaction*): Starting from formaldehyde in water solution at high pH, this reaction leads to the formation of a large variety of sugars. Its importance in prebiotic chemistry remains an open question. It is also called Butlerow reaction.

Fossil (*geology*): All kinds of trace of passed life (bone, shell, cast, biomolecule, track, footprint, etc.).

Frasil: Ice disks with a diameter of a few mm that are observed in water as soon as surfusion occurs. Frasil is common in Artic and Antarctic rivers but also below the huge ice platforms moving forward in the Antarctic Ocean.

Free-fall time: Time required for an object of mass m, initially at rest, to reach an object of mass M ($M > m$) under effect of gravitation alone. It gives a good approximation of the time required for an accretion disk to collapse during the protostar stage (typically 10^5 years).

Furanose: A "furanose ring" is a cyclic ose formed of 4 carbons and an oxygen atom.

G

Ga: Giga annum = one billion years (= Gyr). (Ga is more frequently used in geology, whereas Gyr is preferred in astronomy.)

Gabbro: Plutonic magmatic rock. It has a granular texture and mainly consists of pyroxenes and plagioclase (± olivine). Basalt is its effusive equivalent.

Gaia: Ambitious project of ESA to measure the position of one billion stars with a precision of one microarcsecond. Gaia is essentially devoted to the search for extraterrestrial planets.

Galactic "open" cluster (*astronomy*): Cluster that can contain from a dozen to a few thousand stars ("I" population) and are younger than globular clusters.

Garnet: Mineral. Nesosilicate (isolated SiO_4 tetrahedron). Its general composition is $Y_2^{+++}X_3^{++}(SiO_4)_3$ (Y = Al, Fe^{+++}, Cr and X = Ca, Mg, Fe^{++}, Mn). In the mantle, garnet is stable at high pressure (> 70 km depth) where it is the only aluminium-bearing phase.

GC (*gas chromatography*): Chromatographic method using gas as the moving phase. It can be used in analytical and preparative chemistry.

Gene: Segment of DNA, containing hundreds to thousands of nucleotides, found in a chromosome. A gene codes for a specific protein.

Genome: Ensemble of genes of an organism.

Genotype: Ensemble of the genetic characters of an organism.

Geographic pole: Point where the rotation axis (instantaneous) of a planet intersects the globe surface.

Geothermal gradient: Thermal gradient corresponding to the temperature increase with depth. In the Earth crust, geothermal gradient is $\sim 30\,°C\ km^{-1}$.

Giant planets: Large-size planets of low density, such as Jupiter, Saturn, Uranus and Neptune. One can distinguish two groups: gaseous giants, Jupiter and Saturn, mainly made up of gas (H_2, He) coming from the protosolar nebula and icy giants, Uranus and Neptune, rich in ice (H_2O, NH_3, and CH_4). They all have a core made of heavy elements and were formed in the outer part of the solar nebula, beyond the ice line.

Glaciation: Cold period in the Earth history characterized by the presence of a large cryosphere (ice). Frozen water accumulates and forms ice caps on the continents; consequently, it becomes unable to return to ocean whose level decreases. The main glaciations occurred during Precambrian, Early Cambrian, end of Tertiary and during Quaternary.

Glass inclusion: 1 to 100 µm-sized cavities in minerals that contain magma trapped during mineral crystallization.

Glass: Amorphous material. In volcanic rocks, it can result from the rapid cooling of the magma.

Globular cluster (*astronomy*): Large spherical cluster containing from a few thousand to several million old stars ("II" population).

Glutamic acid (*Glu*): Alpha AA with a side chain containing an acidic COOH function. Described as a hydrophilic AA.

Glutamine (*Glu*): Amino acid containing 5 C atoms with a NH_2 group in the side chain; it is considered as hydrophilic.

Glycan: Synonymous of polysaccharides.

Glyceraldehyde: (HO–CH_2–CHOH–CHO); the simplest aldose, containing only one chirotopic carbon atom. The two enantiomers play an historical role in stereochemistry because they are at the origin of the D, L nomenclature that describes absolute configuration (Fisher). Glyceraldehyde is the biochemical precursor of other oses.

Glycerol: (HO–CH$_2$–CHOH–CH$_2$OH); 1,2,3-propanetriol also called glycerine, component of many membrane phospholipids (which are esters of glycerol).

Glycine (*Gly*): The simplest amino acid and the only one that is achiral.

Glycolic acid: HO–CH$_2$–COOH.

Gneiss: Metamorphic rock made up of quartz, feldspars and micas. All mica crystals show the same orientation thus defining a surface of preferential cutting up called "foliation plane".

Gondwana (*Gondwanaland*): Palaeozoic super-continent formed by convergence and agglomeration of continents (Peninsular India, Madagascar, Africa, Australia, South America and Antarctica) due to plate tectonic activity. It was mainly located in the South hemisphere.

Gram +: Bacteria previously coloured during a Gram test and that does not lose the colour after a treatment with ethanol ("positive" response).

Granite: Plutonic magmatic rock. It has a granular texture and mainly consists of quartz, alkali feldspar and plagioclase feldspar; mica can be present whereas amphibole is rare. Rhyolite is its effusive equivalent.

Granitoid: Family of quartz-bearing plutonic magmatic rocks including granites, granodiorites, tonalites and trondhjemites.

Granodiorite: Plutonic magmatic rock. It is similar to granite but contains no more than 10% alkali feldspar.

Green bacteria: Micro-organisms of the Bacteria domain able to perform anoxic photosynthesis.

Greenhouse effect: Warming of a planet surface due to the trapping by the planet's atmosphere of the electromagnetic waves received and radiated by the planet.

Greenstone belts: Volcanic (basalts and komatiites) and volcano-sedimentary formation widespread in Archaean terrains. It generally presents an elongated shape ($\sim$ 100 km long and few tens of km wide). Its green colour is due to metamorphism of basalts and komatiites.

Guanine (*G*): Nucleic base with purine structure.

Gyr: Giga year = one billion years (= Ga).

H

Hadean (*Aeon*): Period of time (Aeon) ranging from 4.55 Ga (Earth formation) to 4.0 Ga (oldest known rock: Acasta gneisses). Hadean aeon belongs to Precambrian.

Harzburgite: Peridotite made up of olivine and orthopyroxene. It generally corresponds to residual mantle after lherzolite melting and extraction of basaltic magma.

HD 209458b: First exoplanet, whose previous detection by radial velocimetry method has been confirmed by the observation of a transit in front of its parent star (HD 209458).

Heavy element (*astronomy*): Any element other than hydrogen and helium.

Helium: Rare gas whose ^{3}He isotope is used in geology as a marker of a recent degassing process from the deep mantle.

Hertzsprung–Russel diagram (*HR diagram*): In astronomy, a two-dimensional diagram with star temperature (or spectral type) as abscissa and star luminosity (or absolute magnitude) as coordinate. Temperatures decrease from left to right and the spectral types sequence is OBAFGKM.

Heterocycle: Cyclic organic molecule containing heteroatoms (i.e. atoms other than C) as constituents of the cyclic structure.

Heterotrophous: Organism that uses reduced organic molecules as principal carbon source for its biosynthesis. Nowadays, these reduced organic molecules are generally produced by other organisms. At the beginning of life, these reduced molecules were probably found in the environment.

Histidine (*His*): Proteinic amino acid containing an imidazole group in its side chain. Histidine residues (hystidyl) are frequently found in active sites of enzymes.

HMT: ($C_6H_{12}N_4$); hexamethylenetetramine, could be a minor component of a comet nucleus.

Homeostasis: Property of a living organism to maintain unchanged some of its physicochemical characteristics even in the presence of a change in the environment. Homeostasis requires autoregulation.

Homoacetogens: Micro-organisms of the Bacteria domain producing acetate from H_2 and CO_2.

Homochirality: Of the same chirality. All proteinic amino acids are L, while ribose in RNA or ATP is always D. The origin of homochirality for the large majority of the chiral constituents of organisms remains an active research subject.

Homology (*biology*): Two structures in two different species are homologous (and therefore comparable) irrespective of their forms, if they are connected in the same way to identical structures.

Hornblende: Mineral. Inosilicate (double-chain silicate). [$Na_{0-1}Ca_2$ $(Fe^{++},Mg)_{3-5}(Al, Fe^{+++})_{0-2}$ $Si_{8-6}Al_{0-2}O_{22}$ $(OH,F)_2$]. It belongs to the amphibole group.

Hot Jupiter: Jupiter massive-like exoplanet, orbiting close to a star. Most of the extrasolar planets so far discovered belong to this type.

Hot spot: see mantle plume.

HPLC (*high-performance liquid chromatography*): Very efficient liquid-phase chromatography performed under high pressure up to 100–400 bars.

Hydrocarbons: Molecules containing only C and H atoms. If the hydrocarbon contains an aromatic system, it is called aromatic. If the hydrocarbon contains only tetracoordinated C atoms, the hydrocarbon is called aliphatic. Some hydrocarbons result from the polymerization of isoprene (2-methylbutene); they are called isoprenoid hydrocarbons. Latex contains isoprenoid hydrocarbons.

Hydrogen bond: Intermolecular but sometimes intramolecular low energy bond (about 20 kJ/mol) involving generally an H atom linked to an electronegative atom like O, N, S and an atom bearing nonbonding electron pairs such as O or N. The H bond implies an H donor and an H acceptor.

Hydrogen cyanide: H–CN.

Hydrogen sulfide: H_2S.

Hydrolysis: Cleavage of a molecule due to reaction with H_2O.

Hydrothermal vent: See black smoker.

Hydroxy acid: Carboxylic acid containing also an alcohol function. Glycolic acid is the simplest hydroxy acid.

Hydroxyl (*group*): –OH or alcohol group.

Hypoxanthine (*6-hydroxypurine*): Purine base and biological precursor of adenine and guanine.

I

Ice shelves: Ice platforms generated by glaciers whose ice progresses from Antarctica land over ocean. Ross shelf is about 1000 km wide and extents over 600 km into the ocean. They are the source of very large tabular icebergs.

Ices (*astronomy*): Solid form (crystalline or amorphous) of volatile molecules like water, carbon dioxide or ammonia.

IDP: Interplanetary dust particle.

Igneous rocks: Magmatic rocks = due to magma crystallization.

Impactite: Heterogeneous breccia generated at depth by the impact of a large-size stellar body ($> 10\,000$ tons). It is made up of fragments of the rock substrate included in a vitreous matrix.

Inclination (1) (*astronomy*): Angle between the orbital plane of a solar object and the ecliptic plane in degrees (always lower than 90°).

Inclination (2) (*astronomy*): Angle between the orbital plane of an interstellar object and the "sky plane" (plane perpendicular to the "line of sight", i.e. the straight line joining the observer and the stellar object).

Indels: Acronym for insertions/deletions. Phylogenetic analysis on several DNA or protein sequences requires sequences with same length. During the process of alignment, alignment gaps (indels) must be introduced in sequences that have undergone deletions or insertions.

Interplanetary dust: Small grains left behind by asteroids and comets, and dispersed in a cloud including the whole Solar System.

Interstellar cloud: Cloud of gas (98%) and dust (2%). The gas is mainly H (diffuse cloud) or H_2 (molecular cloud). Molecular clouds are called dense clouds ($n(H_2) > 10^3$ molecules cm^{-3}) or dark clouds if dense and cold (10–20K).

Intertidal environment: Environment between high and low tide. Also called tide range.

Intron: Noncoding sequence of nucleotides that separates exons. Introns are removed during the maturation processes of the three types of RNA by splicing.

Iodine (*I*): Halogen element. As iodine is incorporated in living organisms, including marine organisms, it provides information on interactions between living world and marine sediments.

Ion–molecule (*reaction*): Reaction between two gaseous reactants and initiated by ionizing cosmic rays, X-rays or UV radiation. They are important in interstellar clouds and in planetary ionospheres.

Iridium (*Ir*): Element belonging to platinum element family. Its concentration in Earth crust is extremely low. A local Ir enrichment as at the Cretaceous–Cainozoic (K/T) boundary is interpreted as a strong argument in favour of meteoritic impact or of deep originated volcanic activity.

Isochron (geology): Rectangular diagram plotting isotopic ratio of a disintegration system (abscissa = parental isotope; ordinate = daughter isotope) (e.g., $^{87}Sr/^{86}Sr$ versus $^{87}Rb/^{86}Sr$). In this diagram, cogenetic rocks of the same age plot along a straight line whose slope is proportional to age. This method of age determination is widely used in geology.

Isocyanhydric acid: H–N=C.

Isocyanic acid: HN=C=O.

Isoleucine (*Ile*): Proteinic amino acid containing six carbon atoms and described as hydrophobic. Ile is considered as one of the prebiotic AA.

Isotopic ratio: Concentration ratio of two isotopes or concentration ratio of two isotopomers of a molecule (like H_2O and D_2O). Isotopic ratio can provide information on the age of a sample (when used in isochron calculation) as well as on its origin and source.

Isovaline: Hydrophobic nonproteinaceous amino acid. This constitutional isomer of valine contains five C atoms.

Isua: Region of Greenland where the oldest sediments so far recognized 3.865 Ga (gneiss) are exposed. They contain carbon whose origin could be biogenic. (See also Akilia.)

J

Jeans escape: Process leading to the escape of atomic or molecular species from a planet atmosphere. It happens when the thermal agitation rate is greater than the escape rate. The lighter elements or molecules (like H, H_2 or He) escape faster than the heavier ones.

Jovian planets: Other name for giant planets.

Jupiter: The fifth and largest planet (1400 times the Earth volume, 320 times the Earth mass) of the Solar System. Jupiter is 5.2 AU away from the Sun. Its gaseous envelope mainly made of H_2 and He, surrounds a core of ice and rocks (10–20 Earth mass).

Juvenile gases: Gases produced by or trapped inside the Earth and that reach the surface of Earth for the first time. ^{3}He is an example of juvenile gas detected in submarine geothermal fluids.

K

K/T (Strata): Few centimetre-thick sedimentary layer located at the Cretaceous–Cainozoic boundary. Its iridium-enrichment is interpreted as due to a giant meteoritic impact.

Kepler: Spacecraft NASA mission devoted to detection of Earth-type exoplanets (equipped with a 1-m telescope).

Keplerian rotation: Orbital motion that follows Kepler's laws.

Kerogen: Insoluble organic matter found in terrestrial sediments and in some types of meteorites like carbonaceous chondrites.

Kilo base or kilo base pair (kb): Unit used to measure the number of nucleotides in a gene or a genome: 1000 base pairs of DNA or 1000 bases of RNA.

Komatiite: Ultramafic high-Mg lava. It contains olivine and pyroxene; minerals that sometimes can have needle or dendritic shapes (spinifex texture). Komatiites were abundant before 2.5 Ga and extremely rare after.

Kuiper Belt, or Edgeworth–Kuiper: A large ring-shaped reservoir of comets beyond Neptune at about 30 astronomical units (AU) from the Sun.

L

L/D (*ratio*): For a chiral molecule, the ratio between the L and the D enantiomer concentrations.

Lactic acid: $HO-CH(CH_3)-COOH$.

Lagrange points: The five points determining the equilibrium position of a body of negligible mass in the plane of revolution of two bodies (ex: star-planet couple) in orbit around their common-gravity centre.

Late Heavy Bombardment: Heavy bombardment of the Moon (and certainly also of the Earth and other telluric planets) that happened between 4 and 3.8 Ga ago. It could correspond to either the end of a long period of bombardment by asteroids, meteorites and comets or to a short-term cataclysmic phenomenon.

Laurasia: Palaeozoic supercontinent formed by convergence and agglomeration of continents (Europe, North America and Asia) due to plate tectonic activity. It was mainly located in the northern hemisphere. Continent resulting as Pangaea broke into two parts at the end of Palaeozoic.

Lava: Magma emplaced as flow at Earth surface.

Leucine (*Leu*): Proteinic amino acid containing six C atoms and considered as one of the prebiotic AA.

Lherzolite: Peridotite made up of olivine and pyroxenes (ortho- and clinopyroxenes) as well as of an Al-bearing mineral. Its melting generates basaltic magmas leaving a harzburgite residue.

Ligand: Any atom or group of atoms bonded to the atom we are interested in. As an example, the four ligands of a chirotopic (asymmetric) carbon atom are necessarily different.

Light-year: Measure of distance used in astronomy, it corresponds to the distance that light runs in one year (0.946×10^{16} m).

Liquidus: Line that, in composition vs. temperature or pressure vs. temperature diagrams, separates the domain where crystals and liquid coexist from the field where only liquid exists.

Lithosphere: External rigid shell of the Earth. Its definition is based on the rheological behaviour of rocks. It includes crusts (continental and oceanic) as well as the upper rigid part of the mantle. Its thickness varies between 0 and 250 km and more or less corresponds to the 800°C isotherm.

LUCA: Last universal common ancestor. Hypothetical micro-organism that stood at the root of all lines leading to the present-day living beings. Appeared after a long evolution, it cannot be considered as a primitive form of life.

Lysine (*Lys*): Proteinic amino acid containing six C atoms with an amino group in its side chain and that, therefore, is basic and hydrophilic. Lysyl residues are frequently found in the active site of enzymes.

M

Ma: Mega annum = Mega year = one million years (= Myr).

Mag: See Magnitude.

Magma: Molten rock that can be completely liquid or consists of a mixture of liquid and crystals. It is produced by high-temperature ($> 650\,°C$ for granite; $> 1200\,°C$ for basalt) melting of pre-existing rocks. Mantle melting generates basalts, whereas oceanic crust fusion rather generates adakites or TTG and continental crust gives rise to granites.

Magnetic anomaly: Difference between the measured and the theoretical value of the magnetic field intensity of Earth.

Magnetic pole: Point where the magnetic dipole axis of a planet intersects the globe surface.

Magnetite: Mineral: Iron oxide $[Fe^{++}Fe^{+++}_2O_4]$. Its ferromagnetic properties make it able to record past Earth magnetic field characteristics. It can also exist in some bacteria called "magnetotactic".

Magnitude (Mag) (*astronomy*): Measure of brightness of a stellar object on a logarithmic scale. The difference between two successive magnitudes is a factor 2.512. $Mag = -2.5 \log_{10}(I/I_0)$; The less bright a star, the greater is its magnitude. The magnitude is calculated on a chosen spectral interval (visible, IR) or on the total spectrum (bolometric magnitude).

Major half-axis: For an elliptic orbit, half of the distance aphelia–perihelia.

Mantle plume: Ascending column of hot mantle assumed to be generated near the mantle-core boundary or at the upper/lower mantle boundary, (= hot spot). Close to the surface, this column can melt giving rise to oceanic island magmatism (i.e. Hawaii; La Réunion, etc.).

Mantle: In a planet, the mantle is the shell comprised between the crust and the core. On Earth it represents 82% of the volume and 2/3 of the mass, it is divided into upper mantle (until $\sim 700\,km$ depth) and lower mantle (until $2900\,km$ depth).

Mars Express: ESA space mission towards Mars.

Mass loss rate (*astronomy*): Mass ejected per time unit by a star during its formation. Ejection takes place through stellar winds and bipolar jets (typically 10^{-5} to 10^{-8} solar mass per year).

Maturation (*genetics*): Transformation step of mRNAs leading to their functional form. It occurs by splicing "introns sequences".

Megaton: Unit of energy equivalent to the energy released by $10^9\,kg$ of TNT (trinitrotoluene). Corresponds to $4.2 \times 10^{15}\,J$.

Mercaptans: Other name of thiols: sulfur analogues of alcohols.

Mesophase: Matter state exhibiting characteristics of two phases. Liquid crystals have the fluidity of liquids but are characterized by an order at short range similar to what is observed in crystals.

Mesotartric acid: HOOC–CHOH–CHOH–COOH, alpha-beta-dihydroxy-succinic acid, molecule containing two asymmetric carbon of opposite chirality. This molecule is achiral by internal compensation.

Mesozoic (*Era*): Period of time (Era) ranging from 250 Ma to 65 Ma, it is also called Secondary Era.

Metabolism: All the reactions taking place in a cell or in an organism. Metabolism is divided into two subclasses: anabolism and catabolism. Most metabolic reactions are catalyzed by proteinic enzymes.

Metamorphism: Solid-state transformation of a rock due to change in pressure and/or temperature conditions. New mineral assemblage, stable in new P–T conditions will appear. New minerals crystallise perpendicular to oriented pressure thus defining a new planar structure called foliation. Most often metamorphism corresponds to dehydration of the rock.

Metasediments: Sediments transformed by metamorphism.

Metasomatism: Change in rock composition due to fluid circulation. For instance, in a subduction zone, the fluids (mainly water) released by dehydration of the subducted slab, up-rise through the mantle wedge. These fluids, which also contain dissolved elements, not only rehydrate the mantle peridotite, but also modify its composition.

Meteor Crater: Impact crater in Arizona. It is about 1.2 km in diameter and 170 m deep. The impact, which took place 50 000 years ago, was due to an iron meteorite of about 25 m in diameter. Impacts of this kind generally occur every 25 000 year.

Meteorite: Extraterrestrial object, fragment of an asteroid, of a planet (like Mars) or of the Moon that falls on the Earth's surface.

Methanogen: Archeobacteria producing methane, CH_4, from CO_2 and H_2. Some methanogens are hyperthermophilic.

Methionine (*Met*): Proteinic amino acid containing five carbon atoms with a –SCH_3 group in its side chain.

Methylalanine: Synonymous of alpha aminoisobutyric acid (alpha-AIB).

MGS (*Mars Global Surveyor*): American probe that has carried out a complete cartography of Mars (from September 1997).

MHD: Magnetohydrodynamic.

Mica: Mineral family. Phyllosilicate (water bearing sheet silicate): biotite = black mica $[K(Fe,Mg)_3Si_3AlO_{10}(OH)_2]$; muscovite = white mica $[KAl_2Si_3AlO_{10}(OH)_2]$.

Micrometeorite: Very tiny meteorite (< 1 mm). The 50–400 µm fraction is the most abundant found on Earth. Micrometeorites constitute more than 99% of the extraterrestrial material able to reach the Earth surface (major impacts excepted).

Micro-organism: Organism invisible without a microscope. Includes prokaryotes and unicellular eukaryotes (i.e. yeasts).

Microspheres: Spherical clusters of organic molecules found in Precambrian rocks or produced in laboratory from amino acids polymers (proteinoids). Today, the Fox proteinoids microspheres are not still considered as plausible models of primitive cells.

Microsporidia: Parasitic unicellular eukaryotes that have been shown to be highly derived fungi from the fungi. Microsporidia were thought for some time to be primitive.

Migmatite: High-temperature metamorphic rock affected by partial melting.

Milankovitch (*theory of*): Theory connecting the Earth climate variations to astronomic variations such as changes of Earth's orbit or obliquity with time.

Miller–Urey (*experiment of*): One of the first experimental simulations of what was considered as atmospheric prebiotic chemistry (1953). Synthesis of a large variety of organic molecules including few amino acids from a very simple mixture containing reduced small molecules (H_2, CH_4, NH_3 and H_2O) submitted to an electric discharge.

Mineral: Solid material defined by both its chemical composition and crystalline structure.

Minimal protosolar nebula: Minimal mass of gas, necessary for the formation of the Solar System planets (= mass of all planets + H + He ≈ 0.04 solar mass).

Minor planet: Asteroid or planetoid.

Mitochondria: Organelles in the cytoplasm of all eukaryotic cells where ATP synthesis takes place during aerobic respiration. Mitochondria have their own DNA and could have an endosymbiotic origin.

Mitosis: Nucleus division, cell-division step including cytokinesis.

MM: Micrometeorites.

Moho (*Mohorovicic*): Discontinuity in seismic waves that marks the crust/mantle boundary.

Mole: Amount of substance of a system which contains as many elementary entities as they are atoms in 0.012 kg of carbon -12.

Molecular clouds: See interstellar clouds.

Molecular flow: See bipolar flow.

Monophyly: Term that describes a taxonomic group sharing a single ancestor and all its descendants (i.e. the mammals).

Monosaccharides: See oses.

Montmorillonite: Mineral. Phyllosilicate (water-bearing sheet silicate). Clay mineral belonging to the smectite group.

MORB (*mid-ocean ridge basalt*): Basalt generated in midoceanic ridge systems where oceanic crust is created. Most of ocean floor has a MORB composition.

m-RNA: Messenger RNA. Obtained by transcription of a DNA segment and able to orient the synthesis of a specific protein in the ribosome.

MS (*mass spectrometry*): Analytical method involving a preliminary ionization of atoms or ionization and fragmentation of molecules followed by measurement of atomic or molecular masses. These measurements can be carried out from precise study of ion trajectories or time of flight in an electric and/or magnetic field.

MSR (*Mars Sample Return*): NASA-CNES space mission project for the return of Martian soil samples (~ 1 kg) extracted by automatic probes. Launch expected between 2009 and 2014, and samples return three to five years later.

Murchison: Carbonaceous chondrite (CM) that fell in Australia in 1969. Fragments recovered immediately after the fall were (and still are) subjected to many analyses, mainly chemical. More than 500 organic compounds were identified, including amino acids and nucleic bases.

Muscovite: Mineral. Phyllosilicate (water-bearing sheet silicate). [$KAl_2Si_3AlO_{10}(OH)_2$]. It belongs to the mica group and is also called white mica.

Mutation: Any change of the genetic material, transmitted to the descendants.

My: Mega year = one million years (= Ma).

Mycoplasma: The simplest and the smallest known micro-organisms, they live as parasites in animal or vegetal cells. They have long been considered as possible analogues of the first cells; now considered as Gram+ bacteria that lack their rigid cell wall and evolved by reduction.

N

Nanobacteria: Hypothetical bacteria, whose size could be around few nanometres, smaller than any known bacteria. Their existence is very much debated.

N-carbamoyl-amino acids: A molecule showing many similarities with aminoacids except that one of the H atoms of the amino group is substituted by the carbamoyl polyatomic group ($-CO-NH_2$). N-carbamoyl-amino acids could have been prebiotic precursors of some amino acids.

Neutral (mutations) (*genetics*): Term coming from the neutral theory of molecular evolution proposed by Kimura. A neutral mutation is a mutation leading to sequences selectively and functionally equivalent. They are said to be neutral with regard to evolution.

NGST (*Next-Generation Space Telescope*): NASA project of space telescope (4m), it must succeed the HST (Hubble Space Telescope).

Nitrification: Microbial oxidation, autotrophic or heterotrophic, of ammonium to nitrate.

Nitrile: R–CN where the CN group is the cyano group (cyano as prefix but nitrile as suffix).

Nonreducing atmosphere: Atmosphere of CO_2, N_2, H_2O where hydrogen is absent or in low quantity, either in the form of free H_2 or hydrogen-containing compounds, such as methane or ammonia. Also named oxidized or neutral atmosphere according to its composition.

Nonsense (*codon*): When a codon (triplet of nucleotides) does not specify an amino acid but corresponds to a termination codon (Term.). In the "universal code", these codons are UAA, UAG and UGA.

Normative (*rock composition*): Rock mineralogical composition recalculated from its chemical composition.

Nuclear pores: Complex structures, highly specialized, embedded in the nuclear membrane. They allow the transfer of macromolecules between nucleoplasm and cytoplasm.

Nucleic acid: Long-chain polymeric molecule obtained by condensation of nucleotides. DNA and RNAs are nucleic acids.

Nucleic base: Linked to ribose or desoxyribose by a hemi-acetal bond, it gives nucleosides. Nucleic bases are purine bases or pyrimidine bases. Nucleosides together with a phosphate group are the subunits of nucleotides. By condensation, nucleotides give the polynucleotides (including DNA and all RNAs).

Nucleides: Constituents of atom nucleus, i.e. protons and neutrons.

Nucleon: Common name for proton or neutron.

Nucleoplasm: Protoplasm within the nucleus of eukaryotes.

Nucleotide: Molecule made by condensation of a base (purine or pyrimidine), an ose and a phosphate group linked to the ose. Nucleotides are ribonucleotides when the ose is ribose or deoxyribonucleotide when the ose is deoxyribose. DNA is a polydeoxyribonucleotide while RNAs are polyribonucleotides. The symbol of a nucleotide is determined by the base (A for adenine, C for cytosine, G for guanine, T for thymine, U for uracil).

Nucleus (*Biology*): Eukaryote cell substructure that contains the chromosomes.

O

Obduction: Mechanism leading to the thrusting of oceanic lithosphere onto continental crust.

Obliquity: Angle between the ecliptic and the celestial equator, actually, 23.3 degrees for the Earth. This angle is $> 90°$ if the planet has a retrograde rotation.

Oceanic rift: Central depression in a midocean ridge, this is the place where oceanic plates are created.

Oligomer: Small polymer, generally containing less than 25 monomeric units.

Oligomerization: Polymerization involving a small number of monomers.

Oligopeptides: Small polypeptide (less than 25 AA residues even if the definition is not so strict).

Olivine: Mineral. Nesosilicate (isolated SiO_4 tetrahedron). [(Fe, Mg)$_2$SiO$_4$]. This mineral which belongs to the peridot family, is silica-poor and is one of the main components of the terrestrial upper mantle. It is also common in meteorites.

Oort cloud: Huge spherical collection of comets, orbiting the Sun between 10 000 and 100 000 AU.

Ophiolite: Part of oceanic lithosphere tectonically emplaced (obducted) on a continental margin.

Optical activity: Orientation change of the linearly polarized plane of light after its passage through a chiral medium.

Optical rotatory dispersion: Change of the optical power of a chiral medium with the wavelength of the linearly polarized light.

Orbital Migration: Change, in the course of time, of a planet–star distance; this hypothesis is proposed in order to explain the presence of massive planets close to their star.

Organic molecule: Until the beginning of the 19th century, organic molecules were molecules extracted from plants or animals. Today, any molecule containing carbon atoms is called organic. Carbonates, CO and CO_2 are borderline cases. Organic molecules generally contain C atoms with an oxidation number lower than 4.

Organometallic: Organic molecule containing one or more metallic atoms bonded by covalent bonds or by coordination to the organic moiety of the molecules. Metallic salts of organic acids are not considered as organometallic compounds.

Orgueil: Large CI carbonaceous chondrite (very primitive), without chondres, which fell in France in 1864, near Montauban.

Orogenesis: Mountain-chain genesis.

Orthoclase: Mineral. Tectosilicate (3D silicate). Alkali feldspar $KAlSi_3O_8$.

Oses (*or saccharides*): Large group of molecules of primeval importance in the living world. Some of them are monomers like glucose ($C_6(H_2O)_6$) or ribose ($C_5(H_2O)_5$). Some of them are dimers like lactose or saccharose; some of them are polymers (polysaccharides) like starch or cellulose.

Outer membrane: membrane surrounding the plasma membrane in Gram negative bacteria (i.e. *Escherichia coli*).

P

P4: Certification for laboratories accredited to analyze high-risk infectious agents. Such laboratories must be protected against any risk of contamination by viruses and micro-organisms, from inside to outside and from outside to inside. These two conditions must absolutely be realised for extraterrestrial sample analysis.

PAH (*polycyclic aromatic hydrocarbons*): Organic molecules like naphthalene or anthracene containing several fused aromatic rings.

Paleosoil: Fossil soil. These formations are able to have recorded O_2 and CO_2 concentration of the primitive terrestrial atmosphere.

Paleozoic or Palaeozoic(*Era*): Period of time (Era) ranging from 540 Ma to 250 Ma, it is also called Primary Era.

Pangaea: Supercontinent that existed at the end of Paleozoic era (–225 Ma) and that later broke into two parts: Laurasia (N) and Gondwana (S).

Paralogous: Paralogous genes originate in gene-duplication events, in contrast to the standard orthologous genes, which originate via speciation events.

Parent molecule (*in comet*): Molecule present in the cometary nucleus.

Parity (*violation of*): Characterizes any physical property that changes when space is inverted or, in other words, which is not the same in the mirror world. Parity violation is observed in several phenomena related to the weak intranuclear interactions like the beta radioactivity. Parity violation is at the origin of the very small energy difference between enantiomers (PEVD for parity-violation energy difference).

Parsec (*secpar*): Unit of astronomical length of 3×10^{18} cm (about 200 000 AU or 3.26 light-years); it is based on the distance from the Earth at which the stellar parallax is 1 second of arc. Average distance between stars in the Sun vicinity is around 1 parsec.

Pathfinder: American probe that landed on Mars on July 4th, 1997, formally named the Mars Environmental Survey (MESUR). It contained a rover "Sojourner" that explored Ares Vallis for several months. For instance, an "alpha proton X-ray spectrometer" was used to analyse soil and rock samples in order to determine their mineralogy.

PCR (*polymerase chain reaction*): Experimental method that, by successive molecular duplications, leads to a dramatic increase of a small initial amount of DNA. This result is obtained with an enzyme called DNA-polymerase isolated from thermophile bacteria.

Peptide: Polymer obtained by condensation of amino acids. In a peptide (or polypeptide), the number of AA residues is generally lower than 60. With a higher number of residues, the polymer generally adopts a well-defined conformation and is called a protein.

Peridot: Mineral family, olivine is a peridot.

Peridotite: Rock made up of olivine and pyroxenes (ortho- and clino-pyroxenes) as well as of an Al-bearing mineral (spinel at low pressure and garnet at high pressure. Earth mantle is made up of peridotite.

Perihelia: In the case of an object in elliptical motion around a star, the point that corresponds to the shortest distance with respect to the star.

Permafost: In arctic regions, permanently frozen soil or subsoil.

Petrogenesis: Mechanism(s) of formation of rocks.

pH: Measure of the acidity of an aqueous solution. $pH = -\log_{10} (H^+)$ where (H^+) is the molar concentration of hydroxonium ion in solution. $pH = 7$ corresponds to the neutrality while a solution with $pH > 7$ is basic and a solution with $pH < 7$ is acidic.

Phanerozoic (*Aeon*): Period of time (Aeon) ranging from 0.54 Ga to today; it followed Precambrian and was characterized by metazoa development.

Phenetic: Taxonomic system for living beings based on overall or observable similarities (phenotype) rather than on their phylogenetic or evolutionary relationships (genotype).

Phenotype: The observable characteristics of an organism, i.e. the outward, physical manifestation of an organism.

Phenylalanine (*Phe*): Amino acid containing an aromatic phenyl group in its side chain. Phenylalanine contains nine amino acid residues.

Phosphoric acid: H_3PO_4, dihydrogen phosphate. Molecule that plays an important role in the living world: nucleotides are esters of phosphoric acid.

Photochemistry: Chemistry involving energy supply coming from "light" (from IR to UV). When electromagnetic frequency is greater (X-rays, beta-rays or gamma rays), the term radiochemistry is generally used. Specific processes like photoactivation, photoionisation, photodissociation (including the production of free radicals), and photolysis are various aspects of photochemistry.

Photosynthesis: Synthesis using photons as energy supply. Photosynthesis can be performed at laboratory or industrial level as a subdomain of photochemistry. In biochemistry, the term "photosynthesis" describes the different biosynthetical pathways leading to the synthesis of molecules under the influence of light. The photons are absorbed by cell pigments and their energy is converted into chemical energy stored in chemical bonds of complex molecules, the starting material for the synthesis of these complex molecules being small molecules like CO_2 and H_2O. It is important to make a clear distinction between the aerobic photosynthesis (also called oxygenic photosynthesis) and the anaerobic photosynthesis. In the first case, water is the reductive chemical species and O_2 is a by-product of the reaction. It must be kept in mind that atmospheric dioxygen as well as oxygen atoms of many oxidized molecules on Earth surface have a biosynthetical origin. Green plants, algae and cyanobacteria are able to perform aerobic photosynthesis; their pigments are chlorophylls. The anoxygenic photosynthesis is not based on H_2O but on reductive species like H_2S. Finally, some halophile archaeas contain a pigment called bacteriorhodopsin and are able to use light energy supply for ATP synthesis. So, light energy is converted into chemical energy or better, in chemical free energy.

Phototroph: Organism whose energy source is light (photosynthesis).

Phylogenesis: History of the evolution of a group of organisms.

Phylogenic tree: Schematic representation describing the evolutive relationships between organisms. It gives an image of the evolution pattern.

Phylotype: Environmental sequence, representing an organism. Sequence of a clone obtained from environment and representing an organism.

Phylum (*pl. phyla*): Group of organisms evolutionary connected at high taxonomic rank.

Pillow lava: Lava extruded under water (ocean, lake) that produces its typical rounded pillow shape. Pillow lava form the upper part of oceanic crust.

PIXE (*proton-induced X-ray emission*): Device that allows the detection and identification of many elements (metals) in proteins.

Plagioclase: Mineral. Tectosilicate (3D silicate). Calco-sodic feldspar whose composition ranges between a sodic ($NaAlSi_3O_8$ = Albite) and a calcic ($CaAl_2Si_2O_8$ = Anorthite) poles. They represent about 40% of the Earth crust minerals.

Planet: Body formed in circumstellar disks by accretion of planetesimals and may be of gas.

Planetesimals: Small solid bodies ($\sim$ 1–100 km) formed in the protosolar nebula, probably similar to asteroids and comets. Their collision and accretion built the planets.

Plankton: The whole organisms living in water and drifting along ocean and lake currents. It includes zooplankton (small animals) and phytoplankton (plants). Fishes and sea mammals, able to swim independently of current flow, constitute the nekton. Neston refers to organisms drifting at the air/water interface, whereas benthos designates organisms living in/on the aquatic ground.

Plasma membrane (*or cell membrane*): Semipermeable membrane that surrounds contemporary cells. It consists of a double layer of amphiphilic molecules (hydrophobic tail with hydrophilic head), mainly phospholipids, with proteins embedded in it. It may also contain some molecules, such as cholesterol or triterpenes, able to rigidify the whole.

Plasmid: Extrachromosomic DNA found in bacteria and yeasts, able to replicate independently.

Plasts: Organites found in phototrophic Eukaryotes and containing photosynthetic pigments such as chlorophylls.

Plate (*lithospheric*): Piece of the rigid lithosphere that moves over the ductile asthenosphere affected by convection.

Plate tectonic: Theory that describes and explains the rigid lithospheric plate motion.

Plutonic: Magmatic rock, resulting of the slow cooling and crystallisation of magma at depth. Its texture is granular (big crystals), e.g., granite.

PMS Star: Newborn star such that the internal temperature is still too low ($< 10^7$ K) to initiate nuclear fusion of hydrogen into helium (i.e. T-Tauri stars) and characterized by a low mass.

PNA (*peptide nucleic acids*): Synthetic analogues of nucleic acids such that the ose-phosphate strand is replaced by a polypeptide backbone to which the bases are linked.

Polarized light: It is important to differentiate two limit cases. A linearly polarized light (or plane polarized light) is an electromagnetic radiation, whatever is its frequency, such that the electric vector and thus also the magnetic vector oscillate in a plane. If this radiation travels through a chiral medium, the plane of polarization is deviated (optical rotation). A circularly polarized light is an electromagnetic radiation such that its electric vector and thus also its magnetic vector describe, in space, a helix around the propagation direction. This helix can be right or left corresponding to the two possible circularly polarized lights of a definite frequency. The linearly polarized light corresponds to the superposition of two circularly polarized lights of the same frequency.

Pole motion: Geographic pole motion at the Earth surface. This motion has low amplitude, since the pole only moves a few metres.

Polymerization: Chemical reaction such that molecules called monomers are covalently linked together and form long-chain molecules, with or without branching. The polymerization can be the result of an addition of polymers like in polyethylene or polystyrene (industrial polymers) but the polymerization can also be the result of condensation reactions involving at each step, the elimination of a small molecule (generally water). Nylon is an industrial condensation polymer and most biological macromolecules such as polypeptides, polynucleotides and polysaccharides are condensation polymers.

Polynucleotide: Polymer resulting from condensation of nucleotides.

Polypeptide: Polymer resulting from condensation of amino acids.

Polysaccharide: Polymer resulting from condensation of oses (in the past, oses were also called monosaccharides).

POM (*polyoxymethylene*): Polymer of formaldehyde.

Population I stars: Stars enriched in heavy elements (O, C, N...). They have been formed from the interstellar gas enriched by the previous generation(s) of stars formed over the billions of years of lifetime of the Galaxy. They are predominantly born inside the galactic disk. The Sun belongs to population I stars.

Massive, hot stars are necessarily young and are therefore always population I stars.

Population II stars: Stars poor in heavy elements (O, C, N...). They have been formed from low-metallicity gas that has not been enriched by successive generations of stars. They are believed to be born in the early ages of the Galaxy. They are actually distributed predominantly in the halo of the Galaxy thus confirming that they are probable remnants of the infancy of our Galaxy.

Poynting–Robertson (effect): Effect of stellar light on a small orbiting particle. This causes the particle to fall slowly towards the star. Small particles (below 1 cm) are more affected because the effect varies as the reciprocal of particle size.

ppb (*part per billion*): Relative concentration in mass = nanogram/gram.

ppbv (*part per billion in volume*): Relative concentration in volume = nanolitre/litre).

ppm (*part per million*): Relative concentration in mass = microgram/gram.

ppmv (*part per million in volume*): Relative concentration in volume = microlitre/litre).

Prebiotic chemistry: All chemical reactions that have contributed to the emergence of life.

Precambrian: Group of aeons ranging from Earth genesis (4.55 Ga) until the beginning of Palaeozoic era (0.54 Ga). It includes: Hadean, Achaean and Proterozoic aeons.

Precession (*of equinoxes*): Motion of Earth's rotation axis with respect to the celestial sphere due to the other bodies of the Solar System and more particularly to the Moon. The terrestrial pole describes a circle on the celestial sphere in 26 000 years.

Primary structure (*of a protein*): Sequence of the amino acid residues in the proteinic chain.

Primitive Earth: The young Earth from its formation until 2.5 Ga.

Primitive nebula: See protosolar nebula.

Prion: Protein able to induce a pathological state because its conformation is modified.

p-RNA: Synthetic RNA molecule in which the sugar is a pyranose instead of a furanose.

Prokaryote: Micro-organism in which chromosomes are not separated from the cytoplasm by a membrane. Bacteria and Archaea are prokaryotes.

Proline (*Pro*): Amino acid containing six C atoms with a unique characteristic. Its side chain links together the alpha atom and the N atom to give a five-membered ring containing one N atom and four C atoms; the amino group is no longer a $-NH_2$ group but a -NH- group. Proline is frequently observed in protein secondary structures called beta turns.

Proper motion (*astronomy*): Apparent angular movement of a star on the celestial sphere during a year (perpendicular to the line of sight). Proper motion analysis can lead to detection of planets in orbit around this star.

Propionaldehyde: CH_3–CH_2–CHO or propanal.

Proteins: Long-chain biological polymers obtained by condensation of amino acids. The degree of polymerization (number of residues) ranges between 60 and 4000. Some proteins form aggregates. For example, haemoglobin is a tetramer containing two proteinic chains of one type and two proteins chains of another type; in each chain, a heme molecule containing a ferrous cation is settled in without being covalently bonded. Structural proteins are components of muscles or flagella; most enzymes are proteins and proteins are also carriers of other molecules like dioxygen or carbon dioxide. The activity of proteins is extremely dependent on their molecular conformation.

Proteobacteria: Group of bacteria including *Escherichia coli* and purple bacteria (photosynthetic bacteria). Mithochondria are relics of proteobacteria.

Proterozoic (*Aeon*): Period of time (Aeon) ranging from 2.5 to 0.54 Ga. This aeon belongs to Precambrian and follows Achaean. Apparition of oxygen in atmosphere and of metazoa.

Proton motive force: Free-energy difference (measured in volts) associated with proton translocation across a membrane. It depends on the electrical membrane potential and on the pH difference between the two reservoirs separated by the membrane. The proton motive force provides energy required for ATP synthesis.

Protoplanetary disks: Disk around a newborn star where accretion of planets is supposed to take place.

Protosolar nebula: Rotating disk of gas, dust and ice, from which the Solar System is originated.

Protostar: several similar definitions exist. 1) newborn star such that half of its luminosity is due to accretion. 2) Body involved in an accretion process that will bring it on the main sequence. 3) Collapsing interstellar cloud. 4) Young object that is not yet optically visible. Protostars are rare because their time life is short (10^4 to 10^5 years): "Protostars are the Holy Grail of IR and submillimetric astronomy".

Psychrophilic organism: Organism that lives optimally at a temperature lower than $10\,°C$. Some psychrophilic organisms live at temperatures lower than $0\,°C$.

Pulsar: Small neutron star in very fast rotation (one rotation in less than 1 s) emitting a highly focalized radiation, circularly polarized and detected as very regular pulses. Pulsars are remnants of supernovae.

Purine bases: Guanine and adenine are examples of purine bases because their molecular skeleton corresponds to purine. These bases are found in DNA as in RNAs.

Purple bacteria: Micro-organisms of the bacteria domain able to perform anoxic photosynthesis.

PVED (*parity-violation energy difference*): Energy difference between enantiomers due to parity violation at the level of the weak forces.

Pyranose: A "pyranose ring" is a cyclic ose formed of 5 carbons and an oxygen atom. It is the more stable form of oses.

Pyrimidine bases: Thymine, cytosine and uracil are examples of pyrimidic bases because their molecular skeleton corresponds to pyrimidine. Cytosine is found in DNA as in the RNAs while thymine is specific of DNA and uracil is specific of RNAs.

Pyrite: Mineral. Sulfide [FeS_2].

Pyrolysis: Thermal degradation of a molecule.

Pyroxene: Mineral family. Inosilicate (simple chain silicate). Divided in two families: 1) Orthopyroxene [$(Fe,Mg)_2Si_2O_6$] (e.g., enstatite) and 2) Clinopyroxenes [$(Ca,Fe,Mg)_2 Si_2O_6$] (e.g., augite, diopside).

Q

Q: Orbital parameter of a planet orbiting a star; distance between planet aphelia and star.

q: Orbital parameter; distance between planet perihelia and star.

Quartz: Mineral. Tectosilicate (3D silicate). It crystallizes in hexagonal or rhombohedric systems. In magma it characterizes silica sursaturation. The rhombohedric crystals are chiral: quartz can therefore exist as D- or L-quartz depending on the helicity of the –O-Si-O-Si- chain. A chiral quartz crystal can induce an enantioselectivity during a chemical reaction between achiral reactants via a catalytic effect.

Quaternary structure (*of a protein*): In the case of protein-forming supramolecular aggregates like haemoglobins, the quaternary structure corresponds to the arrangements in space of the subunits.

R

R: solar radius (0.69×10^6 km) or $1/200\,AU$; approximately 10 times the Jupiter radius and 100 times the Earth radius.

Racemate (*crystal*): Crystalline form of a chiral substance such that each unit cell of the crystal contains an equal number of molecules of opposite chirality.

Racemic (*mixture*): Mixture of enantiomers containing an equal number of the two enantiomers. Such a mixture is described as achiral by external compensation.

Racemisation: Diminution of the initial enantiomeric excess of a homochiral ensemble of molecules or of a nonracemic mixture of enantiomers. It corresponds to an equilibration reaction: the system evolves spontaneously towards a state characterized by higher entropy and therefore, lower free energy. The highest entropy and lower free energy corresponds to the racemic mixture. Racemisation can be a very slow process: this is why enantiomers of amino acids, sugars and many other components of living species can be separated in many cases.

Radial velocity: Star velocity component parallel to the view line. It causes the frequency shift observed in spectral emission lines (Doppler effect). Periodical change in radial velocity can be an indirect proof that a planet orbits around the star (reflex motion).

Radiation pressure: Pressure applied by electromagnetic waves on any atom, molecule or particle.

Radioactivity (*long-lived species*): Radioactive elements with long period (10^9 to 10^{11} years); they are still present in the Solar System.

Radioactivity (*short-lived species*): Radioactive elements with short period ($< 10^7$ years). Nowadays they have totally disappeared in the Solar System.

Raman (*spectroscopy*): Physical method used for molecular structural analysis. Raman spectroscopy is based on inelastic diffusion of visible or UV light and gives information about the vibration modes of diffusing molecules. Raman spectroscopy must be considered as a complementary method with respect to IR spectroscopy.

Rare Earth Elements (*REE*): Chemical elements with very similar chemical properties. This family (lanthanides) ranges from lanthanum ($Z = 57$) to lutetium ($Z = 71$). In geochemistry, they are commonly used as geological tracers of magmatic processes. Indeed they are poorly sensitive to weathering or metamorphism, but on the contrary they are excellent markers of magmatic processes such as melting or crystallization.

Rare gases (*He, Ne, Kr, Ar, Xe, Rn*): Monatomic gases corresponding to the (VIIIA) period of Mendeleyev periodic table (see "Astrobiological data"). Their isotopes can be used to trace some geological events of Earth differentiation (e.g., atmosphere and ocean formation).

Red giant (*astronomy*): Old star (spectral type K or M) still performing fusion of hydrogen but already having a helium core.

Reducing atmosphere: Atmosphere with high hydrogen content. Carbon, oxygen and nitrogen mainly exist as CH_4, H_2O and NH_3.

Reduction potential: Measure of the tendency for a molecule or an ion to give an electron to an electron acceptor that, itself, can be a molecule, an ion or an electrode. A conventional reduction potential scale for molecules in water allows determination, a priori, of what chemical species will be the electron acceptor and what will be the electron donor when they are mixed together at a well-defined concentration.

Refractory: Substance which remains solid in all temperature conditions available in a particular body of the Solar System (ex. dust particles in a comet). If not refractory a substance is said to be volatile.

Region (*HII region*): Interstellar cloud such that H exists essentially as H^+. Ionization is due to intense UV radiation coming from OB stars. HII is the old name used by spectroscopists to describe lines coming from H^+ recombination.

Replication: Biochemical process by which a DNA strand or a RNA molecule is copied into the complementary molecule. Replication is different from transcription and from translation.

Retrotransposons: DNA sequences able to move from one site to another along the chromosome. Retroposons belong to a transposons family that requires RNA molecule as intermediate. The more frequent retrotransposons are the Alu sequences; around one million of such sequences have been identified in the human genome.

Ribose: Aldopentose of major importance in the living world; part of RNA nucleotides.

Ribosomes: Intracellular structures containing many rRNA (ribosomal RNA) molecules together with a complex of 60–80 proteins. Synthesis of proteins takes place in ribosomes by condensation of amino acids; information about the correct sequence is given by a mRNA (messenger RNA), while each amino acid is linked to a specific tRNA (transfer RNA).

Ribozyme: RNA molecule acting as an enzyme.

Ridge (*geology*): Submarine mountain chain located at divergent lithospheric plate margins. On the Earth the total length of ridges is 60 000 km.

Rift: Rift valley limited by faults = graben.

RNA world: Often considered as an early hypothetical stage of life evolution, based on a life without protein and such that RNAs would have played a double role: catalysis and support of genetic information. This theory is mainly based on the discovery of catalytic RNAs (ribozymes) and on an increasing knowledge about the importance of RNAs in contemporary life. For other scientists, the RNA world is the stage of evolution that preceded the DNA emergence.

RNA: Ribonucleic acid, a class of nucleic acids containing ribose as a building block of its nucleotides. RNAs themselves are divided into subgroups like messenger-RNA (m-RNA), transfer-RNA (t-RNA), ribosomal-RNA (r-RNA).

Rock: Solid material made up of mineral assemblage. It constitutes telluric planets, asteroids and probably the core of giant planets.

Rocky planets: Other name for telluric planets.

Rodinia: Super continent that formed at about 750 Ma ago.

Root (*genetics*): for a particular genetic tree, the last common ancestor of all the organisms of the tree.

Rosetta: ESA mission launched at the beginning of 2004; it will reach the P/Cheryumov-Gerasimenko comet after a 10-year trip. An orbiter will follow the comet during one year, and a lander will perform "in situ" analyses on comet nucleus surface.

Rotatory power: Deviation of the polarization plane of a radiation of well-defined frequency by a chiral medium.

rRNA: Ribosomal RNA.

r-RNA: Ribosomal RNA. r-RNAs are the major components of ribosomes. Some of them have catalytic properties.

Rubisco (*or RuBisCo*): Ribulose-1,5 biphosphate carboxylase-oxygenase; enzyme that catalyses CO_2 metabolism. It is the most abundant enzyme on Earth.

Runaway greenhouse effect: Amplification of a greenhouse effect due to vaporization of molecules able to absorb infrared radiation emitted or received by the planet and that, themselves, contribute to greenhouse effect (i.e. Venus).

Runaway growth (*astronomy*): Increasing rate of planetary accretion with planet growth, it leads to an increasing size difference between the small and large bodies.

Runoff channels: Kind of channels observed on Mars and that seems to be originated by large water flows over a long period of time.

S

Sagduction: Gravity-driven rock deformation. When high-density rocks (e.g., komatiites) emplace over low-density rocks (e.g., TTG), they create an inverse density gradient that results in the vertical sinking of dense rocks in the lighter ones. Sagduction was widespread before 2.5 Ga.

Sarcosine: $(CH_3)NH–CH_2–COOH$, nonproteinic N-methyl amino acid.

Schist: Fine-grain sedimentary rock characterized by a cleavage (slate). It results from fine sediment (i.e. mud) metamorphism.

Secondary structure (*of a protein*): Spatial arrangement of the main chain of a protein. Alpha-helices, beta-sheets, beta-turns are examples of secondary structures.

Sedimentary rock: Rock generated on Earth surface. It can consist of the accumulation of rock particles (detrital) or of organic matter (oil, petrol, coal). It can also be produced by physicochemical or biogenic precipitation of dissolved ions. Detrital particles and ions derived of alteration and erosion of pre-existing rocks.

Selenocysteine: Frequently described as the 21st proteinic amino acid because sometimes coded by the genetic code. It has the structure of the cysteine with a Se atom replacing the S atom.

Semi-minor axis: Orbital parameter. Small axis of orbital parameter.

Sense: DNA strand that is not copied into mRNA. Therefore, the sequence of the sense strand corresponds to the mRNA sequence that, itself, corresponds to the transcription of the antisense strand. Sense and antisense DNA sequences are strictly complementary.

Sequence (*genetics*): Series of directly linked nucleotides in a DNA or a RNA strand; series of directly linked amino acids residues in proteins.

Sequence (main sequence) (*astronomy*): Stage in star evolution when it performs hydrogen fusion in its core. In a HR diagram, main sequence stars draw a straight line.

Sequencing: Experimental determination of a DNA, RNA or protein sequence.

Serine (*Ser*): Proteinic amino acid with a $-CH_2-OH$ side chain. Being hydrophilic, serine is generally present in the external part of the skin and can be used as the proof for human contamination of meteoritic samples. Serine is produced in very small amounts during simulation experiments considered as experimental models of interstellar chemistry.

Serpentine: Mineral family. Phyllosilicate (water-bearing sheet silicate) (i.e. Antigorite $[Si_4O_{10}Mg_6(OH)_8]$). They are generated by olivine (and sometimes pyroxene) alteration. They play an important role in internal water cycle.

SETH (*search for extraterrestrial homochirality*): Search for proof of enantiomeric excesses in extraterrestrial objects.

SETI (*search for extraterrestrial intelligence*): Search for electromagnetic signals (essentially in the radiowave domain) that should be intentionally emitted by some living organisms and coming from sources located outside the Solar System.

Shield (*geology*): Huge block of old (often Precambrian in age) and very stable continental crust, e.g., Baltic shield (includes Finland, Sweden, Norway, and Western part of Russia).

Shocked quartz: Quartz crystal whose structure contains defaults characteristic of the high pressures realised by meteorite impacts.

Siderite: Mineral, iron carbonate $[FeCO_3]$.

Siderophile: Elements frequently associated with iron (like Au, Pt, Pd, Ni...).

SIDP (*stratospheric IDP*): IDP collected in the stratosphere by captors placed on airplanes.

Silicate: Wide mineral family of silicium oxides. The structure is based on a tetrahedron $(SiO_4)^{4-}$.

Site (*active site*): For an enzyme, specific locus where the substrate is fixed, ready to react with the reactant.

SL (*astronomy*): Solar luminosity (3.826×10^{24} W) or 3.826×10^{33} erg s^{-1}.

SM: Solar mass (2×10^{30} kg).

Small bodies: Comets and asteroids.

SMOW (*standard mean ocean water*): Standard reference sample for H and O isotopic abundance measurements.

SNC: Family of about 30 meteorites that, based on several experimental observations, are considered as having a Martian origin. SNC is the abbreviation for Shergotty, Nakhla and Chassigny, three meteorites of this family.

Solar constant: Total energy delivered each second by the Sun and measured in $W\,m^{-2}$, the surface being placed at 1 AU of the Sun and perpendicular to the light rays. The value of the solar constant is equal to $1360\,W\,m^{-2}$.

Solar type star: Star of G type, similar to the Sun.

Solidus: Line that, in composition vs. temperature or pressure vs. temperature diagrams, separates the domain where crystals and liquid coexist from the domain where only crystals (solid) exists.

Spallation: Atomic nuclei breaking due to the collision between two atomic nuclei with energy greater than the coulombic barrier; it leads to the formation of new elements, stable or radioactive.

Spectral type: Star-classification procedure based on electromagnetic spectrum that, itself, depends on star surface temperature. The OBAFGKM classification ranges from surface temperature of about $50\,000\,K$ (O type) to $3500\,K$ (M type = Sun).

Spinel: Mineral. Oxide [$MgAl_2O_4$]. Stable at low pressure (depth $< 70\,km$). In the mantle, at shallow depth, spinel is the only aluminium-bearing phase.

Spore (*biology*): a) Resistant, dormant, encapsulated body formed by certain bacteria in response to adverse environmental conditions. b) Small, usually single-cell body, highly resistant to desiccation and heat and able to develop into a new organism.

Star: Celestial object, generally spherical, where thermonuclear reactions take place (e.g., the Sun) or will take place in the future (e.g. the PMS stars) or has taken place in the past (neutron star).

Stellar cluster: Group of a few hundreds to several millions of stars. The smallest groups are named associations. Most of the stars are formed in open clusters.

Stereoisomers (*or stereomers*): Isomers (molecules with identical atomic composition, i.e. the same number of the same atoms) such that the atoms are identically interconnected by covalent bonds. If two stereoisomers are different in the same way that a left hand is different from a right hand, the stereoisomers are enantiomers. In all other cases, stereoisomers are called diastereoisomers (or diastereomers).

Stereoselectivity: When a chemical reaction leads to the formation of stereoisomers and when these latter are not exactly produced in the same amount, the reaction is called stereoselective. Similarly, when two stereoisomers react at a different rate with a reactant, the reaction is stereoselective. In the chemical literature, some authors have introduced subtle differences between stereoselectivity and stereospecificity.

Steric effect: One of the multiple "effects" introduced by organic chemists to explain the relative stability or the relative reactivity inside a group of molecules. The steric effect takes into account the size of each atom or groups of atoms. For instance, the steric effect of a $–C(CH_3)_3$ group is larger than the steric effect of a CH_3 group. The steric effect can be explained on the basis of repulsive term in the Van der Waals forces.

Stony meteorites: Mainly made up of silicates, they can contain from 0 to 30% of metal grains and several per cent of sulfides. They can be differentiated (achondrite) or undifferentiated (chondrite) even if CI undifferentiated chondrites do not contain chondrules.

Stop codon (*genetics*): Codon that does not code for amino acids but that indicates the translation end. For the Universal code, these codons are UAA, UAG and UGA (synonym: termination codons).

Stratopause: Atmospheric boundary between stratosphere and mesosphere.

Stratosphere: Atmospheric layer located above the troposphere and below the mesosphere, between 9–17 km and 50 km. In the stratosphere, temperature slightly increases with altitude that prevents it from convective movements.

Strecker (*synthesis*): Synthetic method producing amino acids from aldehyde, HCN, NH_3 and water. Frequently considered as important prebiotic reaction.

Stromatolite: Sedimentary structure consisting of laminated carbonate or silicate rocks and formed by the trapping, binding, or precipitation of sediment by colonies of micro-organisms (bacteria and cyanobacteria).

Strong force: One of the four fundamental forces in physics that contribute to the stability of the atomic nucleus.

Subduction: Plate-tectonic mechanism where an oceanic plate sinks under an other lithospheric plate (generally a continental plate, but sometimes also an oceanic plate).

Sublimation: Direct phase change from solid to gas state.

Succinic acid: $HOOC–CH_2–CH_2–COOH$.

Sun: Star belonging to the main sequence (in H-R diagram); it is 4.56 Ga old. Sun–Earth distance corresponds to 8 light-minutes or 1 astronomic unit (1 AU).

Supernova: Exploding star that, before explosion, was either a binary star (type I) or a massive star (type II). After explosion, the remnant becomes a neutron star.

Symbiosis: Prolonged association between two (or more) organisms that may benefit each other. In the case of endosymbiosis, one organism lives within another one; their relationship is irreversible and implies a complete interdependence, such as the two become a single functional organism. Mitochondria and chloroplasts are remnants of the endosymbiosis of photosynthetic bacteria.

Synchrotron radiation: Electromagnetic waves covering a large frequency domain and emitted under vacuum by high-velocity electrons or ions, when their trajectory is altered, as by a magnetic field. Synchrotron radiation is naturally polarized.

Synonymous (*codons*): Different codons that specify the same amino acid; e.g., AAA and AAG specify lysine.

T

T Tauri star: Newborn low-mass star (lower than two solar masses) that starts to become optically observable. Classical T-Tauri are very young (less than 1 Ma old) and are not yet on the main sequence (PMS stars). They are surrounded by a disk, their luminosity is variable with an excess of IR with respect to UV. T-Tauri stars we observe today are probably similar to the young Sun.

Taxonomy: Science that classifies living species. Similar species belong to the same taxon.

Tectonic: Study of structure (and deformation) of rocks and Earth crust.

Tektites: Natural glasses formed at very high temperature ($> 2000\,°C$) by meteoritic impacts creating craters greater than 10 km in diameter.

Telluric planets: Small and dense rocky planets (density: 3 to $5.5\,\mathrm{g\,cm^{-3}}$). These planets (namely Mercury, Venus, Earth and Mars) are silicate rich and were formed in the inner part of the protosolar nebula, beyond the dust line.

Terraforming: Voluntary transformation of a planetary atmosphere in order to allow colonization by plants and animals (including man).

Tertiary structure (*of a protein*): Arrangement of the side chains of the residues in space, overall conformation of a protein.

Theia: Name given by some authors to Mars-sized object, which, after impacting the young Earth, led to the Moon formation. This cataclysmic event took place 4.5 to 4.45 years ago.

Thermolysis: Thermal degradation of a molecule into smaller fragments (atoms, radicals, molecules and, rarely, ions).

Thermonuclear reaction: Reaction leading to formation of heavier nuclei by fusion of lighter nuclei. Thermonuclear reactions require high-T and high-P conditions. In natural conditions, these reactions occur spontaneously in cores of main sequence stars or of heavier stars like giant stars. It occurs also during supernovae explosions.

Thermophile: Organism living optimally at high temperatures. They can be divided into "moderate thermophiles", living optimally between 40 and $60\,°C$, extreme thermophiles, between 60 and 80 °C, and hyperthermophiles, living optimally at temperatures higher than 80 °C (and up to ~ 120 °C).

Thin section (*geology*): Rock slice generally 30-µm thick. At such a thickness most minerals are transparent and can be observed by transmitted light with a polarizing microscope.

Thiocyanate: R–S–CN.

Thioester: R–S–CO–R'.

Thiol: R–SH.

Tholins: Solid mixture of complex organic molecules obtained by irradiation of reduced gases like CH_4, NH_3. Could be present on Titan.

Threonine (*Thr*): Proteinic amino acid containing four C atoms and a –OH group in its side chain. Threonine is described as a hydrophilic amino acid.

Thymine (*T*): Purinic nucleic base purine and specific of DNA.

Titan: The biggest satellite of Saturn (~ 5000 km in diameter; which is approximately the same size as Mercury). In the Solar System, this is the only satellite that possesses a dense atmosphere. Very probably, organic reactions are taking place in its multicomponent hydrocarbon-bearing atmosphere.

Titus–Bode law: Empirical law giving approximately the planet–Sun distance d as a function of the planet ranking n ($d = (4 + 3.2^{(n-2)})/10$). Mercury is an exception to this law ($d = 0.4$ AU).

Tonalite: Plutonic magmatic rock (granitoid), made up of quartz and calcic plagioclase feldspar; biotite and sometimes amphibole are minor mineral phases. Tonalite does not contain alkali feldspar. Dacite is its effusive equivalent.

TPF (*Terrestrial Planet Finder*): NASA project with a similar goal as the ESA Darwin project, i.e. discovery of extrasolar terrestrial planets.

Transcription: Synthesis of an m-RNA as a copy of an antisens DNA single strand.

Transduction: Transfer of genetic material from one bacteria to another through viral infection.

Transfer of genes (*horizontal transfer*): Transfer of a gene from one organism to another that does not belong to the same species. Such transfer can occur through viral infection or by direct inclusion of genetic material present in the external medium. Horizontal transfer is different from vertical transfer from parents to children.

Transferase: Enzyme that catalyses transfer of a chemical group from one substrate to another.

Transform fault: Boundary between two lithospheric plates that slide without any crust creation or destruction.

Transit: Motion of a planet in front of the disk of its star.

Translation: Sequence enzymatic reactions such that the genetic information coded into a messenger RNA (m-RNA) leads to the synthesis of a specific protein.

Triple point: In a P–T phase diagram of a pure compound, it corresponds to the unique P, T value where the three phases (gas, liquid, solid) coexist at equilibrium. For water, $P = 6.11\,\text{mbar}$ and $T = 273.16\,\text{K}$.

t-RNA synthetase: Enzyme that catalyzes, in a very specific way, bond formation between an amino acid and its t-RNA.

t-RNA: Transfer RNA. Polymer containing 70 to 80 ribonucleotides and specific of each amino acid to which it is linked. Able to recognize a triplet of nucleotides of m-RNA (codon) by a specific molecular-recognition process involving a triplet of nucleotides of the t-RNA (anti-codon). The t-RNA's play a fundamental role for proteinic synthesis.

Trojans: Family of asteroids located at the Lagrange point on the Jupiter orbit. Their position together with the Sun and Jupiter positions determines an equilateral triangle.

Trondhjemite: Plutonic magmatic rock (granitoid), made up of quartz and sodic plagioclase feldspar; biotite is a minor mineral phase. Tonalite and trondhjemite are similar rocks except that in tonalite plagioclase is calcic, whereas it is sodic in trondhjemite.

Tropopause: Atmospheric boundary between troposphere and stratosphere.

Troposphere: Lowest part of Earth's atmosphere, as the temperature at its base is greater than at its top, it is the place of active convection. On Earth, the troposphere thickness ranges between 9 km (pole) and 17 km (equator). Most meteorological phenomena take place in troposphere.

Tryptophane (*Trp*): Proteinic amino acid containing eleven C atoms; tryptophane contains a heterocycle in its side chain. It is described as an aromatic amino acid.

TTG: Tonalite, Trondhjemite, Granodiorite. Rock association typical of the continental crust generated during the first half of Earth's history.

Tunguska event: Explosion that took place in 1908 (June 30th) in Siberia and devastated $2000\,\text{km}^2$ of forest. It was probably due to the impact of a comet or of an asteroid of a size of a few tens of metres.

Tunnel effect: Description of tunnel effect requires the use of quantum mechanics because it is a direct consequence of the wave properties of particles. When a system A gives another system B, while the internal energy of A is lower than the energy barrier (activation energy) required to cross the barrier from A to B, it can be said that A gives B by passing "through the barrier" by a tunnel effect.

Tyrosine (*Tyr*): Proteinic amino acid containing nine C atoms. Its side chain contains a phenolic group.

U

Uracil (*U*): Nucleic base belonging to pyrimidine family and specific of RNAs.

Urea ($H_2N-CO-NH_2$): First organic molecule that has been synthesised from a mixture of inorganic molecules (Wohler).

UV radiation: Electromagnetic radiation characterized by wavelengths ranging from 0.01 to 0.4 micrometres (energies from 124 to 3.1 eV).

V

Valine (Val): Hydrophobic proteinic amino acid containing five carbon atoms.

Van der Waals forces (*VdW*): Interatomic forces acting between nonbonded atoms at the intramolecular level but also at the intermolecular levels. Repulsive VdW forces are responsible for the less than infinite compressibility of matter and for the fact that atoms have sizes. Attractive VdW forces are responsible for matter cohesion. VdW forces play a major role in biochemistry: together with H bonds and electrostatic interactions, they determine the preferred conformations of molecules and they contribute to molecular-recognition phenomena.

Vernal point: Sun location on the celestial sphere at the vernal equinox (spring equinox). It is the origin of coordinates in the equatorial system.

Vertical tectonic: See sagduction.

Viking: NASA mission to Mars that started in 1976. The two landers (Chryse Planitia and Utopia Planitia) performed a series of very ambitious experiments to detect the presence of life on Mars. Unfortunately, many results were ambiguous.

Virus: System containing DNA or RNA surrounded by a proteinic envelope (capside). When introduced in a living cell, a virus is able to replicate its genetic material by using the host-cell machinery.

VLBI (*Very Long Baseline Interferometry*): Technique that allows a very accurate determination (50 microarcsec) of the position of astronomical sources of radiowaves. This method is based on interferometry measurements using very distant radiotelescopes (large base) located on the same continent or on different continents or even on Earth and on a satellite.

VLT (*Very Large Telescope*): Group of four large telescopes (4 to 8 metres) and several smaller telescopes, able to work as interferometers and located in Chile. VLT is managed by ESO.

Volatile: see refractory.

Volcanoclastic sediment: Material due to sedimentation of volcanic products (i.e. ashes) in the sea or in a lake.

W

Wall (*of a cell*): Extracellular membrane. In bacteria, cell-wall structure is complex: the walls of gram-positive and gram-negative bacteria are different.

Water triple point: See triple point.

Watson–Crick: Canonical model of DNA (double helix) involving the pairing of two polynucleotide strands via H bonds between A and T or G and C. RNA is generally single-stranded but Watson–Crick pairing can occur locally within a single strand. When happens, it involves A–U and G–C pairing.

Weak bonds: Intermolecular or intramolecular bonds involving nonbonded atoms (atoms not bonded by covalent, coordination or electrostatic bonds). H bonds are well-known examples of weak bonds but Van der Waals forces and electrostatic interactions also contribute to weak bonds. The weak bonds play a fundamental role in the living world: they determine the conformation of molecules and more particularly the conformation(s) of biopolymers; they are responsible for the specificity of molecular recognition. The intermolecular association due to weak bonds is generally reversible.

Weak force: One of the four fundamental forces of physics. Parity is violated for this force. The coupling between weak forces and electromagnetic forces is at the origin of the very small energy difference between enantiomers (PVED for parity-violation energy difference).

Weathering: See alteration.

Wind (*solar or stellar*): Flow of ionized matter ejected at high velocity (around 400 km/s) by a star. Solar wind mainly contains protons.

Wobble (*genetics*): Describes imprecision in base pairing between codons and anticodons. It always involves position 3 of codon, mainly when the base is U.

X

Xenolith: Inclusion or enclave of foreign rock or mineral (xenocrystal), in a magmatic rock.

Y

Young sun paradox: Apparent contradiction between the lower brightness of the young Sun, (70% of the present-day intensity in the visible spectral range) and the early presence of liquid water (–4.4 Ga) on Earth. A strong greenhouse effect due to high concentrations of atmospheric CO_2 could account for this apparent paradox.

YSO (*Young Stellar Object*): Star that has not yet completed the process of star formation. YSO includes objects ranging from dense cores, (that can be detected in the submillimetric IR frequency range) to pre-main sequence stars (T Tauri, Herbig AeBe) and HII regions.

YSO: Young stellar object.

Z

Z (*astronomy*): Abundance of "heavy elements" i.e. all elements except H and He.

ZAMS (*zero-age main sequence*): Ensemble of newborn stars in which H fusion has just started.

Zircon: Mineral. Nesosilicate (isolated SiO_4 tetrahedron). [$ZrSiO_4$]. This mineral which also contains traces of Th and U is extremely resistant to weathering and alteration. This is why it is commonly used to determine rock ages. The oldest zircon crystals so far dated gave an age of 4.4 Ga. They represent the oldest known terrestrial material.

Zodiacal light: Diffuse faint light observed in a clear sky close to the ecliptic. It is due to the diffusion of the solar radiation by the electrons and the interplanetary dusts. Also used for any light diffused by dust particles in a planetary system.

4 Authors (Photos and addresses)

BARBIER Bernard
CNRS Researcher
Center for Molecular Biophysics, Orléans, France
E-mail: barbier@cnrs-orleans.fr

Physico-chemistry/analytical chemistry:
Interstellar chemistry, prebiotic chemistry, evolution, origin of the genetic code

BOITEAU Laurent
CNRS Researcher
Chemistry Department
University of Montpellier 2, France
E-mail: laurent.boiteau@univ-montp2.fr

Chemistry, physical organic chemistry:
Origins of life, origins of amino acids and peptides

BRINKMANN Henner
Research assistant
Biochemistry Department, University of Montréal, Montréal, Québec, Canada
E-mail: Henner.Brinkmann@umontreal.ca

Molecular biology: Evolution and molecular phylogeny, origin and early evolution of life, plastid and mitochondrial endosymbioses, evolution of the main lineages of the three domains

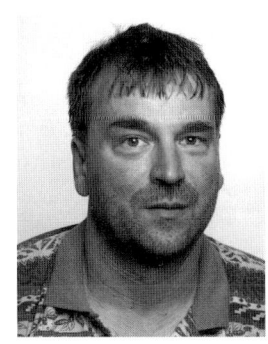

COLL Patrice
Assistant Professor
Laboratoire Interuniversitaire
des Systèmes Atmosphériques,
University Paris 12 and Paris 7, Créteil, France
E-mail: pcoll@lisa.univ-paris12.fr

Chemistry/Planetology: Astro/Exobiology, extraterrestrial organic chemistry and physics (Titan, Mars,...), theoritical and experimental developments

COMMEYRAS Auguste
Professor Emeritus
Chemistry Department,
University of Montpellier 2, France
E-mail: acommeyras@univ-montp2.fr

Chemistry, physical organic chemistry:
Origins of life, origins of amino acids and peptide

CRONIN John
Professor Emeritus,
Arizona State university, Tempe, Arizona (USA)
E-mail: jcronin@asu.edu

Chemistry: Origin of life, evolution of organic matter, organic chemistry of meteorites

FORTERRE Patrick
Professor
Institut de Génétique et Microbiologie,
University Paris Sud, Orsay, France
Head of the Department
of fundamental and medical microbiology
at the Pasteur Institut, Paris
E-mail: forterre@igmors.u-psud.fr
        forterre@pasteur.fr

Biology/exobiology: Hyperthermophiles, archaea, phylogenomics, universal tree of life, LUCA, origin of DNA

GARGAUD Muriel
CNRS Researcher
Observatoire Aquitain des Sciences de l'Univers,
Bordeaux, France
E-mail: muriel@obs.u-bordeaux1.fr

Astrophysics/astrobiology: Interstellar medium physico-chemistry, origins of life

GRIBALDO Simonetta
Post-doc researcher,
Institut de Génétique et Microbiologie,
University Paris-Sud, Orsay, France
E-mail: simonetta.gribaldo@igmors.u-psud.fr

Biology: Early evolution, molecular phylogenetics, comparative genomics

HAENNI Anne-Lise
CNRS Research Director Emeritus
Institut Jacques-Monod,
University Paris 6 and Paris 7, Paris, France
E-mail: haenni@ijm.jussieu.fr

Biochemistry: RNA viruses, protein biosynthesis

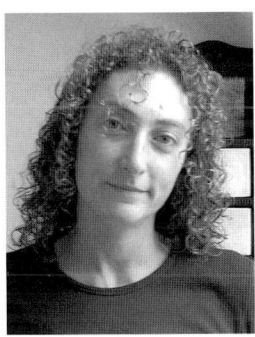

LOPEZ-GARCIA Purificación
CNRS Researcher
Unité d'Ecologie, Systématique et Evolution,
University Paris-Sud, Orsay, France
E-mail: puri.lopez@ese.u-psud.fr

Microbiology: Microbial diversity, phylogeny and evolution, extreme environments

MARTIN Hervé
Professor
Laboratoire Magmas et Volcans,
Blaise Pascal University, Clermont-Ferrand, France
E-mail: H.Martin@opgc.univ-bpclermont.fr

Geochemistry: Geological and geochemical evolution of the primitive Earth. Subduction zone magmatism

MAUREL Marie-Christine
Professor
Biochemistry of Evolution
and Molecular Adaptability
Institut Jacques-Monod,
University Paris 6, Paris, France
E-mail: maurel@ijm.jussieu.fr

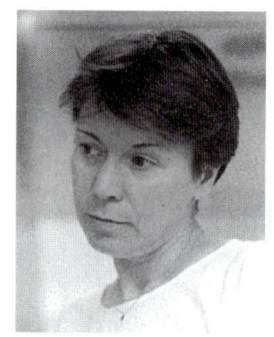

Biology-Biochemistry: Molecular evolution, origins of life, etiology, activity and persistence of RNA

NAVARRO-GONZALEZ Rafael
Research Professor
Universidad Nacional Autónoma de México
Ciudad Universitaeia, Mexico City
E-mail: navarro@nuclecu.unam.mx

Atmospheric Chemistry and Astrobiology:
Lightning chemistry in the atmosphere of Earth and of other planets and satellites: impact in the origin and evolution of the biosphere. Study of terrestrial analogs of Mars and Europa for the identification of biosignatures required for the search of extraterrestrial life.

NAKATANI Yoichi
CNRS Research Director Emeritus
Louis Pasteur University, Strasbourg, France
E-mail: nakatani@chimie.u-strasbg.fr

Chemistry: Membrane chemistry, biochemistry and biophysics

OURISSON Guy
Emeritus Professor,
Louis Pasteur University, Strasbourg, France
Member of the French Academy of Sciences
E-mail: ourisson@chimie.u-strasbg.fr

Organic chemistry: Natural substances, biological and fossil membranes, origin of life

PHILIPPE Hervé
Professor
Biochemistry department, University of Montréal,
Montréal, Québec, Canada
E-mail: herve.philippe@umontreal.ca

Biology: Bio-informatics, universal tree of life, molecular phylogeny, comparative genomics model of sequence evolution

RAULIN François
Professor
Laboratoire Interuniversitaire
des Systèmes Atmosphériques,
University Paris 12 and Paris 7, Créteil, France
E-mail: raulin@lisa.univ-paris12.fr

Chemistry/Planetology: Exo/astrobiology, extraterrestrial organic chemistry (Titan, Giant planets, Comets, Mars,...), experimental, theoretical and observational approaches

REISSE Jacques
Professor Emeritus,
Free University of Brussels, Belgium
Member on the Belgian Academy of Sciences
E-mail: jreisse@ulb.ac.be

Chemistry: Stereochemistry and chirality, NMR, sonochemistry, organic cosmochemistry, study of the liquid phase

SELSIS Franck
CNRS Researcher
Centre de recherchen en astrophysique
Ecole Normale Superieure, Lyon, France
E-mail: selsis@ens-lyon.fr

Planetary Sciences and Astrobiology: Formation and evolution of planetary atmospheres, atmospheres of exoplanets: detection and characterization

VANDENABEELE-TRAMBOUZE Odile
CNRS Researcher
Chemistry Department,
University of Montpellier 2, France
E-mail: o.trambouze@univ-montp2.fr

Analytical chemistry: Development of analytical methodologies for trace analysis of organic compounds in meteorites and complex matrices

5 Index

α-helices, 4
absolute asymmetric synthesis, 98
Acasta, 151
adenine, 185
Akilia, 4
alanine, 75
algae, 269
altritol nucleic acid (ANA), 182
amino acids, 51, 85, 119, 125
   α, 120, 128
   N-carbarmoyl (CAA), 147
   non-α amino acid, 123
   protein, 121
amino diacids
   β, 122
   γ, 122
   δ, 122
amphiphiles, 29, 30, 37
anaerobic organism, 263
aptamer, 187
archaea, 204, 205, 236, 238, 261
   archaebacteria, 217, 221
   glycerolipid, 200
   lipid, 38
Archaean, 272
archaeon, 257
Archezoa, 223
asthenosphere, 286
asymmetric autocatalysis, 95
atmosphere, 52
   primitive, 52, 53, 152
autocatalytic systems, 10

β-sheet structures, 4
Bacillus subtilis, 270
bacteria, 206, 229, 238, 261
   cyanobacteria, 269
   Gram positive, 270
   green sulfur, 269
   spores, 17
Barberton, 3
barotolerant, 259
Beagle 2, 14
biosignature, 190, 276
biotechnology, 260
biotope, 259
brownian movement, 36
Bücherer–Bergs, 129, 132, 136

Cahn–Ingold–Prelog (CIP) system, 78
Cassini–Huygens, 15, 65, 66
chemo
   selectivity, 161
chirality, 73, 79
   catalysis, 95
   homochirality, 82–84
      analysis, 107
      origin, 91, 101
   magnetochiral, 92, 103
chloroplast, 210
chondrites
   carbonaceous, 119, 121
circular dichroism, 77
circularly polarized light, 91, 102
coenzyme, 183
cofactors, 186
collision (continental), 293
Comet
   Hale–Bopp, 8
   Halley, 8
   Hyakutake, 8
conglomerate, 81
continental rift, 290
core
   inner, 283
   outer, 284

covarion model, 231
CPT theorem, 90
crust
  continental, 286
  oceanic, 285
cryopreservation, 270
crystallization, 81
Curie principle, 91
cyanic acid, 135
cyanobacteria, 269
cyanohydrin, 123, 138

Darwin, 187, 211
Dead Sea, 269
desiccation, 270
DNA, 206, 208

Earth
  primitive, 118, 151
  subsurface, 275
editing mechanism, 175
EETA 79001, 13
enantiomer, 4, 51, 74, 75
  enantiomeric excess, 93, 99, 150
endolithic, 269
endosymbiosis, 210, 217, 223, 241, 243
Eubacteria, 217, 221
Eukarya, 204, 205, 261
eukaryogenesis, 211
Eukaryote, 217, 221, 229, 238, 259
Europa, 14
evaporite, 274
evolution, 177, 183, 260
  biochemical, 179
  chemical, 163
  evolutionary process, 179
  evolutionary rate, 273
  molecular, 187
exon, 175, 206
exoplanet, 16
extreme environment, 261
extremophile, 257, 261, 271
extremotolerant, 261

Fisher (projection), 75, 78, 80
formaldehyde, 52–54, 57, 180
Formose reaction, 57
fossil trace, 184
FOTON 8 and 11, 9

fungi, 269

Galileo, 14
gene
  genetic take-over, 181
  genotype, 186
  lateral/horizontal transfer, 224, 235
genomic (comparative), 204
Giotto, 8
greenhouse effect, 153

hairpin, 172
halophile, 11, 257, 274
halotolerant, 259
heterotachy, 228, 230
hexitol nucleic acid (HNA), 182
homochirality, see chirality
hot spot, 293, 294
hydrogen cyanide, 6, 50, 52, 55–57, 63
hydrothermal
  system, 6
hypersaline, 269
hyperthermophile, 11, 197, 200, 201, 257

imidazole, 184
impact, 154
indel, 232
interaction weak, 89
intron, 175, 206

K-T (Cretaceous Tertiary boundary)
  impact, 155

lichen, 269
light diffusion, 34
lipophilic, 33
liposome, 31
lithosphere, 286
long-branch attraction, 228
longevity, 270
LUCA, 86, 195, 201, 204, 209, 211, 223, 245
lyophilisation, 270

magnetic
  circular dichroism, 92
  rotatory dispersion, 92
magnetochiral
  dispersion, 92
  photochemistry, 103
mantle
  lower, 284

temperature, 284, 285
upper, 285
margin, 287, 288
Mariana Trench, 269
marker (molecular), 261
Mars, 12
    Express, 14
    Global Surveyor, 12
meso compound, 81
mesophile, 200, 258
metabolism, 275
meteor, 154
meteorite
    compound, 124
    Murchison, 7, 119, 126
    SNC, 12, 13
methanogen, 270
Methanopyrus kandlerii, 270
mevalonic pathway, 44
micelle, 31
microfossil, 3
microsporidia, 197
mid oceanic ridge, 269, 288, 289
Miller, 6, 50, 128, 271
MIR station, 9
mitochondria, 210
montmorillonite, 37, 179
Murray, 126
mutation, 273
mycoplasma, 196

N-carbamoylation, 158
N-carboxy anhydride (NCA), 149, 157
neutron star, 7, 92
nitric oxide, 128, 152, 155, 156
nitrogen
    liquid, 270
    oxide, 152
nitrosation, 158
Nuclear Magnetic Resonance (NMR), 34
nucleic acid, 36
nucleoside, 180
    nucleus, 209
nucleotide (modified), 171

ocean
    crust, 285
    primitive, 156
oligotrophic, 259

one-handedness, 4
Oparin, 6
optical rotation (rotatory dispersion), 77
organic
    matter, 128
    molecule, 5, 118
Orion, 7
osmotic pressure, 261

paralog, 221, 235
parity, 89, 99
    parity-violating energy difference
        (PVED), 90
Pasteur, 79
peptide, 117, 127, 147
    emergence, 147
    nucleic acid (PNA), 86, 181
    synthesis, 159
    tag, 175
permafrost, 270
phenetic, 236
phenotype, 186
phosphate, 42
    phytyl phosphate, 39
    polyprenyl phosphate, 39
phospholipid, 37, 38
phosphorimidazolide, 179
phosphorylation, 43
photochemistry, 91
    magnetochiral, 92, 103
photodegradation, 127
photolysis, 103
photosynthesiser, 269
phylogeny, 260
    eukaryotic, 241
    phylotype, 261
    prokaryote, 237
    reconstruction, 273
    tree, 269
Pilbara, 3
plate tectonic, 287–295
PNA world, 181
polarized synchrotron radiation, 7
polymerase, 189
polyterpene, 39
post-transcriptional modification, 185
prebiotic
    chemistry, 49
    synthesis, 271

primary pump, 147, 156, 157, 163
primitive soup, 271
prokaryote, 198, 273
protein world, 185
protocell, 29
psychrophile, 273
pyranosyl-RNA, 10
pyrite, 6
Pyrolobus fumarii, 269

racemate, 81
racemization, 9, 82
redox reaction, 273
resistance form, 270
reverse gyrase, 198
ribonucleoprotein
   small nuclear ribonucleoprotein
      particle snRNP, 175
   small nucleolar (snoRNP), 174
ribose, 180
ribozyme, 174, 184
RNA, 208
   helices, 172
   messenger, 172
   pre-mRNAs, 175
   pre-tRNAs, 173
   pyranosyl RNAs (p-RNAs), 181
   ribosomal RNAs (rRNAs), 174
   RNAi, 177
   signal recognition particle (SrpRNA), 175
   small interfering, 177
   small nuclear RNA (snRNA), 175
   small nucleolar RNA (snoRNAs), 174
   thermodegradation, 203
   transfer (tRNA), 173
   transfer-messenger (tmRNA), 175
   tRNA synthetase, 222, 232, 238
   vault RNA, 175
   world, 147, 171

selection
   non-natural, 187
SELEX, VII, 187
self
   complexification, 29
   organisation, 29, 33
serpentinisation, 6
SETH (search for extraterrestrial homochirality), 107

SETI program, 17
small ribosomal subunit, 260
snorposome, 174
splicing (spliceosome), 206
spore, 270
STONE experiment, 13
strain, 269
Strecker reaction, 54, 55, 120, 129, 132, 133, 136
stromatolite, 273
subduction, 291–293
sugar, 180
Sulfolobus acidocaldarius, 257
superturn, 240
survival rate, 270
symbiosis, 269
symmetry breaking, 82, 86
synchrotron radiation, 102

telomerase, 175
terpenoid, 41
tetraether lipid, 200
thermophilic
   Eukarya, 204
   organisms, 197
Thermus aquaticus, 257
thioester world, 7
tholin, 52, 58, 63, 64
Titan, 15, 61–63
trans-translation, 175
tree of life, 221, 237, 273
Tunguska, 155

variable rate of evolution, 228
vesicle, 31, 35, 36
Viking, 12
viroid, 176
virus, 208
volcanic lightning, 156
Voyager, 14

water
   availability, 261
   liquid water, 5
Wilson cycle, 294
Woese, 221

xenologous, 235

zircon, 151